Lecture Notes in Mathematics

Volume 2375

This series reports on new developments in all areas of mathematics and their applications - quickly, informally and at a high level. Mathematical texts analysing new developments in modelling and numerical simulation are welcome. The type of material considered for publication includes:

1. Research monographs
2. Lectures on a new field or presentations of a new angle in a classical field
3. Summer schools and intensive courses on topics of current research.

Texts which are out of print but still in demand may also be considered if they fall within these categories. The timeliness of a manuscript is sometimes more important than its form, which may be preliminary or tentative. Please visit the LNM Editorial Policy (https://drive.google.com/file/d/1MOg4TbwOSokRnFJ3ZR3ciEeKs9hOnNX_/view?usp=sharing)

Titles from this series are indexed by Scopus, Web of Science, Mathematical Reviews, and zbMATH.

Joseph Cho • Kosuke Naokawa • Yuta Ogata •
Mason Pember • Wayne Rossman •
Masashi Yasumoto

Discrete Isothermic Surfaces in Lie Sphere Geometry

 Springer

Joseph Cho
Global Leadership School
Handong Global University
Pohang, Korea (Republic of)

Yuta Ogata
Department of Mathematics, Faculty of
Science
Kyoto Sangyo University
Kyoto, Japan

Wayne Rossman
Department of Mathematics
Kobe University
Kobe, Japan

Kosuke Naokawa
Faculty of Applied Information Science
Hiroshima Institute of Technology
Hiroshima, Japan

Mason Pember
Department of Mathematical Sciences
University of Bath
Bath, UK

Masashi Yasumoto
Faculty of Science and Technology
Tokushima University
Tokushima, Japan

ISSN 0075-8434   ISSN 1617-9692   (electronic)
Lecture Notes in Mathematics
ISBN 978-3-031-95591-4   ISBN 978-3-031-95592-1   (eBook)
https://doi.org/10.1007/978-3-031-95592-1

This work was supported by Japan Society for the Promotion of Science London, Austrian Science Fund and Ministry of Education, University and Research, Italy (MIUR).

This Springer imprint is published by the registered company Springer Nature Switzerland AG
The registered company address is: Gewerbestrasse 11, 6330 Cham, Switzerland

If disposing of this product, please recycle the paper.

Preface

This text is the outgrowth of a differential geometry seminar which has been running at Kobe University in Japan for over two decades, and its goal is to describe how the notion of isothermicity allows for a discretization of surface theory that preserves certain mathematical structures enjoyed by smooth surfaces.

The impetus for creating this text stems from conversations with two highly regarded experts in the field, Francis E. Burstall and Udo Hertrich-Jeromin, who explained the finer points of the theory surrounding isothermic surfaces to the authors: beginning with Rossman, the seminar's senior member, and gradually including all other authors over time. In fact, they were Pember's doctoral thesis advisors, and Cho has recently conducted joint research with them. They have both regularly visited Kobe University and participated in the geometry seminar, and have been in regular contact with the authors via conferences and research meetings in Germany, Austria, England, and Japan.

Influenced by discussions with discrete differential geometry experts such as Francis E. Burstall, Udo Hertrich-Jeromin, Tim Hoffmann, and Konrad Polthier, the seminar commenced to study discretization that relies fundamentally on isothermicity, that is, existence of conformal curvature line coordinates, and so isothermicity started playing a most central role. As isothermicity is invariant under Möbius transformations, it became clear that the seminar should study it in the context of Möbius geometry. Then, through discussions with various mathematicians, occurring primarily when they would join conferences in Japan, or when the authors would travel to Europe, and TU Wien in Vienna in particular, it also became clear that extending Möbius geometry to Lie sphere geometry is desirable, as it allows for the consideration of a more general class of surfaces, Ω-surfaces, which are not necessarily isothermic themselves, but do have isothermic sphere congruences, and therefore for which the techniques of isothermicity still apply. In particular, the study of Ω-surfaces clarifies and simplifies certain properties and leads to elegant generalizations, for example, a concise expansion of properties of constant mean curvature surfaces to those of the larger class of linear Weingarten surfaces.

In the seminar, a direct route to specific smooth and discrete surfaces was considered, after which surrounding mathematical tools were steadily introduced on a need-to-know basis. The value in this approach was evident when the authors repeatedly witnessed the seminar's students naturally developing strong desires to understand those tools via experience with surface theory applications, wishing to learn about topics such as integrable systems, Riemann surface theory, Legendre manifolds, exterior algebras, Lie groups and Lie algebras, curvature notions in differential geometry, and transformation theory with associated permutabilities. The authors then appreciated how a text like this present one could not only explain a piece of the discrete differential geometry field but also instill further interest in the fundamental tools of geometry in young aspiring mathematicians.

It is worth mentioning that the works of Josef Dorfmeister, Franz Pedit, and Hongyou Wu on harmonic maps into symmetric spaces and its application to the theory of constant mean curvature surfaces, and of Masaaki Umehara and Kotaro Yamada on the utilization of complex analytical methods for the examination of submanifold theory and the integration of singularity theory, had a pervasive influence on the conception of these notes, as advantages could be seen in viewing these concepts through the lens of Lie sphere geometry. Some aspects of this are described in detail in this text, while other very recently explored aspects are alluded to in the sections on further readings here.

Through both direct visits and advice via conversations, each mathematician named here influenced the seminar, often advocating hands-on application to construction of particular surfaces of interest, and only later elucidating the needed background theory. These approachable methods proved effective for students of the seminar, and it was then that the above-mentioned underlying approach of this text first began to take hold: to focus initially on application to surface theory, and bring in needed mathematical tools only when required and only to the degree needed.

Each of the authors made specific contributions to the text: Rossman initially provided the bulk of the material via lecture notes used in the seminar; Naokawa and Yasumoto covered aspects of discrete surfaces, with Yasumoto especially considering discretizations of cases where complex analysis is critical and often taking on the under-appreciated task of bringing mathematicians together through organizing conferences; Ogata contributed more analytic aspects along with a healthy dose of raw computing power; Cho and Pember highlighted and clarified what is most essential in Lie sphere geometry. Cho in particular took on the critical and rather daunting task of turning what were only loosely connected lecture notes into a proper contiguous textbook with clear elucidation of essential principles.

It is the authors' genuine hope that this book will be a suitable introduction to discrete differential geometry for young mathematicians, and that the hands-on approach taken here will leave them well equipped to then explore the broad collection of interesting results and elegant material in this field.

In addition to the mathematicians mentioned above, the authors wish to express their gratitude to others as well for useful discussions and assistance in various ways: Alexander Bobenko, David Brander, David Calderbank, Shoichi Fujimori, Masaya Hara, Atsufumi Honda, Jun-ichi Inoguchi, Yu Kawakami, Callum Kemp, Martin Kilian, Shimpei Kobayashi, Miyuki Koiso, Masatoshi Kokubu, Wai Yeung Lam, Katrin Leschke, Nozomu Matsuura, Reiko Miyaoka, Katsuhiro Moriya, Christian

Müller, Yoshihiro Ohnita, Ulrich Pinkall, Denis Polly, Helmut Pottmann, Thomas Raujouan, Kentaro Saji, Takeshi Sasaki, Wolfgang Schief, Nicholas Schmitt, Yoshihiko Suyama, Gudrun Szewieczek, Keisuke Teramoto, Johannes Wallner, Seong-Deog Yang, Masaaki Yoshida, the referees of this manuscript, and the many students in the Kobe University differential geometry seminar over the last 20 years.

Pohang, Korea (Republic of)   Joseph Cho
Hiroshima, Japan   Kosuke Naokawa
Kyoto, Japan   Yuta Ogata
Bath, UK   Mason Pember
Kobe, Japan   Wayne Rossman
Tokushima, Japan   Masashi Yasumoto

Competing Interests The authors have no competing interests to declare that are relevant to the content of this manuscript.

Contents

Chapter 1
Introduction

1.1 Purpose

The primary purpose of these notes is to understand the underlying structure behind a mathematically rich class of surfaces, and introduce how this comprehension leads to the principles of *structure-preserving discretizations*. We are particularly interested in the *integrable structure* they possess, as the modern surface theory is marked by the influence of integrable systems, and these properties extend to suitable discretizations of such surfaces. These lecture notes will focus on the case of isothermic surfaces in the context of Lie sphere geometry to achieve this goal.

Thus, it is of paramount importance to understand the *integrable structure of isothermicity*, including their *transformations* and *permutability*. In the modern surface theory scene with its influence from soliton theory, integrability is represented by *zero-curvature formulations*: the zero-curvature formulation of isothermicity in this manuscript is based on the gauge-theoretic approach exemplified by [38] for the case of conformal geometry.

In the gauge-theoretic approach, integrability is represented by the existence of a certain 1-parameter family of *flat connections* on vector bundles. Under this paradigm, transformations of surfaces are given by gauge transformations of the flat connections: the spectral deformation is found via the locally trivializing gauge transformations while the transformations subject to a parameter correspond to gauge transformations similar to simple factor dressings of Terng and Uhlenbeck [209, 211] (cf. [36]). Thus, permutabilities between transformations are given in terms of the compositions of gauge transformations.

This approach is especially suitable for the process of structure-preserving discretization: under a carefully constructed theory of discrete vector bundles, the gauge transformations themselves become discrete connections, while the permutability is translated to a notion of *discrete flat connections*, arriving at the comprehensive structure-preserving discretization of the integrable structure of isothermicity.

© The Author(s), under exclusive license to Springer Nature Switzerland AG 2025
J. Cho et al., *Discrete Isothermic Surfaces in Lie Sphere Geometry*, Lecture Notes in Mathematics 2375, https://doi.org/10.1007/978-3-031-95592-1_1

1.2 Background

To understand why we focus on the case of isothermic surfaces in Lie sphere geometry to achieve our primary purpose, we must first consider the historical and mathematical background of three important key concepts:

1. discretization,
2. isothermic surfaces, and
3. Lie sphere geometry.

Discretization The central theme of these notes is the ***structure-preserving discretization*** of surfaces, which lie at the cross roads between the fields of differential geometry and integrable systems. The proper introduction to the concept of structure-preserving discretizations requires one to examine the starting point of integrable systems, which takes us back to the latter half of the nineteenth century. This period saw the key ingredients for integrable systems identified via the development in the transformation theory of surfaces, with the seminal example being the class of surfaces with constant negative Gaussian curvature, also referred to as K-surfaces or *pseudospherical surfaces*, whose history we briefly review below.

In the examination of pseudospherical surfaces, Bour showed that the integrability condition for pseudospherical surfaces amounts to the angle between asymptotic coordinates satisfying the now celebrated *sine-Gordon equation* [31, Equation (12)]. Then applying Bonnet's result on constant mean curvature (cmc) surfaces [28], Lie mentioned in [154, p. 510] that pseudospherical surfaces also admit a 1-parameter family of associated surfaces, also known as *Lie transformation*.

On the other hand, Bianchi constructed a method to create a new pseudospherical surface from a given one in [9], referred to as Bianchi transformation, with Darboux providing the related partial differential equations (PDEs) in terms of the solutions to the sine-Gordon equation in[1] [83, Equation (5)]. Bäcklund then generalized this method in [6] introducing the *spectral parameter* of the transformation, defining *Bäcklund transformations* for pseudospherical surfaces. It was Bianchi this time who derived the *Bianchi PDEs* for Bäcklund transformations in terms of the solutions to sine-Gordon equations in [11, Equation (33)].

The interpretation of the transformations in terms of the solutions to the sine-Gordon equation was crucial for obtaining the *permutability* between the transformations. In [155], Lie showed that the three transformations, Lie transformation, the transformation of Bianchi, and Bäcklund transformation, are intertwined: Symbolically representing the three representations by L_σ, B_0 and B_σ, respectively following Bianchi, Lie noticed that

$$B_\sigma = L_\sigma B_0 L_{-\sigma}.$$

---

[1] In the same work, Darboux also mentions that the construction of Bianchi is a special case of cyclic systems of Ribaucour [187].

Bianchi then discovered that the Bäcklund transformations *permute* in [12, Equation (11)]; namely, starting with a pseudospherical surface f and two Bäcklund transforms f_1 and f_2 with respective spectral parameter σ_1 and σ_2, Bianchi showed that there exists a fourth pseudospherical surface f_{12} that is simultaneously a Bäcklund transform of f_1 and f_2 with switched spectral parameter. Symbolically, this amounts to

$$B_{\sigma_1} B_{\sigma_2} = B_{\sigma_2} B_{\sigma_1},$$

and this phenomenon is now often referred to as *Bianchi permutability*. Remarkably, the permutability removes the need to solve the PDEs to obtain the fourth surface; instead, the fourth surface can be found *algebraically* from the first three surfaces.

Thus the works of classical geometers on pseudospherical surfaces laid the blueprint for the following ingredients of a system (with the corresponding elements of the pseudospherical surface theory written in parentheses):

- *non-linear PDEs* (sine-Gordon equation),
- *spectral deformation* (Lie transformation),
- *transformation with a parameter* (Bäcklund transformation), and
- *permutability* (Bianchi permutability),

features we now commonly associate with *integrable systems*.

Many of the systems arising in various physical applications governed by certain non-linear PDEs share the common ingredients, and exhibit some *integrable structure*: Korteweg-de Vries (KdV) equation, modified KdV equation, non-linear Schrödinger equation, and Painlevé equations, to name a few. For example, the KdV equation was originally given in [32, 33, 143] to model solitary waves in fluid dynamics as first observed by Russell [192], and appeared again in the context of magnetohydrodynamic waves [224] where, due to the behavior of the solitary waves resembling that of a particle, the term *soliton* was coined. Due to their close ties with physical models, methods for finding solutions to these systems have received voluminous interest, fueling the modernization of the classical methods in the field of integrable systems, including the *Lax pair* [150], *inverse scattering transformations* [107, 108], and *dressing transformations* [225].

It was also in the field of integrable systems where the ethos of integrable discretization was first conceived. Discretizing methods and equations of integrable systems had received numerous interest, as evidenced by works that discretize the inverse scattering method [61, 62], KdV equation [122], and sine-Gordon equation [123], for example. However, it was the work [152] on the multichannel Schrödinger spectral problem that proposed a simple yet beautiful method for discretizations. Noting that the process of transformations is *discrete* in nature so that the solutions obtained by successive transformations can be viewed as a system with smooth parameters of the object and a discrete parameter of the transformation, called a *semi-discrete system*, the work suggested that the transformations of a system should yield a semi-discrete analogue of the system. Building on this approach, the work [169] suggested that the permutability of transformations, resulting in multi-dimensional lattices, be interpreted as a discrete analogue of the system. Since then, these principles have been heavily used in various settings, and have cemented themselves as the central tenet of *integrable discretization: the semi-discrete and discrete analogues of an integrable system stem from the transformations and permutabilities of the smooth system, respectively.*

Let us explain this in a bit more detail, schematically, using the example of pseudospherical surfaces. Representing a Bäcklund transformation f_1 of a pseudospherical surface f with spectral parameter σ_1 with an arrow as

$$f \xrightarrow{\ \sigma_1\ } f_1,$$

the permutability says

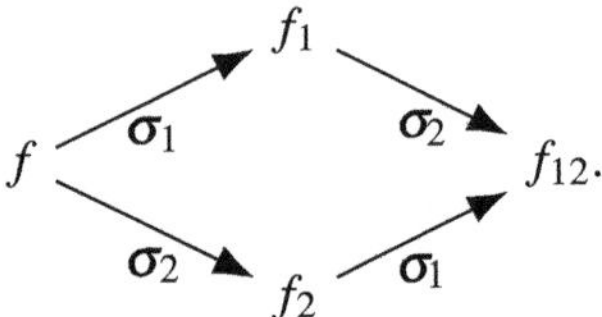

Therefore, one can consider the following lattice given by permutabilities:

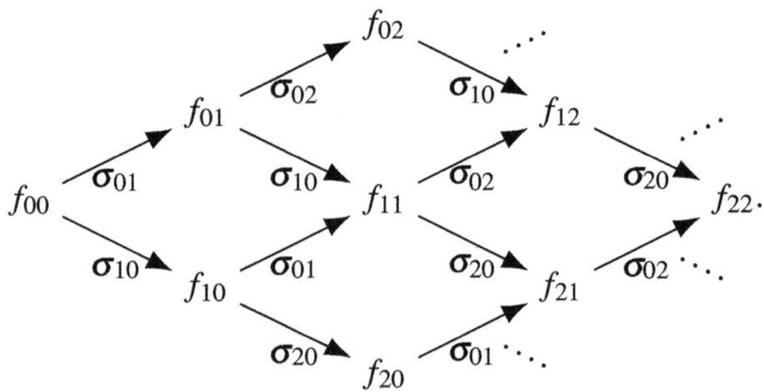

Then the corresponding points on the surfaces in the lattice constitute a discrete surface comprised of polygons, represented here schematically:

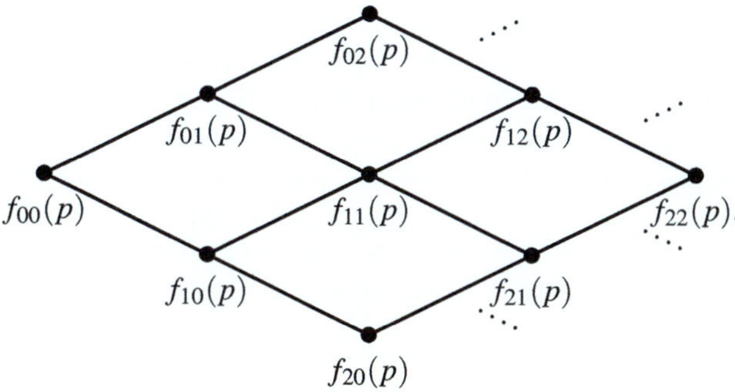

It is easily shown that this discrete surface coincides with the geometrically constructed discrete pseudospherical surface of Sauer and Wunderlich [196, 220]. In the pioneering work [23], it is further shown that the angle between the discrete edges must satisfy the previously found discrete sine-Gordon equation [123] as in the smooth case, showing the compatibility between discrete surface theory and integrable systems.

The work on discrete pseudospherical surfaces [23] is one of the first two examples exhibiting the characteristics of *discrete differential geometry*; namely, discretized surfaces should possess a mathematical structure rivaling that of the smooth counterpart. The other early example of integrable discretization of surfaces is given in the class of **discrete isothermic surfaces** [22].

Isothermic Surfaces It is well-known that locally, a surface always admits conformal coordinates or curvature line coordinates away from umbilics. However, only a certain class of surfaces admit conformal curvature line coordinates away from umbilics; such surfaces are called **isothermic surfaces**. First examined by Bour in [31, §54], many well-known classes of surfaces such as surfaces of revolution, quadrics, minimal surfaces, or constant mean curvature (cmc) surfaces in Euclidean space belong to the class of isothermic surfaces.

Thus, classical geometers were deeply interested in the class of isothermic surfaces, developing a rich transformation theory rivaling that of the pseudospherical surfaces. En route to studying minimal surfaces, Christoffel characterized isothermic surfaces via the existence of a certain dual surface called the *Christoffel dual* or the *Christoffel transformation* in [76, p. 221]. In [84, 85], Darboux showed that one can obtain a new isothermic surface from a given one so that they are conformally equivalent, and envelop a common sphere congruence[2] along curvature lines. This process involves a choice of a spectral parameter, and the transformation is now commonly referred to as the *Darboux transformation* of isothermic surfaces. On the other hand, Bianchi [7, 10] and Calapso [57, 58] recovered another type of transformation for isothermic surfaces, now referred to as *T-transformation* or *Calapso transformation*.

---

[2] We will explain this notion shortly.

Various permutability theorems were also found by the classical geometers: Bianchi showed in [10, §5] that Darboux transformations admit permutability akin to the permutability of Bäcklund transformations. Namely, given an isothermic surface f, and two Darboux transforms f_1 and f_2 with respective spectral parameters m_1 and m_2, there exists a fourth isothermic surface f_{12} that is a common Darboux transform of both f_1 and f_2 with the spectral parameters switched. Demoulin then provided a simple relationship between the four surfaces and the spectral parameters in terms of cross-ratios in [90]. Then Bianchi further showed in [7, 10] that the Calapso transformations intertwine the Christoffel transformations and Darboux transformations.

Therefore, the classical geometers uncovered the following features of the class of isothermic surfaces:

- spectral deformation via Calapso transformation,
- transformation with a parameter via Darboux transformation, and
- permutability of transformations.

The work [77] noticed these classical results, and showed that the class of isothermic surfaces constitute an integrable system, rekindling a deep interest in that class. Following the ethos of integrable discretizations, *discrete isothermic surfaces* were defined via the permutability of Darboux transformations of isothermic surfaces in [22, Definition 1] using the cross ratios condition found by Demoulin [90]. Remarkably, so defined discrete isothermic surfaces exhibit many of the properties enjoyed by the smooth counterpart; a characterization via Christoffel duality [22, Theorem 6], existence of a spectral deformation [113, Definition 3.11], and Darboux transformation involving a spectral parameter [114, § 4].

A notable approach to isothermic surfaces is provided via the observation that conformal curvature line coordinates of a surface are invariant under Möbius transformations, the transformation group generated by inversions with respect to spheres. Thus, Möbius transformations map spheres to spheres, where planes are regarded as spheres with infinite radius. In fact, Cartan showed in [60] that isothermic surfaces are the second order deformable class of surfaces in *Möbius geometry*, or *conformal geometry*, the geometry defined by the Möbius transformations. Therefore, Möbius geometry presents itself as the natural setting for the examination of the integrable structure of isothermic surfaces. Many modern works including [36, 46, 111, 116, 117] reinterpret and expand upon the classical results on the integrable structure of isothermic surfaces using Möbius geometry.

An interesting perk of using Möbius geometry is the fact that the well-studied Riemannian geometries including Euclidean geometry, spherical geometry, and hyperbolic geometry are all subgeometries of Möbius geometry. Otherwise said, the well-studied Riemannian space forms with constant sectional curvature, including Euclidean 3-space, 3-sphere, and hyperbolic 3-space, are all identified as the *conformal 3-sphere* in Möbius geometry, paving the way for uniformly treating all such space forms. Thus, the study of isothermic surfaces in Möbius geometry allows us to treat many well-known surface classes including minimal and cmc surfaces in all Riemannian space forms with constant sectional curvature in a uniform manner.

Here lies the advantage of discretizing isothermic surfaces in the context of Möbius geometry. By obtaining the integrable structure of isothermic surfaces in Möbius geometry, one can use the aforementioned permutability of Darboux transforms to define *discrete isothermic surfaces* in the conformal 3-sphere, which naturally subsumes the definitions of discrete isothermic surfaces in each Riemannian space form. Furthermore, well-known surface classes including minimal and cmc surfaces in a Riemannian space form are contained in the class of isothermic surfaces, so that the definitions of discrete minimal and cmc surfaces may be obtained via integrable reductions within the class of discrete isothermic surfaces.

In discretization, *it is often advantageous to discretize a larger class of surfaces (such as isothermic surfaces) in a higher geometry (such as Möbius geometry).*

Lie Sphere Geometry How, then, does Lie sphere geometry fit into the theory of isothermic surfaces? Recall that Möbius transformations maps spheres to spheres (viewing planes as infinite radius spheres), and points to points. Thus, points are preserved under Möbius transformations, making them distinguished geometric objects in Möbius geometry, and a surface in the conformal 3-sphere can be viewed as a map into the set of points.

Lie sphere geometry is a higher geometry than Möbius geometry, that is, Möbius geometry is a subgeometry of Lie sphere geometry. Thus every Möbius transformation is a Lie sphere transformation, but not every Lie sphere transformation is a Möbius transformation, as Lie sphere transformations merely map spheres to spheres (including points and planes), while preserving oriented contact. For example, let us take a sphere as a surface (in Euclidean 3-space) with inward pointing normals. Parallel surfaces of the sphere then consist of concentric spheres of varying radii; in fact, the point of the center is also a parallel surface of a sphere. Parallel surfaces are examples of Lie sphere transforms; thus, as opposed to Möbius transformations, Lie sphere transformations can map spheres into points.

Lie sphere transformations make no distinction between points (now viewed as zero radius spheres) and spheres (including planes); thus, when we refer to the notion of "spheres" in Lie sphere geometry, it is understood that spheres includes the notions of "points" and "planes" as well. Now that spheres are the distinguished geometric object in Lie sphere geometry, surfaces in Lie sphere geometry are now viewed as maps into the set of spheres. Such maps are called *sphere congruences*. As such, one can also consider sphere congruences that are *isothermic* in Lie sphere geometry; most notably, isothermic sphere congruences in Lie sphere geometry exhibit many integrable features of isothermic surfaces in Möbius geometry, including

- spectral deformations via Calapso transformations,
- transformations subject to a parameter via Darboux transformations, and
- permutability of transformations.

To consider surfaces in space form geometries in the context of Lie sphere geometry without the notion of points, we recall the fact that Lie sphere transformations preserve the oriented contact between two spheres. Now, given two

Fig. 1.1 Two spheres in oriented contact and a contact element

Fig. 1.2 Contact elements as Legendre lift of a surface (on the left); an enveloping sphere congruence of a surface (on the right)

spheres in oriented contact, one can recover a 1-parameter family of spheres that are pairwise in oriented contact (see Fig. 1.1); such pencil of spheres is called a *parabolic sphere pencil*. Thus, Lie sphere transformations map parabolic sphere pencils to parabolic sphere pencils. Once a certain space form geometry is chosen, one can then identify within a given parabolic sphere pencil the point of contact and the (geodesic) plane tangent to all the spheres in the parabolic sphere pencil. In other words, the information regarding contact is encoded in a parabolic sphere pencil; for this reason, a parabolic sphere pencil is referred to as a *contact element*.

Therefore, given a surface in some space form geometry, one can use a point on the surface and the (geodesic) plane tangent to that point to recover the contact element; this way, a surface in a space form geometry can be viewed as a map into the set of contact elements in Lie sphere geometry, called a *Legendre lift* of the given surface (see Fig. 1.2). In every contact element is a 1-parameter family of spheres; thus, choosing a sphere from each contact element, one can obtain a sphere congruence. In particular, since every sphere is chosen from a contact element, each sphere is in contact with the surface; such sphere congruences are said to be *enveloping* a surface (see Fig. 1.2).

Special attention is reserved for those surfaces in space forms that admit enveloping isothermic sphere congruences: these surfaces are called Ω-*surfaces*, first introduced by Demoulin in [87–89]. The class of Ω-surfaces includes a wide range of well-studied surface classes such as:

- minimal and constant mean curvature surfaces (examples of isothermic surfaces),
- pseudospherical surfaces (examples of Guichard surfaces), and
- linear Weingarten surfaces of Bryant-type (examples of L-isothermic surfaces).

Ω-surfaces constitute an integrable class of surfaces and exhibit many characteristics of an integrable system; remarkably, their integrable structure is inherited by those of the enveloping isothermic sphere congruences. Thus, the well-known features of an integrable system,

- spectral deformations,
- transformations subject to a parameter, and
- permutability of transformations

of Ω-surfaces are induced by those of the enveloping isothermic sphere congruences. Therefore, the consideration of Ω-surfaces allows the rich theory of isothermicity to be used to study the integrable structure of a wide-range of surfaces in a uniform fashion, including but not limited to isothermic surfaces. In particular, the examination of integrable structure of Ω-surfaces leads to the discretization of Ω-surfaces, allowing for a uniform discretization of a wide-range of surfaces classes, including:

- discrete minimal and discrete constant mean curvature surfaces,
- discrete pseudospherical surfaces, and
- discrete linear Weingarten surfaces of Bryant-type.

1.3 Approach

As the primary purpose of this text is to provide an accessible and approachable introduction to the concept of isothermicity and their discretizations within the context of sphere geometries, these notes are created to imitate the learning process of a student: starting from the known and discovering the unknown.

In the case of sphere geometries under consideration, we start from the well-known notions of 3-dimensional space form geometry (including Euclidean, spherical, and hyperbolic geometry) and explicitly calculate how these concepts translate to the notions of sphere geometry. As we are fully committed to this philosophy, we introduce geometric concepts reaching beyond the scopes of surface theory in space forms only when necessary; even then, we will cover the bare minimum of these advanced concepts required to comprehend the machinery behind isothermicity in sphere geometries.

Our choice of approach may come at a price of sacrificing brevity and, in some cases, elegance of arguments; we hope that this text will provide the readers with a comprehensive grasp of the concepts, and ultimately lead to a deeper appreciation of those works that employ advanced machinery to demonstrate the mathematical beauty of sphere geometries. To assist the readers in finding these works, we have provided a selection of recommended further readings at the end of selected sections with short descriptions of what each work does, where the lists include both historically important texts as well as cutting-edge modern day research articles.

These notes have been structured to achieve the stated purpose; this can be seen in the considerable amount of attention given to Möbius geometry. Möbius geometry stands as a bridge between space form geometries and Lie sphere geometry, as it is a geometry of spheres where points are the distinguished objects. Thus close examination of Möbius geometry allows us to introduce sphere geometric notions in tandem with space form geometric analogues.

Möbius geometry is especially apt for introducing isothermicity and its integrable structure. Isothermic surfaces can be defined in space forms as those surfaces admitting conformal curvature line coordinates; however, their most natural consideration comes within the context of Möbius geometry, as isothermicity is a Möbius invariant notion. Therefore, we will build from the space form geometric characterizations of isothermic surfaces to recover a Möbius geometric characterization of isothermicity through the gauge-theoretic approach, where the tools of Möbius geometry such as the notion of cross-ratios play a crucial role. Careful consideration of isothermicity in the context of Möbius geometry will allow us to reconcile the gauge-theoretic characterizations with space form geometric characterizations of isothermicity. The gauge-theoretic approach to isothermicity will then allow for a rapid generalization to the case of isothermic sphere congruences in Lie sphere geometry.

As the structure-preserving discretization process comes directly from integrability, we will also devote a large part of the notes to considering the class of discrete isothermic surfaces in Möbius geometry. Many of the advantages of introducing isothermicity in the context of Möbius geometry naturally apply to the examination of discrete isothermic surfaces, as the familiar space form notions will serve to strengthen the gauge-theoretic definitions of discrete isothermicity. Again, the gauge-theoretic description of discrete isothermicity in Möbius geometry will naturally extend to that in Lie sphere geometry.

Notably, we will exclude umbilics in our consideration of isothemicity and focus on the local theory, because umbilics introduce complications in the transformation theory. By avoiding the additional difficulties posed by umbilics, we can thoroughly introduce the fundamental aspects of the transformation theory of isothermic surfaces, laying the groundwork for the notion of discrete isothermicity.

In short, these notes should be regarded as a first dictionary on the integrable structure of isothermicity in the context of sphere geometries, translating space form geometric notions to truly sphere geometric notions. A thorough understanding of the local theory of isothermicity and its integrability will leave the readers able to explore the frontiers of the field, including global theory and generalizations to broader contexts.

1.4 Prerequisites and Intended Audience

As reflected in the approach, the prerequisites to read these notes is a good understanding of surface theory, typically covered by one or two semesters of differential geometry in an undergraduate curriculum. Differential geometers with seminal interest in sphere geometries will find the text helpful; any graduate student and even highly motivated undergraduate students with particular interest in differential geometry will have little to no difficulty following this text.

Chapter 2
Isothermic Surfaces in Möbius Geometry

2.1 3-dimensional Space Forms

We first describe the ambient spaces that we will be using in this text.

Euclidean 3-space $\mathbb{R}^3$ Euclidean 3-space is

$$\mathbb{R}^3 = \{(x_1, x_2, x_3)^t \mid x_j \in \mathbb{R}\},$$

where the superscript "t" denotes transposition. The standard Euclidean metric g with respect to these standard rectangular coordinates (x_1, x_2, x_3) is

$$g_3 = \mathrm{d}x_1^2 + \mathrm{d}x_2^2 + \mathrm{d}x_3^2,$$

or equivalently, in matrix form g_3 is the 3×3 identity matrix. Because the entries g_3 are all constant, the Christoffel symbols are all zero, and so the sectional curvature of $\mathbb{R}^3$ is identically zero. So $\mathbb{R}^3$ is a simply-connected complete 3-dimensional manifold with constant sectional curvature zero, and in fact it is the unique such manifold. (Here, completeness means that all geodesics extend to infinite length within the manifold.)

The set of isometries of $\mathbb{R}^3$ is a group under the composition operation, and is generated by

1. translations,
2. reflections across planes containing the origin $(0, 0, 0)^t$, and
3. rotations fixing the origin.

Isometries of $\mathbb{R}^3$ are also called rigid motions of $\mathbb{R}^3$. The reflections and rotations fixing the origin can be described by left multiplication (i.e. multiplication on the left) of $(x_1, x_2, x_3)^t$ by matrices in the orthogonal group

$$\mathrm{O}_3 = \{A \in M_{3\times 3} \mid A^t A = I_3\} = \{A \in M_{3\times 3} \mid x \cdot x = Ax \cdot Ax \text{ for all } x \in \mathbb{R}^3\},$$

© The Author(s), under exclusive license to Springer Nature Switzerland AG 2025
J. Cho et al., *Discrete Isothermic Surfaces in Lie Sphere Geometry*, Lecture Notes in Mathematics 2375, https://doi.org/10.1007/978-3-031-95592-1_2

where $\cdot$ denotes the standard inner product of $\mathbb{R}^3$, $M_{n \times n}$ denotes the set of all $n \times n$ matrices, and I_n is the $n \times n$ identity matrix.

Spherical 3-space $\mathbb{S}^3$ We define the *spherical 3-space* $\mathbb{S}^3$ as

$$\mathbb{S}^3 = \left\{ (x_1, x_2, x_3, x_4)^t \in \mathbb{R}^4 \,\middle|\, \sum_{j=1}^{4} x_j^2 = 1 \right\}.$$

The metric $g_{\mathbb{S}^3}$ on $\mathbb{S}^3$ can be defined by restricting the metric $g_4 = \mathrm{d}x_1^2 + \mathrm{d}x_2^2 + \mathrm{d}x_3^2 + \mathrm{d}x_4^2$ to the 3-dimensional tangent spaces of $\mathbb{S}^3$, and then the form of $g_{\mathbb{S}^3}$ (but not $g_{\mathbb{S}^3}$ itself) will depend on the choice of coordinates we make for $\mathbb{S}^3$. One of the ways to achieve this is to use the *stereographic projection*.

In this book, we will consider stereographic projection from $(0, 0, 0, -1)^t$, which is a one-to-one mapping of $\mathbb{S}^3 \setminus \{(0, 0, 0, -1)^t\}$ to the 3-dimensional vector space

$$\{(x_1, x_2, x_3, 0)^t \in \mathbb{R}^4\}$$

lying in $\mathbb{R}^4$. Since the last coordinate is identically zero, we can remove it and then view this 3-dimensional subspace of $\mathbb{R}^4$ simply as a 3-dimensional Euclidean space. More explicitly, stereographic projection $\mathrm{Pr} : \mathbb{S}^3 \setminus \{(-1, 0, 0, 0)^t\} \to \mathbb{R}^3$ is given by

$$\mathrm{Pr}((x_1, x_2, x_3, x_4)^t) = \frac{1}{1 + x_4}\, (x_1, x_2, x_3)^t,$$

and its inverse map $\sigma_1 := \mathrm{Pr}^{-1}$ is

$$\sigma_1((x_1, x_2, x_3)^t) = \frac{1}{1 + x_1^2 + x_2^2 + x_3^2} (2x_1, 2x_2, 2x_3, 1 - x_1^2 - x_2^2 - x_3^2)^t. \qquad (2.1)$$

Exercise 2.1.1 By viewing σ_1 as a coordinate patch, compute that the metric $g_{\mathbb{S}^3}$ of $\mathbb{S}^3$ is

$$g_{\mathbb{S}^3} = \left(\frac{2}{1 + x_1^2 + x_2^2 + x_3^2} \right)^2 (\mathrm{d}x_1^2 + \mathrm{d}x_2^2 + \mathrm{d}x_3^2).$$

Note that the metric $g_{\mathbb{S}^3}$ of $\mathbb{S}^3$ is certainly not the Euclidean metric g_3, but is conformally equivalent to the Euclidean metric. Therefore, the angles between vectors in the tangent spaces are the same from the viewpoints of both the spherical and Euclidean metrics. From this metric, one can compute directly that the sectional curvature must be 1. (See, for example, [29].)

The isometries of $\mathbb{S}^3$ are the restrictions of the rotations and reflections of $\mathbb{R}^4$ fixing the origin $(0, 0, 0, 0)^t$ to $\mathbb{S}^3$. Thus the isometry group of $\mathbb{S}^3$ is represented by

$$\mathrm{O}_4 = \{A \in M_{4\times4} \mid A^t A = I_4\} = \{A \in M_{4\times4} \mid \langle x, x\rangle_4$$

$$= \langle Ax, Ax\rangle_4 \text{ for any } x \in \mathbb{R}^4\},$$

where $\langle \ , \ \rangle_4$ denotes the standard Euclidean inner product of $\mathbb{R}^4$.

Spheres of $\mathbb{S}^3$ are "hyperplanar slices" of $\mathbb{S}^3$ in $\mathbb{R}^4$, which we formulate as follows: recall that any point $(x_1, x_2, x_3, x_4)^t$ in a 3-dimensional hyperplane in $\mathbb{R}^4$ can be described via the equation

$$m_1 x_1 + m_2 x_2 + m_3 x_3 + m_4 x_4 = q$$

for some $m_1, m_2, m_3, m_4, q \in \mathbb{R}$. Rewriting the above equation, we have

$$\langle x, m\rangle_4 = q$$

for $m = (m_1, m_2, m_3, m_4)^t$ and $x = (x_1, x_2, x_3, x_4)^t$. Therefore, we may consider spheres in the 3-sphere as

$$\tilde{S}_{\mathbb{S}^3}[m, q] := \{x \in \mathbb{S}^3 \subset \mathbb{R}^4 \mid \langle x, m\rangle_4 = q\}, \tag{2.2}$$

provided that the set is non-empty, a condition we can characterize as follows.

Exercise 2.1.2 $\tilde{S}_{\mathbb{S}^3}[m, q]$ is non-empty (and includes more than one point) if and only if $\langle m, m\rangle_4 > q^2$.

Remark 2.1.3 In the case of $q = 0$, we have that $\tilde{S}_{\mathbb{S}^3}[m, 0]$ is a geodesic plane. Note that in the choice of m and q, we have the freedom of scaling, i.e. they are only projectively defined.

Minkowski 3-space $\mathbb{R}^{2,1}$ and 4-space $\mathbb{R}^{3,1}$ Minkowski 3-space $\mathbb{R}^{2,1}$ is defined as

$$\{(x_0, x_1, x_2)^t \mid x_j \in \mathbb{R}\},$$

endowed with the non-Euclidean metric

$$g_{2,1} = -\mathrm{d}x_0^2 + \mathrm{d}x_1^2 + \mathrm{d}x_2^2,$$

or equivalently, in matrix form

$$g_{2,1} = \begin{pmatrix} -1 & 0 & 0 \\ 0 & 1 & 0 \\ 0 & 0 & 1 \end{pmatrix}.$$

With this metric we have that the Minkowski 3-space is a Lorentzian manifold.

Similarly, Minkowski 4-space is the 4-dimensional vector space

$$\{(x_0, x_1, x_2, x_3)^t \mid x_j \in \mathbb{R}\}$$

endowed with the metric

$$g_{3,1} = -dx_0^2 + dx_1^2 + dx_2^2 + dx_3^2,$$

or equivalently, in matrix form

$$g_{3,1} = \begin{pmatrix} -1 & 0 & 0 & 0 \\ 0 & 1 & 0 & 0 \\ 0 & 0 & 1 & 0 \\ 0 & 0 & 0 & 1 \end{pmatrix}.$$

For these two manifolds, the entries in the metrics g are all constant; hence, $\mathbb{R}^{2,1}$ and $\mathbb{R}^{3,1}$ have constant sectional curvature 0.

Denoting the inner product of $\mathbb{R}^{n,1}$ with respect to the metric $g_{n,1}$ by $\langle \ , \ \rangle_{n,1}$ for $n = 2, 3$, the isometries of $\mathbb{R}^{2,1}$, respectively $\mathbb{R}^{3,1}$, that fix the origin can be described by the matrix group

$$O_{2,1} = \{A \in M_{3\times3} \mid A^t I_{2,1} A = I_{2,1}\}$$

$$= \{A \in M_{3\times3} \mid \langle x, x\rangle_{2,1} = \langle Ax, Ax\rangle_{2,1} \text{ for all } x \in \mathbb{R}^{2,1}\},$$

respectively,

$$O_{3,1} = \{A \in M_{4\times4} \mid A^t I_{3,1} A = I_{3,1}\}$$

$$= \{A \in M_{4\times4} \mid \langle x, x\rangle_{3,1} = \langle Ax, Ax\rangle_{3,1} \text{ for all } x \in \mathbb{R}^{3,1}\},$$

where $I_{n,1}$ is described as

$$n \left\{ \begin{pmatrix} -1 & & \overbrace{}^{n} & \\ & 1 & & \\ & & \ddots & \\ & & & 1 \end{pmatrix} \right. .$$

The group $O_{2,1}$, respectively $O_{3,1}$, is called the orthogonal group of $\mathbb{R}^{2,1}$, respectively $\mathbb{R}^{3,1}$. We are interested in these particular isometries because they can be used to describe the isometries of hyperbolic 3-space $\mathbb{H}^3$, as we are about to see.

Hyperbolic 3-space $\mathbb{H}^3$ We now describe *hyperbolic 3-space*, a 3-dimensional Riemannian manifold with constant sectional curvature -1. However, it can be described by a variety of models, each with their own advantages. Here we describe the models we will need: the Poincaré ball model, and the Hermitian matrix model. We derive these models from the *Minkowski model*.

We define $\mathbb{H}^3$ by way of the Minkowski 4-space $\mathbb{R}^{3,1}$ with its Lorentzian metric $g_{3,1}$. Denote the upper and lower sheet of the two-sheeted hyperboloid by $\mathbb{H}^3_+$ and $\mathbb{H}^3_-$, that is,

$$\mathbb{H}^3_+ = \left\{ x = (x_0, x_1, x_2, x_3)^t \in \mathbb{R}^{3,1} \,\middle|\, \langle x, x \rangle_{3,1} = -1, x_0 > 0 \right\}$$

$$\mathbb{H}^3_- = \left\{ x = (x_0, x_1, x_2, x_3)^t \in \mathbb{R}^{3,1} \,\middle|\, \langle x, x \rangle_{3,1} = -1, x_0 < 0 \right\},$$

and take two copies of the hyperbolic space as $\mathbb{H}^3 := \mathbb{H}^3_+ \cup \mathbb{H}^3_-$. The metric $g_{\mathbb{H}^3}$ is given by the restriction of $g_{3,1}$ to the tangent spaces of this 3-dimensional object. Although the metric $g_{3,1}$ is Lorentzian and therefore not positive definite, the restriction g to each hyperbolic space is positive definite, so $\mathbb{H}^3$ is a Riemannian manifold, as we will see using the Poincaré ball model.

As in the case of $\mathbb{S}^3$, the isometry group of $\mathbb{H}^3$ can be described using the matrix group $O_{3,1}$. We now give two other models for $\mathbb{H}^3$.

The Poincaré Ball Model   By stereographic projection from the point $(-1, 0, 0, 0)^t \in \mathbb{R}^{3,1}$ of the Minkowski model for $\mathbb{H}^3$, one has the *Poincaré ball model* $\mathcal{P}$ for $\mathbb{H}^3$, where the stereographic projection is given by

$$\mathbb{H}^3 \ni (x_0, x_1, x_2, x_3)^t \mapsto \left(\frac{x_1}{x_0 + 1}, \frac{x_2}{x_0 + 1}, \frac{x_3}{x_0 + 1} \right)^t \in \mathcal{P} \tag{2.3}$$

with the inverse σ_{-1} defined as

$$\mathcal{P} \ni (x_1, x_2, x_3)^t \mapsto \frac{1}{1 - x_1^2 - x_2^2 - x_3^2} (1 + x_1^2 + x_2^2 + x_3^2, 2x_1, 2x_2, 2x_3)^t \in \mathbb{H}^3. \tag{2.4}$$

Exercise 2.1.4   Show that we can view the Poincaré ball model of $\mathbb{H}^3_+$ as the Euclidean unit ball

$$B^3 = \{ (x_1, x_2, x_3)^t \in \mathbb{R}^3 \mid x_1^2 + x_2^2 + x_3^2 < 1 \}$$

in $\mathbb{R}^3 = \{ (x_1, x_2, x_3)^t \in \mathbb{R}^3 \}$.

Exercise 2.1.5   Viewing the inverse stereographic projection (2.4) as a coordinate chart for $\mathbb{H}^3$, compute that the metric of $\mathbb{H}^3$ is

$$g_{\mathbb{H}^3} = \left(\frac{2}{1 - x_1^2 - x_2^2 - x_3^2}\right)^2 (dx_1^2 + dx_2^2 + dx_3^2). \tag{2.5}$$

This metric $g_{\mathbb{H}^3}$ in (2.5) is written as a function times the Euclidean metric $dx_1^2 + dx_2^2 + dx_3^2$, thus the Poincaré ball model's metric is conformal to the Euclidean metric.

However, distances are clearly not Euclidean. In fact, the boundary

$$\partial B^3 = \{(x_1, x_2, x_3)^t \in \mathbb{R}^3 \mid x_1^2 + x_2^2 + x_3^2 = 1\}$$

of the Poincaré ball model is infinitely far from any point in B^3 with respect to the hyperbolic metric $g_{\mathbb{H}^3}$ in (2.5). For example, consider the curve

$$c(s) = (s, 0, 0)^t, \quad s \in [0, 1)$$

in the Poincaré ball model. Its length is

$$\int_0^1 \sqrt{g_{\mathbb{H}^3}(c'(s), c'(s))} ds = \int_0^1 \frac{2ds}{1 - s^2} = +\infty.$$

Thus the point $(0, 0, 0)^t$ is infinitely far from the boundary point $(1, 0, 0)^t$ in the Poincaré ball model. For this reason, the boundary ∂B^3 is often called the *ideal boundary at infinity*.

Spheres in $\mathbb{H}^3$ are given similarly to those in $\mathbb{S}^3$, as "hyperplanar slices" in $\mathbb{R}^{3,1}$: given a vector $m = (m_0, m_1, m_2, m_3)^t \in \mathbb{R}^{3,1}$ and $q \in \mathbb{R}$, we let

$$\tilde{S}_{\mathbb{H}^3}[m, q] := \{x \in \mathbb{H}^3 \mid \langle x, m \rangle_{3,1} = -q\} \tag{2.6}$$

be a sphere.

Exercise 2.1.6 Show that $\tilde{S}_{\mathbb{H}^3}[m, q]$ is non-empty (and includes more than one point) if and only if $\langle m, m \rangle_{3,1} + q^2 > 0$.

Spheres in $\mathbb{H}^3$ come in three different types, which can be characterized via m as follows:

- *Bounded spheres* arise when $\langle m, m \rangle_{3,1} < 0$. These appear as genuine spheres within the Poincaré ball model.
- *Horospheres* appear when $\langle m, m \rangle_{3,1} = 0$. These appear as spheres tangent to the ideal boundary in the Poincaré ball model.
- *Unbounded spheres* arise when $\langle m, m \rangle_{3,1} > 0$, and appear as parts of spheres that intersect the ideal boundary transversally in the Poincaré ball model. If $q = 0$ (so that $\langle m, m \rangle_{3,1} > 0$), then we obtain a special type of unbounded spheres, called *geodesic planes*, that intersect the ideal boundary orthogonally in the Poincaré ball model.

Surfaces in 3-dimensional Space Forms Let us now treat surfaces in the 3-dimensional space forms.

Definition 2.1.7 A *surface* in a smooth manifold $\mathcal{N}$ refers to a smooth mapping from a 2-dimensional manifold $\mathcal{M}$ to the manifold $\mathcal{N}$. A surface is called *regular* if its differential map is injective.

We will always assume that a surface is regular throughout unless otherwise noted.

Let $x : \mathcal{M} \to \mathbb{R}^3$ be a surface. We will restrict our attention to a local coordinate patch $\Sigma \subset \mathcal{M}$ with coordinates (u, v). Thus, we assume without loss of generality that $x : \Sigma \to \mathbb{R}^3$. Then the unit normal to the surface x is given by

$$n = \frac{x_u \times x_v}{|x_u \times x_v|},$$

where $x_u = \frac{\partial}{\partial u} x$ and $x_v = \frac{\partial}{\partial v} x$ denote the partial derivatives of x with respect to u and v, respectively, and $|x| = \sqrt{\langle x, x \rangle_3}$. The *first fundamental form* of a surface x is

$$g = \begin{pmatrix} g_{11} & g_{12} \\ g_{21} & g_{22} \end{pmatrix} := \begin{pmatrix} x_u \cdot x_u & x_u \cdot x_v \\ x_v \cdot x_u & x_v \cdot x_v \end{pmatrix}, \tag{2.7}$$

while the *second fundamental form* is

$$b = \begin{pmatrix} b_{11} & b_{12} \\ b_{21} & b_{22} \end{pmatrix} := \begin{pmatrix} x_{uu} \cdot n & x_{uv} \cdot n \\ x_{vu} \cdot n & x_{vv} \cdot n \end{pmatrix}. \tag{2.8}$$

Now let $x : \Sigma \to \mathbb{S}^3$ be a surface in the 3-sphere. The normal $n : \Sigma \to \mathbb{S}^3$ is given by

$$n(u, v) \in T_{x(u,v)}\mathbb{S}^3 \quad \text{and} \quad \langle n, x_u \rangle_4 = 0 = \langle n, x_v \rangle_4$$

at each point $(u, v) \in \Sigma$, so that x, span $\{x_u, x_v\}, n$ are all perpendicular with respect to the inner product $\langle \ , \ \rangle_4$. The first fundamental form g and second fundamental form b for surfaces in the 3-sphere are then calculated the same way as in the Euclidean case, (2.7) and (2.8), respectively, where the inner product is now $\langle \ , \ \rangle_4$.

Finally, for a surface $x : \Sigma \to \mathbb{H}^3$, the normal $n : \Sigma \to \mathbb{S}^{2,1}$ is given similarly to that in the 3-sphere,

$$n(u, v) \in T_{x(u,v)}\mathbb{H}^3 \quad \text{and} \quad \langle n, x_u \rangle_{3,1} = 0 = \langle n, x_v \rangle_{3,1},$$

so that x, span $\{x_u, x_v\}, n$ are all perpendicular with respect to the inner product $\langle \ , \ \rangle_{3,1}$. Here,

$$\mathbb{S}^{2,1} = \{ n \in \mathbb{R}^{3,1} \mid \langle n, n \rangle_{3,1} = 1 \}.$$

Using the normal n, the first fundamental form g and second fundamental form b for surfaces in the 3-dimensional hyperbolic space are again calculated the same way as in the Euclidean case, (2.7) and (2.8), respectively, using the inner product $\langle \; , \; \rangle_{3,1}$.

In all cases, we see that for $n_u = \alpha x_u + \gamma x_v$ and $n_v = \beta x_u + \delta x_v$,

$$b = -\begin{pmatrix} \alpha g_{11} + \gamma g_{12} & \beta g_{11} + \delta g_{12} \\ \alpha g_{12} + \gamma g_{22} & \beta g_{12} + \delta g_{22} \end{pmatrix} = -g \begin{pmatrix} \alpha & \beta \\ \gamma & \delta \end{pmatrix},$$

where we used the fact that $g_{12} = g_{21}$. Therefore, we have

$$\begin{pmatrix} \alpha & \beta \\ \gamma & \delta \end{pmatrix} = -g^{-1}b.$$

The eigenvectors of $g^{-1}b$ gives us the principal curvature directions, and the eigenvalues the principal curvatures. Furthermore, the *mean curvature H* and the *Gaussian curvature K* can be found via

$$H = \frac{1}{2}\frac{g_{22}b_{11} - 2g_{12}b_{12} + g_{11}b_{22}}{g_{11}g_{22} - g_{12}^2} \quad \text{and} \quad K = \frac{b_{11}b_{22} - b_{12}^2}{g_{11}g_{22} - g_{12}^2}. \tag{2.9}$$

Given a surface without umbilics (i.e. a surface with distinct principal curvatures on Σ) in some space form of constant sectional curvature, one can always find *curvature line coordinates* $(u, v) \in \Sigma$, i.e.

$$g_{12} = 0 \quad \text{and} \quad b_{12} = 0,$$

or

$$x_u \perp x_v \quad \text{and} \quad x_{uv} \in \text{span}\{x_u, x_v\}$$

where the perpendicularity depends on the ambient space. On the other hand, one can also always find *conformal coordinates* $(u, v) \in \Sigma$, that is,

$$g_{11} = g_{22} \quad \text{and} \quad g_{12} = 0.$$

2.2   Minkowski Model of the Conformal 3-sphere

We begin by describing the 3-dimensional space forms using the 5-dimensional Minkowski space $\mathbb{R}^{4,1}$. Let

$$\mathbb{R}^{4,1} = \left\{ X = (x_0, x_1, x_2, x_3, x_4)^t \mid x_j \in \mathbb{R} \right\} \tag{2.10}$$

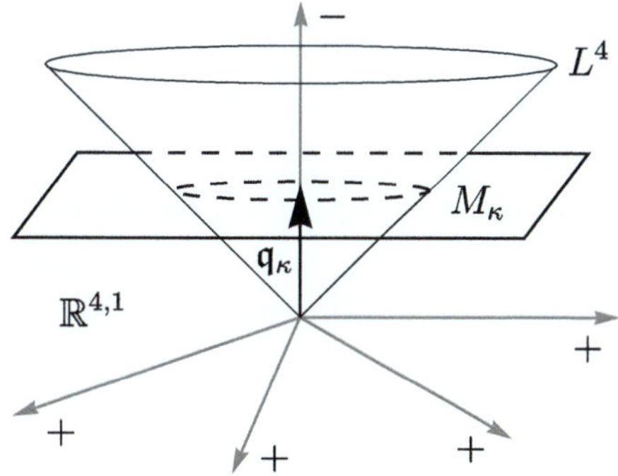

Fig. 2.1 The Möbius geometric model for 3-dimensional space forms M_κ

with indefinite metric $\langle\ ,\ \rangle$ of signature $(-++++)$, that is,

$$\langle X, Y \rangle = -x_0 y_0 + x_1 y_1 + x_2 y_2 + x_3 y_3 + x_4 y_4,$$

for $X = (x_0, x_1, x_2, x_3, x_4)^t$, $Y = (y_0, y_1, y_2, y_3, y_4)^t \in \mathbb{R}^{4,1}$, and we denote $\langle X, X \rangle$ by $\|X\|^2$. The 4-dimensional *light cone* is

$$L^4 = \{X \in \mathbb{R}^{4,1} \mid \|X\|^2 = 0\}.$$

We can make the 3-dimensional space forms as follows: A space form M_κ is

$$M_\kappa = \{X \in L^4 \mid \langle X, \mathfrak{q}_\kappa \rangle = -1\} \tag{2.11}$$

for any nonzero $\mathfrak{q}_\kappa \in \mathbb{R}^{4,1}$, which is often referred to as the *space form vector* (see Fig. 2.1). It will turn out (see the upcoming Lemma 2.2.12) that M_κ has constant sectional curvature κ, where $\kappa = -\|\mathfrak{q}_\kappa\|^2$, so without loss of generality we can obtain any space form by choosing

$$\mathfrak{q}_\kappa = \begin{pmatrix} \frac{1}{2}(\kappa + 1) \\ 0 \\ 0 \\ 0 \\ \frac{1}{2}(\kappa - 1) \end{pmatrix}. \tag{2.12}$$

M_κ is called a *quadric*, because it is the intersection of a hyperplane with the light cone, which is also a (non-singular) quadric (determined by the quadratic form $\|\ \|^2$).

Remark 2.2.8 There is no real computational advantage to restricting $\mathfrak{q}_\kappa$ to the form in (2.12). However, we frequently do this, as it gives us a convenient way to explicitly express 3-dimensional space forms, surfaces, normal vector fields, and such.

Let $\mathbb{R}^3 \cup \{\infty\}$ denote the one point compactification of

$$\mathbb{R}^3 = \{x = (x_1, x_2, x_3)^t \mid x_j \in \mathbb{R}\},$$

with $|x|^2 = x_1^2 + x_2^2 + x_3^2$, and define

$$\mathfrak{R}_\kappa := \{(x_1, x_2, x_3)^t \in \mathbb{R}^3 \cup \{\infty\} \mid |x|^2 \neq -\kappa^{-1}\}.$$

We can write

$$M_\kappa = \left\{ X = \frac{1}{1 + \kappa|x|^2} \begin{pmatrix} 1 + |x|^2 \\ 2x^t \\ 1 - |x|^2 \end{pmatrix} \, \middle| \, x \in \mathfrak{R}_\kappa \right\} \tag{2.13}$$

as the next lemma shows:

Lemma 2.2.9 *The map $\psi_\kappa : \mathfrak{R}_\kappa \to M_\kappa$ defined by*

$$\psi_\kappa(x) = \frac{1}{1 + \kappa|x|^2} \begin{pmatrix} 1 + |x|^2 \\ 2x^t \\ 1 - |x|^2 \end{pmatrix}$$

is a bijection. We often refer to $\psi_\kappa(x)$ as the lift *of x.*

Proof The fact that ψ_κ is well-defined and one-to-one can be proved easily. To show that it is a surjection, write $X = (y_0, y, y_4)^t \in M_\kappa$ with $y = (y_1, y_2, y_3)$.

Suppose that $y_0 \neq -y_4$. Since we have that $\langle X, X \rangle = 0$ and $\langle X, \mathfrak{q}_\kappa \rangle = -1$ for $\mathfrak{q}_\kappa$ as in (2.12), we can deduce

$$2 = (y_0 + y_4) - \kappa(y_4 - y_0)$$

$$2(y_0 + y_4) = (y_0 + y_4)^2 + \kappa|y|^2.$$

Setting $x = (y_0 + y_4)^{-1} y$, one can then directly calculate that

$$\frac{2}{y_0 + y_4} = 1 + \kappa|x|^2,$$

telling us

$$\psi_\kappa(x) = X.$$

If $y_0 = -y_4$, then $\|X\|^2 = 0$ implies $y = (0, 0, 0)^t$, and using $\langle X, \mathfrak{q}_\kappa \rangle = -1$ gives

$$y_0 = -y_4 = \kappa^{-1},$$

so this case occurs only when $\kappa \neq 0$. Set $x = \infty$ to get the desired conclusion. $\square$

Exercise 2.2.10 Since Lemma 2.2.9 shows that ψ_κ is a bijection, there is a well-defined $\phi_\kappa = \psi_\kappa^{-1} : M_\kappa \to \mathfrak{R}_\kappa$. Show that

$$\phi_\kappa(X) = \phi_\kappa((y_0, y_1, y_2, y_3, y_4)^t) = \frac{1}{y_0 + y_4}(y_1, y_2, y_3)^t.$$

Therefore, we see by Lemma 2.2.9 and Exercise 2.2.10 that ϕ_κ induces a coordinate chart for M_κ. Using this coordinate chart, we calculate the tangent space of M_κ as follows: let $X(t) : (-\epsilon, \epsilon) \to M_\kappa$ be a smooth function of a real variable t such that $X(0) = X$, and let $x(t) = \phi_\kappa(X(t)) : (-\epsilon, \epsilon) \to \mathfrak{R}_\kappa$ with $x(0) = x$. For $\prime$ denoting differentiation with respect to t, we can then use the coordinate chart to calculate that

$$X'(0) = \frac{2}{(1 + \kappa|x|^2)^2}\begin{pmatrix} (1 - \kappa)(x \cdot x') \\ (1 + \kappa|x|^2)(x')^t - 2\kappa(x \cdot x')x^t \\ (-1 - \kappa)(x \cdot x') \end{pmatrix},$$

where $x' = x'(0) \in T_x \mathfrak{R}_\kappa \cong \mathbb{R}^3$. Therefore, we express the tangent space of M_κ uniformly as

$$T_X M_\kappa = \left\{ \mathcal{T}_a = \frac{2}{(1 + \kappa|x|^2)^2}\begin{pmatrix} (1 - \kappa)(x \cdot a) \\ (1 + \kappa|x|^2)a^t - 2\kappa(x \cdot a)x^t \\ (-1 - \kappa)(x \cdot a) \end{pmatrix} \,\middle|\, a \in \mathbb{R}^3 \right\}. \tag{2.14}$$

A computation then gives

$$\langle \mathcal{T}_a, \mathcal{T}_b \rangle = \frac{4}{(1 + \kappa|x|^2)^2}(a \cdot b), \tag{2.15}$$

the metric induced on M_κ by the ambient space.

Exercise 2.2.11 Confirm (2.15) by using ψ_κ as a coordinate patch for M_κ and calculating the metric.

We can also calculate that

$$X'' = \mathcal{T}_{\frac{-4\kappa(x \cdot x')}{1 + \kappa|x|^2}x' + x''} - \frac{2|x'|^2}{(1 + \kappa|x|^2)^2}\begin{pmatrix} \kappa - 1 \\ 2\kappa x^t \\ \kappa + 1 \end{pmatrix}.$$

Note that generally X'' is not contained in $T_X M_\kappa$. However, X'' can be split into a part contained in $T_X M_\kappa$ and a part perpendicular to $T_X M_\kappa$ as follows:

$$X'' = \mathcal{T}_{\frac{\kappa(-4(x \cdot x')x' + 2|x'|^2 x)}{1 + \kappa|x|^2} + x''} - \frac{2|x'|^2}{(1 + \kappa|x|^2)^3} \begin{pmatrix} (\kappa - 1)(1 - \kappa|x|^2) \\ 4\kappa x^t \\ (\kappa + 1)(1 - \kappa|x|^2) \end{pmatrix}. \tag{2.16}$$

The following lemma follows from (2.15) (cf. [111, Lemma 1.4.1]).

Lemma 2.2.12 M_κ *as determined by* q_κ *in* (2.12) *has constant sectional curvature* κ.

By Lemma 2.2.12, we also see that $\mathfrak{R}_\kappa$ with given metric as in (2.15) is also a 3-dimensional space form with constant sectional curvature κ. Thus, from now on, we always assume that the metric given to $\mathfrak{R}_\kappa$ is as in (2.15) so that M_κ and $\mathfrak{R}_\kappa$ are isometric. We often refer to $\mathfrak{R}_\kappa$ as the *(space form) projection* of M_κ.

Remark 2.2.13 Note that the metric $\langle \mathcal{T}_a, \mathcal{T}_b \rangle$ of M_0 is 4 times the usual $\mathbb{R}^3$ metric $a \cdot b$ of a and b. We tolerate this difference by a factor of 4 in order to conform with the usual expression of the metric in the cases when $\kappa \neq 0$.

Exercise 2.2.14 One can remedy the difference in the metric for M_0 by correctly scaling the space form vector. Check that for $\mathsf{q} = 2\mathsf{q}_0 = (1, 0, 0, 0, -1)^t$, the space form defined via

$$M = \{X \in L^4 \mid \langle X, \mathsf{q} \rangle = -1\},$$

is isometric to the Euclidean 3-space $\mathbb{R}^3$ with the standard metric.

We see also from (2.15) that the collection of M_κ given by the above choice (2.12) for q_κ, for various κ, are all *conformally equivalent* (or *Möbius equivalent*). In fact, for any κ, we see that for a point $x \in \mathfrak{R}_\kappa$,

$$\psi_\kappa(x) \in \text{span}\{(1 + |x|^2, 2x, 1 - |x|^2)^t\}.$$

Therefore, we consider the *projectivized light cone* $P(L^4)$, defined as

$$P(L^4) := \{\text{span}\{X\} \mid X \in L^4 \setminus \{0\}\} = \{\text{null lines through origin in } \mathbb{R}^{4,1}\},$$

as the model for unifying the 3-dimensional space forms M_κ as the *conformal 3-sphere*.

Remark 2.2.15 In the projective light cone, one can recover the geometric meaning of $\mathsf{q}_0 = (1/2, 0, 0, 0, -1/2)^t$. For ψ_κ as defined in Lemma 2.2.9, $\psi_0(x)$ diverges as x approaches infinity; however, in the projectivized setting, one can consider scalar multiples of $\psi_0(\infty)$ to take the appropriate limits to see where $\text{span}\{\psi_0(\infty)\}$ converges to. Doing this, one can see that

Fig. 2.2 Commutative diagrams representing Exercises 2.2.16 and 2.2.17

$$\mathrm{span}\{\psi_\kappa(\infty)\} = \mathrm{span}\{\mathsf{q}_\kappa\}$$

when $\kappa = 0$. Hence, such q_κ for $\kappa = 0$ is sometimes referred to as "point at infinity".

In this setting, the map $M_\kappa \ni X \mapsto x \in \mathfrak{R}_\kappa$ is stereographic projection when $\kappa \neq 0$, which can be checked in the following exercises when $\kappa = \pm 1$.

Exercise 2.2.16 For $\kappa = 1$, use q_κ as in (2.12) to show that $X = (x_0, \tilde{x}^t)^t \in M_1$ if and only if $\tilde{x} = (\tilde{x}_1, \tilde{x}_2, \tilde{x}_3, \tilde{x}_4)^t \in \mathbb{S}^3$ (Fig. 2.2). Further verify that $\psi_1 : \mathfrak{R}_1 \to M_1$ as defined in Lemma 2.2.9 is the inverse stereographic projection σ_1 of $\mathbb{S}^3$ given in (2.1) by checking that, for $\iota_1 : \mathbb{R}^4 \to \mathbb{R}^{4,1}$ defined as

$$\iota_1(\tilde{x}) = (1, \tilde{x}^t)^t,$$

we have that $\iota_1 \circ \sigma_1 = \psi_1$.

Exercise 2.2.17 For $\kappa = -1$, show that $X = (\tilde{x}, x_4)^t \in M_{-1}$ if and only if $\tilde{x} = (\tilde{x}_0, \tilde{x}_1, \tilde{x}_2, \tilde{x}_3)^t \in \mathbb{H}^3$. Further check that ψ_{-1} as defined in Lemma 2.2.9 is the inverse stereographic projection σ_{-1} of $\mathbb{H}^3$ given in (2.4) using $\iota_{-1} : \mathbb{R}^{3,1} \to \mathbb{R}^{4,1}$ defined as

$$\iota_{-1}(\tilde{x}) = (\tilde{x}^t, 1)^t.$$

2.3 Spheres

A vector

$$\mathcal{S} = \begin{pmatrix} z_0 \\ z^t \\ z_4 \end{pmatrix}, \quad z = (z_1, z_2, z_3),$$

in $\mathbb{R}^{4,1}$ with positive squared norm

$$\|\mathcal{S}\|^2 = -z_0^2 + |z|^2 + z_4^2 > 0$$

Fig. 2.3 The sphere $\tilde{S}$

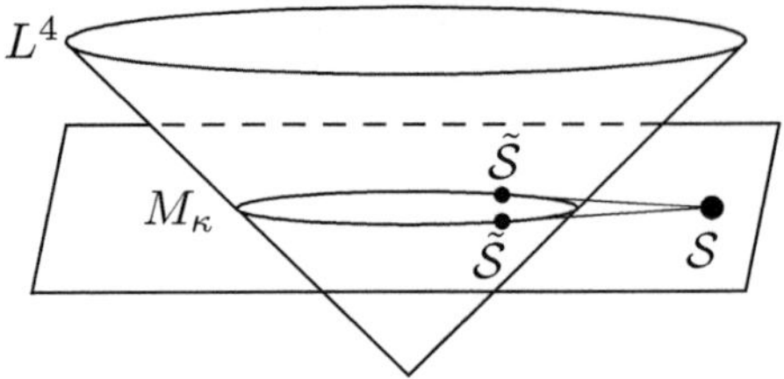

determines a *sphere* $\tilde{S}$ (with orientation) in the space form M_κ as follows (see Fig. 2.3):

$$\tilde{S} = \{Y \in M_\kappa \mid \langle Y, S \rangle = 0\}. \tag{2.17}$$

The above definition of spheres does not depend on the scaling of S; therefore, we assume that $\|S\|^2 = 1$, allowing us to consider

$$\mathbb{S}^{3,1} := \{Y \in \mathbb{R}^{4,1} \mid \|Y\|^2 = 1\} \tag{2.18}$$

as the space of oriented spheres.

In the next examples and exercises, we see how such $\tilde{S}$ project to spheres in some space forms.

Example 2.3.18 For $\kappa = 1$, we have seen that $M_\kappa \cong \mathbb{S}^3$. Recall that a sphere $\tilde{S}_{\mathbb{S}^3}[m, q]$ in the 3-sphere can be written as in (2.2) for some $m \in \mathbb{R}^4$ and $q \in \mathbb{R}$, where $\langle m, m \rangle_4 > q^2$ by Exercise 2.1.2. Setting $S = (q, m^t)^t$, we have that $\|S\|^2 > 0$, and writing $X = (1, \tilde{x}_1, \tilde{x}_2, \tilde{x}_3, \tilde{x}_4)^t = (1, \tilde{x}^t)^t \in M_1$, we see that $X \in \tilde{S}$ if and only if $\tilde{x} \in \tilde{S}_{\mathbb{S}^3}[m, q]$. Therefore, $\tilde{S} \subset M_1$ corresponds to spheres in the 3-sphere.

Exercise 2.3.19 For $\kappa = -1$, show that $\tilde{S} \subset M_{-1}$ corresponds to spheres in the hyperbolic 3-space using (2.6) and Exercise 2.1.6.

Example 2.3.20 Finally, we show that $\tilde{S} \subset M_0$ projects to a sphere (or a plane) in $\mathfrak{R}_0 \cong \mathbb{R}^3$: For $Y \in \tilde{S}$ with some $S \in \mathbb{R}^{4,1}$, we can take

$$Y = (1 + |y|^2, 2y, 1 - |y|^2)^t, \quad y = (y_1, y_2, y_3).$$

Then $\langle S, Y \rangle = 0$ implies

$$2y \cdot z - |y|^2(z_0 + z_4) - (z_0 - z_4) = 0. \tag{2.19}$$

If $z_0 = -z_4$, then (2.19) implies that $\tilde{S}$ projects to a plane with normal vector z. Otherwise, one can use (2.19) to show that

$$\left| y - \frac{z}{z_0 + z_4} \right|^2 = \frac{1}{(z_0 + z_4)^2}. \tag{2.20}$$

Therefore $\tilde{\mathcal{S}}$ projects to a sphere with center $\frac{z}{z_0+z_4}$ and (signed) radius $\frac{2}{z_0+z_4}$.

Remark 2.3.21 In the above Example 2.3.20, one can also note that for $z_0 = -z_4$, $\langle \mathcal{S}, \mathfrak{q}_0 \rangle = 0$; therefore, $\tilde{\mathcal{S}}$ projects to a sphere including the point at infinity, i.e. $\tilde{\mathcal{S}}$ projects to a plane.

Exercise 2.3.22 Use Example 2.3.20 to show that if $\tilde{\mathcal{S}}$ projects to a sphere in $\mathfrak{R}_0 \cong \mathbb{R}^3$ with center $c \in \mathfrak{R}_0$ and (signed) radius $r \in \mathbb{R}$, then

$$\mathcal{S} = \frac{1}{r}\begin{pmatrix} 1 + c \cdot c - \frac{1}{4}r^2 \\ 2c^t \\ 1 - c \cdot c + \frac{1}{4}r^2 \end{pmatrix}. \tag{2.21}$$

(Note that the signature of r, and hence $\mathcal{S}$, changes as the orientation of the sphere changes since r is the signed radius.)

Remark 2.3.23 Assuming that $\kappa < 0$, we have that $\mathfrak{q}_\kappa$ is spacelike, so that one can normalize to define

$$\mathcal{S}_\kappa = \frac{1}{\sqrt{-\kappa}}\mathfrak{q}_\kappa \in \mathbb{S}^{3,1}$$

that determines a sphere $\tilde{\mathcal{S}}_\kappa$. To see what this sphere projects to in space forms, we consider its projection to $\mathfrak{R}_0$. Setting $Y = (1 + |y|^2, 2y, 1 - |y|^2)^t \in M_0$ for $y = (y_1, y_2, y_3) \in \mathfrak{R}_0$, Example 2.3.20 and (2.20) tells us that

$$|y|^2 = -\frac{1}{\kappa}.$$

Thus, we see that $\mathcal{S}_\kappa$, and hence $\mathfrak{q}_\kappa$, determines the ideal boundary of $\mathfrak{R}_\kappa$. This is a generalization of Remark 2.2.15, where we discussed the fact that $\mathfrak{q}_0$ represents the point at infinity of $\mathfrak{R}_0$. Note that we needed to project $\tilde{\mathcal{S}}_\kappa$ to $\mathfrak{R}_0$ since there are no $Y \in M_\kappa$ such that $\langle Y, \mathcal{S}_\kappa \rangle = 0$.

Now suppose that we have a sphere in $\mathfrak{R}_0 \cong \mathbb{R}^3$ with center $c \in \mathbb{R}^3$ and (signed) radius $r \in \mathbb{R}$, and choose a point $x \in \mathbb{R}^3$ on the sphere. Denoting the unit length vector pointing in the direction of $x - c$ by n (unit length here means $4|n|^2 = 1$), we then have $x = -rn + c$. If we let $r \to \infty$, then we obtain the plane containing the point x with normal vector n. Therefore, the vector $\mathcal{S}$ that represents this plane can be found via (2.21) as follows:

$$\mathcal{S} = \lim_{r \to \infty} \frac{1}{r}\begin{pmatrix} 1 + (x + rn) \cdot (x + rn) - \frac{1}{4}r^2 \\ 2(x + rn)^t \\ 1 - (x + rn) \cdot (x + rn) + \frac{1}{4}r^2 \end{pmatrix} = 2\begin{pmatrix} x \cdot n \\ n^t \\ -x \cdot n \end{pmatrix}. \tag{2.22}$$

As in the sphere case, if the signature of the normal vector n changes, then the signature of $\mathcal{S}$ changes as well.

Note that $Y \in \tilde{\mathcal{S}}$ implies Y is perpendicular to $\mathcal{S} - Y$, so $\tilde{\mathcal{S}}$ is the base of the tangent cone from $\mathcal{S}$ to M_κ, once $\mathcal{S}$ is scaled so that $\langle \mathcal{S}, q_\kappa \rangle = -1$. In fact, $Y - \mathcal{S}$ is then a normal vector field to $\tilde{\mathcal{S}}$ within the tangent space of M_κ.

Remark 2.3.24 We have seen above that for a sphere $\tilde{\mathcal{S}}$, if $\langle \mathcal{S}, q_0 \rangle = 0$, then $\mathcal{S}$ represents a plane in M_0. This holds for other space forms as well: if the sphere $\mathcal{S}$ is perpendicular to q_κ, that is, $\langle \mathcal{S}, q_\kappa \rangle = 0$, then $\mathcal{S}$ represents the geodesic plane in M_κ.

Exercise 2.3.25 Show that Remark 2.3.24 holds for $\kappa = \pm 1$.

Remark 2.3.26 Here, we make a few remarks regarding spheres in $M_1 \cong \mathbb{S}^3$.

- If $\|\mathcal{S}\|^2 = 0$, then $\tilde{\mathcal{S}}$ is a point in $\mathbb{S}^3$, and $\tilde{\mathcal{S}}$ consists of just a real scalar multiple of $\mathcal{S}$. Hence $\tilde{\mathcal{S}}$ simply gives back the same point $\mathcal{S}$.
- Let ℓ be the horizontal (i.e. perpendicular to $(1, 0, 0, 0, 0)^t$) line segment from $\mathcal{S}$ to the timelike axis $\{(t, 0, 0, 0, 0)^t \mid t \in \mathbb{R}\}$. Then $m = \ell \cap L^4$ is a single point, which, when considered as being in $\mathbb{S}^3 \cong M_1$, gives the center of $\tilde{\mathcal{S}}$ in $\mathbb{S}^3$.

Lemma 2.3.27 (cf. [111, §1.1.5]) *Let $\tilde{\mathcal{S}}_1, \tilde{\mathcal{S}}_2$ be two intersecting spheres in $\mathbb{S}^3$ produced from $\mathcal{S}_1, \mathcal{S}_2$, respectively. Let α be the intersection angle between $\tilde{\mathcal{S}}_1$ and $\tilde{\mathcal{S}}_2$. Then $\cos\alpha = \langle \mathcal{S}_1, \mathcal{S}_2 \rangle$, where the sign on the right hand side depends on the orientations of $\tilde{\mathcal{S}}_1$ and $\tilde{\mathcal{S}}_2$. (See Fig. 2.4.)*

Proof Let S_1 and S_2 be scales of $\mathcal{S}_1$ and $\mathcal{S}_2$, respectively, so that S_1 and S_2 have inner product -1 with q_1. Take $P \in \tilde{\mathcal{S}}_1 \cap \tilde{\mathcal{S}}_2 \subset M_1$. Then $S_1 - P$ and $S_2 - P$ are normals (in the tangent space of $\mathbb{S}^3$) to $\tilde{\mathcal{S}}_1$ and $\tilde{\mathcal{S}}_2$, respectively, at P. So

$$\cos\alpha = \left\langle \frac{S_1 - P}{\|S_1 - P\|}, \frac{S_2 - P}{\|S_2 - P\|} \right\rangle$$

$$= \frac{1}{\|S_1 - P\|} \frac{1}{\|S_2 - P\|} \langle S_1, S_2 \rangle = \frac{1}{\|S_1\|} \frac{1}{\|S_2\|} \langle S_1, S_2 \rangle = \langle \mathcal{S}_1, \mathcal{S}_2 \rangle,$$

giving us the desired conclusion. $\square$

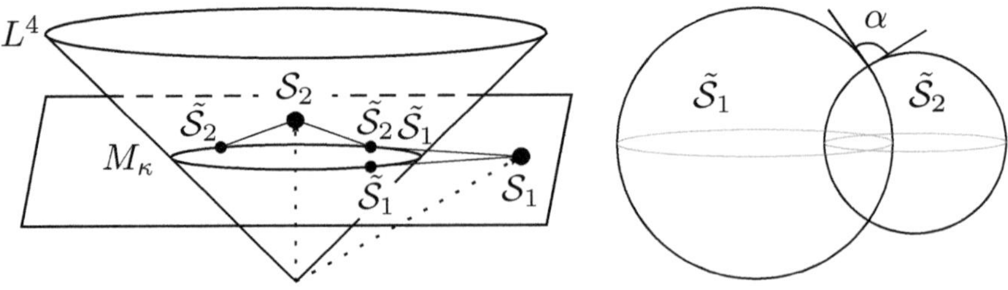

Fig. 2.4 Two depictions of the setting in Lemma 2.3.27

Fig. 2.5 A pencil of spheres

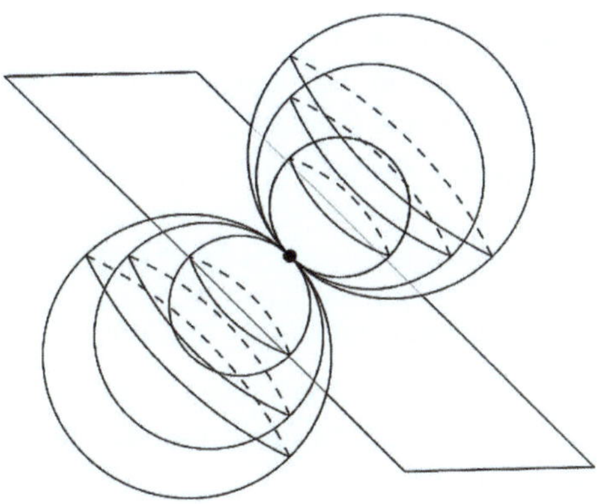

Using Lemma 2.3.27, we can now directly verify the following.

Lemma 2.3.28 *Let S give a sphere $\tilde{S}$ containing $Y \in M_\kappa$. Then $\{S + tY \mid t \in R\}$ gives a (parabolic) pencil of spheres at Y, i.e. the collection of spheres of arbitrary radius through Y and tangent to $\tilde{S}$ sharing the same orientation (see Fig. 2.5).*

Now we explain how one can view inversion through spheres in Möbius geometry.

Lemma 2.3.29 (cf. [111, Lemma 1.4.1]) *Let $P \in M_k$ and $\tilde{S}$ be the sphere given by S. The map given by*

$$f : P \mapsto P - 2\langle P, S\rangle S$$

is inversion through $\tilde{S}$.

Proof First note that $P \in L^4$ implies $P - 2\langle P, S\rangle S \in L^4$. Now let C be a circle that intersects $\tilde{S}$ perpendicularly. We wish to show that $P \in C$ implies $f(P) \in C$. Note that $C = \tilde{S}_1 \cap \tilde{S}_2$ for some spheres $\tilde{S}_1$ and $\tilde{S}_2$. Then $\tilde{S}_1 \perp \tilde{S}$ and $\tilde{S}_2 \perp \tilde{S}$, and so

$$\langle S, S_1\rangle = \langle S, S_2\rangle = 0,$$

by Lemma 2.3.27. Then $P \in C$ implies $P \in \tilde{S}_1 \cap \tilde{S}_2$, which implies $\langle P, S_1\rangle = \langle P, S_2\rangle = 0$. Thus

$$\langle P - 2\langle P, S\rangle S, S_1\rangle = \langle P - 2\langle P, S\rangle S, S_2\rangle = 0,$$

and so $f(P) \in C$. Since this holds for all circles C that intersect $\tilde{S}$ perpendicularly, it follows that the map is inversion through $\tilde{S}$.   □

Viewing a circle as the intersection of two spheres, we can now characterize points lying on a circle via Möbius geometry:

Remark 2.3.30 If $x_1, x_2, x_3, x_4 \in \mathbb{R}^3 \cong \mathfrak{R}_0$ (with associated lifts X_1, X_2, X_3, X_4 $\in M_0$) all lie in the circle of positive radius determined by the intersection of two distinct spheres $\tilde{S}_1$ and $\tilde{S}_2$ given by spacelike vectors S_1 and S_2, then the fact that $X_1, X_2, X_3, X_4 \in \tilde{S}_1 \cap \tilde{S}_2$ is equivalent to

$$X_1, X_2, X_3, X_4 \perp \text{span}\{S_1, S_2\}.$$

This implies, in particular, that $\dim(\text{span}\{X_1, X_2, X_3, X_4\}) \leq 3$.

Exercise 2.3.31 Prove that four points always lie on a sphere. (Hint: Use the fact that the linear span of four null vectors X_1, X_2, X_3, X_4 in the light cone can at most be of dimension 4.)

2.4 Möbius Transformations

Möbius transformations are the diffeomorphisms from $\mathbb{S}^3$ to $\mathbb{S}^3$ that take 2-spheres to 2-spheres, and they are equivalent to the orthogonal transformations $O_{4,1}$ of $\mathbb{R}^{4,1}$. We will not prove that here, and instead just give some examples. Identifying $\mathbb{S}^3$ with $\mathbb{R}^3 \cup \{\infty\}$ (via the conformal 3-sphere), the group of Möbius transformations (including orientation reversing maps), i.e., the group of maps that preserve the collection of spheres and planes in $\mathbb{R}^3 \cup \{\infty\}$, is generated (via repeated composition of maps) by:

$$(y_1, y_2, y_3) \mapsto (r y_1, r y_2, r y_3),$$

$$(y_1, y_2, y_3) \mapsto (y_1, y_2, y_3) + (y_{0,1}, y_{0,2}, y_{0,3}),$$

$$(y_1, y_2, y_3) \mapsto (-y_1, y_2, y_3),$$

$$(y_1, y_2, y_3) \mapsto (y_1 \cos\theta - y_2 \sin\theta, y_1 \sin\theta + y_2 \cos\theta, y_3),$$

$$(y_1, y_2, y_3) \mapsto (y_1 \cos\theta - y_3 \sin\theta, y_2, y_1 \sin\theta + y_3 \cos\theta),$$

$$(y_1, y_2, y_3) \mapsto (y_1, y_2 \cos\theta - y_3 \sin\theta, y_2 \sin\theta + y_3 \cos\theta),$$

$$(y_1, y_2, y_3) \mapsto (y_1, y_2, y_3)/(y_1^2 + y_2^2 + y_3^2),$$

where $\theta, r, y_{0,1}, y_{0,2}, y_{0,3}$ are any real constants. These seven maps are a dilation, a translation, a reflection, three rotations, and an inversion, respectively.

In the conformal 3-sphere, the Möbius transformations are $O_{4,1}$ matrices acting on the light cone via left matrix multiplication. More precisely, for some y in a 3-dimensional space form with sectional curvature κ, we identify y with the null line $L = \text{span}\{\psi_\kappa(y)\} \subset L^4 \subset \mathbb{R}^{4,1}$ via the usual identification. Then, for $A \in O_{4,1}$, we see that AL is another null line, and we project this null line back to the 3-dimensional space form.

The next examples offer the corresponding $O_{4,1}$ matrices for the transformations given at the beginning of the section, for $y = (y_1, y_2, y_3) \in \mathbb{R}^3 \cup \{\infty\}$.

- Given the matrix

$$A_1 = \begin{pmatrix} 1 & 0 & 0 & 0 & 0 \\ 0 & -1 & 0 & 0 & 0 \\ 0 & 0 & 1 & 0 & 0 \\ 0 & 0 & 0 & 1 & 0 \\ 0 & 0 & 0 & 0 & 1 \end{pmatrix} \in O_{4,1},$$

we see that

$$A_1 L = \operatorname{span}\left\{ \left(1 + |y|^2, -2y_1, 2y_2, 2y_3, 1 - |y|^2\right)^t \right\}.$$

To project this back to $\mathbb{R}^3 \cup \{\infty\}$, we need to find $A_1 L \cap M_0$. This amounts to solving for t where

$$\left\langle t\left(1 + |y|^2, -2y_1, 2y_2, 2y_3, 1 - |y|^2\right)^t, \mathfrak{q}_0\right\rangle = -1,$$

telling us that $t = 1$. Now, we project this back to $\mathbb{R}^3 \cup \{\infty\}$ via ϕ_0 to see that left multiplication of A_1 corresponds to the map

$$(y_1, y_2, y_3) \mapsto (-y_1, y_2, y_3).$$

- For the matrix

$$A_2 = \begin{pmatrix} 1 & 0 & 0 & 0 & 0 \\ 0 & 1 & 0 & 0 & 0 \\ 0 & 0 & 1 & 0 & 0 \\ 0 & 0 & 0 & 1 & 0 \\ 0 & 0 & 0 & 0 & -1 \end{pmatrix} \in O_{4,1},$$

we have

$$A_2 L = \operatorname{span}\left\{ \left(1 + |y|^2, 2y, -1 + |y|^2\right)^t \right\}.$$

Similarly, solving for t where

$$\left\langle t\left(1 + |y|^2, 2y, -1 + |y|^2\right)^t, \mathfrak{q}_0\right\rangle = -1,$$

we obtain that $t = \frac{1}{|y|^2}$. Therefore, application of A_2 corresponds to the map

$$(y_1, y_2, y_3) \mapsto (y_1, y_2, y_3)/(y_1^2 + y_2^2 + y_3^2).$$

- Left multiplication by

$$A_3 = \begin{pmatrix} \cosh\varphi & 0 & 0 & 0 & \sinh\varphi \\ 0 & 1 & 0 & 0 & 0 \\ 0 & 0 & 1 & 0 & 0 \\ 0 & 0 & 0 & 1 & 0 \\ \sinh\varphi & 0 & 0 & 0 & \cosh\varphi \end{pmatrix} \in O_{4,1}$$

corresponds to the map

$$(y_1, y_2, y_3) \mapsto (ry_1, ry_2, ry_3)$$

where $r = e^{\varphi}$. (Checking the correspondence is left as an exercise.)
- Applying the matrix

$$A_4 = \begin{pmatrix} 1 + \frac{1}{2}|y_0|^2 & y_{0,1} & y_{0,2} & y_{0,3} & \frac{1}{2}|y_0|^2 \\ y_{0,1} & 1 & 0 & 0 & y_{0,1} \\ y_{0,2} & 0 & 1 & 0 & y_{0,2} \\ y_{0,3} & 0 & 0 & 1 & y_{0,3} \\ -\frac{1}{2}|y_0|^2 & -y_{0,1} & -y_{0,2} & -y_{0,3} & 1 - \frac{1}{2}|y_0|^2 \end{pmatrix} \in O_{4,1}$$

corresponds to the map

$$(y_1, y_2, y_3) \mapsto (y_1, y_2, y_3) + (y_{0,1}, y_{0,2}, y_{0,3}).$$

(Checking the correspondence is left as an exercise.)
- Finally, the left matrix multiplication by

$$A_5 = \begin{pmatrix} 1 & 0 & 0 & 0 & 0 \\ 0 & \cos\theta & -\sin\theta & 0 & 0 \\ 0 & \sin\theta & \cos\theta & 0 & 0 \\ 0 & 0 & 0 & 1 & 0 \\ 0 & 0 & 0 & 0 & 1 \end{pmatrix} \in O_{4,1}$$

corresponds to

$$(y_1, y_2, y_3) \mapsto (y_1\cos\theta - y_2\sin\theta, y_1\sin\theta + y_2\cos\theta, y_3).$$

(Checking the correspondence is left as an exercise.)

Exercise 2.4.1 Check the above correspondences for $A_3, A_4, A_5 \in O_{4,1}$.

When $\kappa \neq 0$, i.e. when q_κ as in (2.12) is not null, M_κ has a particular Möbius transformation called the antipodal map, which we now describe: A point X in M_κ can be decomposed as

$$X = \mathcal{A} + \kappa^{-1} q_\kappa,$$

where $\mathcal{A} \perp q_\kappa$. The antipodal map is

$$\mathcal{A} + \kappa^{-1} q_\kappa \mapsto -\mathcal{A} + \kappa^{-1} q_\kappa,$$

that is, the map is a reflection across the line spanned by q_κ. In detail, X as in (2.13) is

$$\frac{1}{1+\kappa|x|^2} \begin{pmatrix} \frac{1}{2}(\kappa^{-1}-1)(\kappa|x|^2-1) \\ 2x^t \\ \frac{1}{2}(-\kappa^{-1}-1)(\kappa|x|^2-1) \end{pmatrix} + \kappa^{-1} q_\kappa$$

and is mapped to

$$\frac{-1}{1+\kappa|x|^2} \begin{pmatrix} \frac{1}{2}(\kappa^{-1}-1)(\kappa|x|^2-1) \\ 2x^t \\ \frac{1}{2}(-\kappa^{-1}-1)(\kappa|x|^2-1) \end{pmatrix} + \kappa^{-1} q_\kappa = \frac{1}{1+\kappa|y|^2} \begin{pmatrix} 1+|y|^2 \\ 2y^t \\ 1-|y|^2 \end{pmatrix},$$

where $y = -x/(\kappa|x|^2)$. Hence this map is represented by

$$x \mapsto -\frac{x}{\kappa|x|^2}.$$

In light of viewing the set of spheres as the set of spacelike vectors in $\mathbb{R}^{4,1}$ (2.18), we obtain an immediate advantage of viewing Möbius transformations as $O_{4,1}$ matrices acting on the null lines:

Proposition 2.4.33 *Möbius transformations preserve spheres.*

2.5 Cross Ratios

Cross ratios are projectively invariant, and their natural treatment comes from the context of projective geometry. The fact that we have a projective model of Möbius geometry alludes to the fact that cross ratios are Möbius invariant, since Möbius transformations induce projective transformations. However, in this text, we show the Möbius invariance of the cross ratios with the tools we have at hand. (For a projective geometric treatment of cross ratios, we refer the readers to excellent textbooks treating projective geometry such as [205].)

Definition 2.5.34 (cf. [111, §6.5.4]) Let $X_1, X_2, X_3, X_4 \in L^4$. Then the *cross ratio* $\mathrm{cr}(X_1, X_2, X_3, X_4)$ of the four points is defined as

$$\mathrm{cr}(X_1, X_2, X_3, X_4) := \frac{s_{12}s_{34} + s_{14}s_{23} - s_{13}s_{24} \pm \sqrt{\mathcal{E}}}{2s_{14}s_{23}} \qquad (2.23)$$

where $s_{ij} = \langle X_i, X_j \rangle$, and

$$\mathcal{E} = s_{12}^2 s_{34}^2 + s_{13}^2 s_{24}^2 + s_{14}^2 s_{23}^2 - 2s_{13}s_{14}s_{23}s_{24} - 2s_{12}s_{14}s_{23}s_{34} - 2s_{12}s_{13}s_{24}s_{34} \leq 0.$$

Remark 2.5.35 To see that $\mathcal{E} \leq 0$, we consider the following argument: Because the four points X_1, X_2, X_3, X_4 all lie in the light cone, we can deduce that $\mathrm{span}\{X_1, X_2, X_3, X_4\}$ is a subspace of $V \cong \mathbb{R}^{3,1}$. Choose a basis e_1, e_2, e_3, e_4 of this subspace V so that

$$\|e_1\|^2 = \|e_2\|^2 = \|e_3\|^2 = -\|e_4\|^2 = 1 \quad \text{and} \quad \langle e_i, e_j \rangle = 0 \, (i \neq j).$$

Writing $X_\beta = a_{1\beta}e_1 + a_{2\beta}e_2 + a_{3\beta}e_3 + a_{4\beta}e_4$ in terms of the basis e_1, e_2, e_3, e_4, we have

$$\mathcal{E} = \det(\langle X_\beta, X_\gamma \rangle_{\beta,\gamma=1,2,3,4})$$

$$= \det\left(\begin{pmatrix} a_{11} & a_{12} & a_{13} & a_{14} \\ a_{21} & a_{22} & a_{23} & a_{24} \\ a_{31} & a_{32} & a_{33} & a_{34} \\ a_{41} & a_{42} & a_{43} & a_{44} \end{pmatrix}^t \begin{pmatrix} 1 & 0 & 0 & 0 \\ 0 & 1 & 0 & 0 \\ 0 & 0 & 1 & 0 \\ 0 & 0 & 0 & -1 \end{pmatrix} \begin{pmatrix} a_{11} & a_{12} & a_{13} & a_{14} \\ a_{21} & a_{22} & a_{23} & a_{24} \\ a_{31} & a_{32} & a_{33} & a_{34} \\ a_{41} & a_{42} & a_{43} & a_{44} \end{pmatrix} \right).$$

Thus $\mathcal{E} \leq 0$.

Note that this definition is well-defined only up to complex conjugation; however, as in [22], we will refer to the cross ratios as if they are complex numbers, and we ask the readers to remember that either member of the complex conjugate pair can be the cross ratio of the given four points.

First, we check that this definition coincides with the cross ratio of four points in the complex plane $\mathbb{C}$. Recall that for $z_1, z_2, z_3, z_4 \in \mathbb{C}$, we have that the cross ratio of four points z_1, z_2, z_3, z_4 is defined as

$$\mathrm{cr}(z_1, z_2, z_3, z_4) = \frac{z_1 - z_2}{z_2 - z_3} \frac{z_3 - z_4}{z_4 - z_1}.$$

Using the correspondence $\mathbb{C} \cong \mathbb{R}^2 \subset \mathbb{R}^3 \cong \mathfrak{R}_0 \cong M_0$ via

$$\mathbb{C} \ni a + \sqrt{-1}b \sim x = (a, b, 0) \in \mathfrak{R}_0$$

$$\sim X = (1 + |x|^2, 2x, 1 - |x|^2)^t \in M_0,$$

we can directly calculate and verify the following lemma using the relation

$$s_{\beta\gamma} = \langle X_\beta, X_\gamma \rangle = -2|x_\beta - x_\gamma|^2, \tag{2.24}$$

where $\beta, \gamma = 1, 2, 3, 4$.

Lemma 2.5.36 *For $z_1, z_2, z_3, z_4 \in \mathbb{C}$ and corresponding $X_1, X_2, X_3, X_4 \in M_0$, we have that*

$$\mathrm{cr}(z_1, z_2, z_3, z_4) = \mathrm{cr}(X_1, X_2, X_3, X_4).$$

The form of the cross ratio defined in (2.23) tells us two things: first, since the expression is written completely in terms of the inner product of lightlike vectors in $\mathbb{R}^{4,1}$, applying any Möbius transformation to the four lightlike vectors will not change the value under consideration, that is, for any $A \in O_{4,1}$,

$$\mathrm{cr}(X_1, X_2, X_3, X_4) = \mathrm{cr}(AX_1, AX_2, AX_3, AX_4),$$

giving us the Möbius invariance of cross ratios.

Furthermore, scaling $X_1, X_2, X_3, X_4 \in L^4$ by $\alpha_1, \alpha_2, \alpha_3, \alpha_4 \in \mathbb{R}^\times = \mathbb{R} \setminus \{0\}$, respectively, does not change the value of the cross ratio, i.e.

$$\mathrm{cr}(X_1, X_2, X_3, X_4) = \mathrm{cr}(\alpha_1 X_1, \alpha_2 X_2, \alpha_3 X_3, \alpha_4 X_4).$$

Thus, cross ratios are notions that are well-defined not only for four points in each choice of space form, but also for four points in the conformal 3-sphere $P(L^4)$ (or four null lines in L^4), allowing us to also write

$$\mathrm{cr}(L_1, L_2, L_3, L_4) = \mathrm{cr}(X_1, X_2, X_3, X_4),$$

for $L_i = \mathrm{span}\{X_i\}$ for $i = 1, 2, 3, 4$.

Remark 2.5.37 Let $x_1, x_2, x_3, x_4 \in \mathbb{R}^3 \cong \mathfrak{R}_0$ with lifts $X_1, X_2, X_3, X_4 \in M_0$. One can also calculate the cross ratio of these four points in the following way: Since four points always lie on a sphere, we can find a Möbius transformation mapping the sphere to a plane in $\mathbb{R}^3$. Identifying the plane with the complex plane, x_1, x_2, x_3, x_4 can be identified with $z_1, z_2, z_3, z_4 \in \mathbb{C}$, respectively. Then the Möbius invariance of cross ratios and Lemma 2.5.36 guarantees us that

$$\mathrm{cr}(X_1, X_2, X_3, X_4) = \mathrm{cr}(z_1, z_2, z_3, z_4) = \frac{z_1 - z_2}{z_2 - z_3} \frac{z_3 - z_4}{z_4 - z_1}.$$

Exercise 2.5.38 Show that $L_1, L_2, L_3, L_4 \in P(L^4)$ are concircular when projected to some space form if and only if $\mathrm{cr}(L_1, L_2, L_3, L_4) \in \mathbb{R}$. (Hint: One can use Remark 2.3.30 to show that $\mathcal{E} = 0$. One could also show this for four points in the complex plane, and use Remark 2.5.37.)

Remark 2.5.39 Exercise 2.5.38 can be interpreted as follows: let $\mathrm{cr}(X_1, X_2, X_3, X_4) = t \in \mathbb{R}$ for $X_1, X_2, X_3, X_4 \in L^4$. Fixing X_1, X_2, X_4, one can view cross ratio as parametrizing the circle determined by X_1, X_2, X_4 via

$$X_3 = X_1 + \frac{1}{\langle X_2, X_4 \rangle} \{(t - 1)\langle X_1, X_4 \rangle X_2 + (t^{-1} - 1)\langle X_1, X_2 \rangle X_4\} \qquad (2.25)$$

where the point X_3 is seen as moving through the circle as the value of t changes.

We finish this section with four comments on how we can interpret cross ratios.

Comment 1 Given four points $p_1 < p_2 < p_3 < p_4$ along the real line $\mathbb{R}$ in the complex plane $\mathbb{C}$, we can consider the two half circles from p_1 to p_3 and from p_2 to p_4, respectively, in the upper half plane $\mathbb{C}^{i+} = \{x + \sqrt{-1}y \mid y > 0\}$. These are images of geodesics (straight lines, and "diagonals" of the "quadrilateral" $\{p_1, p_2, p_3, p_4\}$) in $\mathbb{C}^{i+}$ when $\mathbb{C}^{i+}$ is given the hyperbolic metric. At the point where these two circles intersect, we can consider the angle θ between the two circles with respect to the adjacent region as in Fig. 2.6. This angle θ can be regarded as a geometric quantity, since it is a notion that is invariant under Möbius transformations of $\mathbb{C} \cup \{\infty\}$ that preserve the set $\mathbb{R} \cup \{\infty\}$. Let

$$\mathrm{cr}(p_1, p_2, p_3, p_4) = \frac{p_2 - p_1}{p_3 - p_2} \frac{p_4 - p_3}{p_1 - p_4} \qquad (2.26)$$

be the cross ratio of p_1, p_2, p_3 and p_4. Here we determine the precise relationship between θ and the cross ratio, thus giving it a geometric meaning.

Lemma 2.5.40 *For four points $p_1 < p_2 < p_3 < p_4$ in the real line $\mathbb{R}$ in $\mathbb{C}$, and for θ as defined above,*

$$\mathrm{cr}(p_1, p_2, p_3, p_4) = \frac{\cos \theta - 1}{\cos \theta + 1} = -\tan^2 \frac{\theta}{2}.$$

Fig. 2.6 The situation in Comment 1

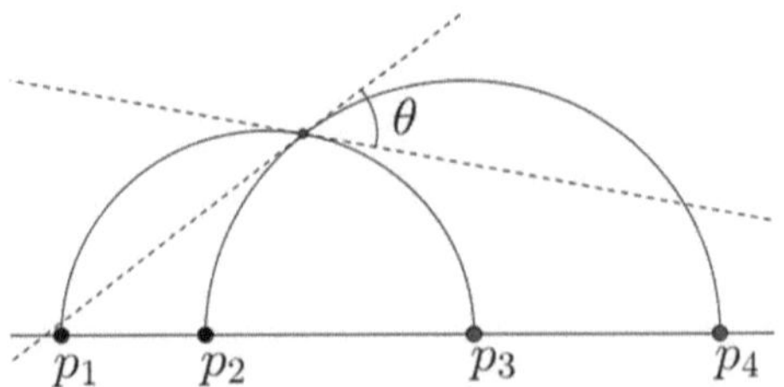

Proof Without loss of generality, applying a Möbius transformation fixing $\mathbb{R} \cup \{\infty\}$ as a set if necessary, we may assume that

$$p_4 = \infty, \; p_3 = 1 > p_2 > p_1 = -1.$$

Then one half-circle is

$$\left\{ e^{\sqrt{-1}t} \mid t \in [0, \pi] \right\},$$

and the other is the half-line

$$\{p_2 + \sqrt{-1}t \mid t \geq 0\}.$$

Then θ is the angle between these two half-circles with respect to the adjacent region to the upper right. It can be immediately checked that $\cos \theta = -p_2$ and that

$$\mathrm{cr}(p_1, p_2, p_3, p_4) = \frac{p_2 + 1}{1 - p_2} \frac{\infty - 1}{-1 - \infty} = \frac{p_2 + 1}{p_2 - 1}.$$

The result follows. $\square$

In fact, using the cross ratios on the real line, we can prove a handy lemma that will be useful later:

Lemma 2.5.41 *Let* $\mathrm{cr}(X_1, X_2, X_3, X_4) = t \in \mathbb{R}$ *for* $X_1, X_2, X_3, X_4 \in L^4$. *Then,*

$$\mathrm{cr}(X_2, X_1, X_3, X_4) = \frac{t}{t - 1}, \quad \text{and} \quad \mathrm{cr}(X_4, X_1, X_3, X_2) = \frac{1}{1 - t}.$$

Proof Since we have that X_1, X_2, X_3, X_4 are concircular by Exercise 2.5.38, there is a (unique) Möbius transformation A that maps the circle to $\mathbb{R} \cup \{\infty\}$, and

$$A(X_1) = 1, \; A(X_2) = \infty, \; A(X_3) = t, \; A(X_4) = 0.$$

Then we can check that

$$\mathrm{cr}(X_2, X_1, X_3, X_4) = \mathrm{cr}(A(X_2), A(X_1), A(X_3), A(X_4)) = \frac{\infty - 1}{1 - t} \frac{t - 0}{0 - \infty} = \frac{t}{t - 1},$$

$$\mathrm{cr}(X_4, X_1, X_3, X_2) = \mathrm{cr}(A(X_4), A(X_1), A(X_3), A(X_2)) = \frac{0 - 1}{1 - t} \frac{t - \infty}{\infty - 0} = \frac{1}{1 - t},$$

giving us the claim. $\square$

Exercise 2.5.42 Show that if $\mathrm{cr}(X_1, X_2, X_3, X_4) = t \in \mathbb{R}$, then $\mathrm{cr}(X_1, X_4, X_3, X_2) = \frac{1}{t}$.

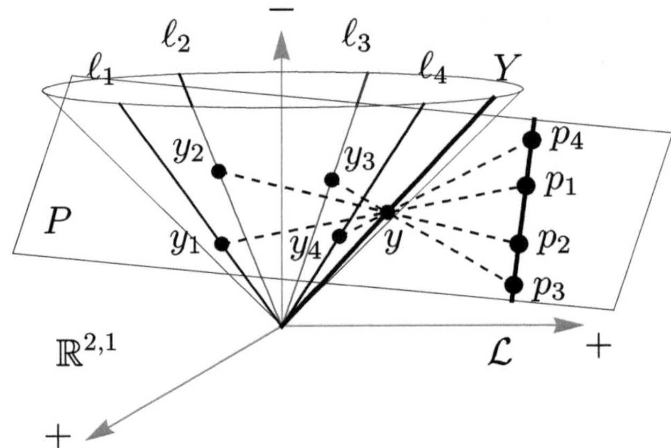

Fig. 2.7 The situation in Comment 2

Comment 2 Consider $\mathbb{R}^{2,1}$ and its 2-dimensional light cone L^2. Let ℓ_1, ℓ_2, ℓ_3, ℓ_4 be four lines in L^2 that are pairwise distinct.

Let P be a spacelike plane in $\mathbb{R}^{2,1}$ that does not contain the origin. Then $C = L^2 \cap P$ will be a conic section, with four intersection points $y_j = \ell_j \cap C$. Let $\mathcal{L}$ be a line in P and Y a line in L^2. Then one can stereographically project y_j within P though the point $Y \cap P \in C$ to a point $p_j \in \mathcal{L}$ (see Fig. 2.7). We have the following fact:

Lemma 2.5.43 *Regarding $\mathcal{L}$ as the real line, and then computing the cross ratio of the four points* $\mathrm{cr}(p_1, p_2, p_3, p_4)$ *as in (2.26), the cross ratio is independent of the choices of P and Y and $\mathcal{L}$.*

Proof Elementary calculations show that the cross ratio does not depend on the choices of Y and $\mathcal{L}$. It remains only to show that it is independent of choice of P.

Suppose we have chosen one line Y in L^2 and have made two different choices P_1 and P_2 for the plane. Choosing the line $\mathcal{L}$ to be $\mathcal{L} = P_1 \cap P_2$, stereographic projection of either

$$\ell_j \cap P_1 \text{ to } \mathcal{L} \text{ through } Y \cap P_1 \text{ within } P_1$$

or

$$\ell_j \cap P_2 \text{ to } \mathcal{L} \text{ through } Y \cap P_2 \text{ within } P_2$$

will produce the same four points p_j in $\mathcal{L}$, and so the value of the cross ratio does not depend on whether P_1 or P_2 was used.   □

Now, choosing the plane P as

$$P = \{(1, x_1, x_2)^t \mid x_1, x_2 \in \mathbb{R}\}, \tag{2.27}$$

one can clearly regard it as a plane in the standard Euclidean $\mathbb{R}^3$, and it follows that the cross ratio can be computed using the definition (2.23). More explicitly, for

$$y_j = (1, y_{1,j}, y_{2,j})^t \in L^2,$$

consider P as a plane in $\mathbb{R}^3$, and translate these points to $(0, y_{1,j}, y_{2,j})^t \in \mathbb{R}^3$ with $y_{1,j}^2 + y_{2,j}^2 = 1$, which is just a Möbius transformation. Now, each y_j lifts to

$$Y_j = (2, 0, 2y_{1,j}, 2y_{2,j}, 0)^t.$$

Then, we can use the definition (2.23) to determine the cross ratio. In fact, one could also obtain the same cross ratio viewing $\mathbb{R}^{2,1}$ as the subspace of $\mathbb{R}^{4,1}$ generated by $\{(a, 0, b, c, 0)^t \mid a, b, c \in \mathbb{R}\}$, and considering

$$\mathbb{R}^{2,1} \ni y_j = (y_{0,j}, y_{1,j}, y_{2,j})^t \sim (y_{0,j}, 0, y_{1,j}, y_{2,j}, 0)^t = \tilde{Y}_j \in \mathbb{R}^{4,1}.$$

Finally, by the homogeneous nature of the definition, we know that the formula applies regardless of which points

$$\hat{y}_j = r_j y_j \quad (r_j \in \mathbb{R} \setminus \{0\})$$

in ℓ_j are chosen.

Comment 3 Here we give a way to compute the cross ratios using matrices in the Lie algebra $\mathfrak{su}_{1,1}$. Identifying $\mathbb{R}^{2,1}$ with $\mathfrak{su}_{1,1}$ via

$$x_\beta = (x_0, x_1, x_2)^t \sim \begin{pmatrix} \sqrt{-1}x_0 & x_1 + \sqrt{-1}x_2 \\ x_1 - \sqrt{-1}x_2 & -\sqrt{-1}x_0 \end{pmatrix} =: X_\beta,$$

and then taking $X_1, X_2, X_3, X_4 \in \mathfrak{su}_{1,1} \cong \mathbb{R}^{2,1}$, let us denote the eigenvalues of

$$(X_1 - X_2)(X_2 - X_3)^{-1}(X_3 - X_4)(X_4 - X_1)^{-1} \tag{2.28}$$

by λ_1, λ_2.

Lemma 2.5.44 λ_1, λ_2 *are invariant under isometries and homotheties of* $\mathbb{R}^{2,1}$.

Proof It is evident that a homothety $X_\beta \mapsto r X_\beta$ for some $r \in \mathbb{R} \setminus \{0\}$ will not change λ_1, λ_2. A rotation of $\mathbb{R}^{2,1}$ is represented by

$$\mathfrak{su}_{1,1} \ni X \mapsto FXF^{-1} \in \mathfrak{su}_{1,1}$$

for some $F \in \mathrm{SU}_{1,1}$. This transformation also will not change λ_1, λ_2. $\qquad\square$

One can also check the following:

Lemma 2.5.45 *The λ_1 and λ_2 in Lemma 2.5.44 are either real or complex conjugate, that is,*

$$\lambda_1, \lambda_2 \in \mathbb{R} \quad or \quad \overline{\lambda_1} = \lambda_2.$$

Furthermore, when $X_\beta \in L^2$, i.e. when $\det X_\beta = 0$ for $\beta = p, q, r, s$, then

$$\lambda := \lambda_1 = \lambda_2 \in \mathbb{R}.$$

Lemma 2.5.46 *When the four points $X_1, X_2, X_3, X_4 \in L^2$ lie in the plane P as in (2.27), then λ is equal to $\mathrm{cr}(x_1, x_2, x_3, x_4)$.*

Proof Rewriting corresponding x_j as $(1, x_{1,j}, x_{2,j})$, we have

$$(X_1 - X_2)(X_2 - X_3)^{-1}(X_3 - X_4)(X_4 - X_1)^{-1}$$

$$= \prod_{j=1}^{4} \begin{pmatrix} 0 & x_{1,j}-x_{1,j+1}+\sqrt{-1}(x_{2,j}-x_{2,j+1}) \\ x_{1,j}-x_{1,j+1}-\sqrt{-1}(x_{2,j}-x_{2,j+1}) & 0 \end{pmatrix}^{(-1)^{j+1}}$$

$$= \begin{pmatrix} cr & 0 \\ 0 & \overline{cr} \end{pmatrix}$$

with $x_5 := x_1$, where

$$cr = \frac{x_{1,1} - x_{1,2} + \sqrt{-1}(x_{2,1} - x_{2,2})}{x_{1,2} - x_{1,3} + \sqrt{-1}(x_{2,2} - x_{2,3})} \frac{x_{1,3} - x_{1,4} + \sqrt{-1}(x_{2,3} - x_{2,4})}{x_{1,4} - x_{1,1} + \sqrt{-1}(x_{2,4} - x_{2,1})},$$

giving us the desired conclusion. $\qquad\qquad\qquad\qquad\qquad\qquad\qquad\qquad\qquad$ $\square$

Further computation shows:

Lemma 2.5.47 *When $X_1, X_2, X_3, X_4 \in L^2$, for any $r_j \in \mathbb{R} \setminus \{0\}$,*

$$(r_1 X_1 - r_2 X_2)(r_2 X_2 - r_3 X_3)^{-1}(r_3 X_3 - r_4 X_4)(r_4 X_4 - r_1 X_1)^{-1}$$

and (2.28) will have the same eigenvalue $\lambda = \mathrm{cr}(x_1, x_2, x_3, x_4)$.

This also shows that the eigenvalue of (2.28), and hence the cross ratio, does not depend on each scaling of $X_j \in L^2$.

Comment 4 Similarly to Comment 3, we can use the Lie algebra $\mathfrak{su}_2$ to compute the cross ratio in the case of four points $(x_{1,j}, x_{2,j}, x_{3,j}) \in \mathbb{R}^3$ ($j = 1, ..., 4$) by using the identification

$$\mathbb{R}^3 \ni (x_1, x_2, x_3) \sim \begin{pmatrix} \sqrt{-1}x_3 & x_1 + \sqrt{-1}x_2 \\ -x_1 + \sqrt{-1}x_2 & -\sqrt{-1}x_3 \end{pmatrix} =: X \in \mathfrak{su}_2.$$

Then the cross ratio will again be the eigenvalues of (2.28), with the X_j now in $\mathfrak{su}_2$, and the analogous result of Lemma 2.5.44 will still hold. However, there will be no analogue of Lemma 2.5.47 here, as $\mathbb{R}^3$ has no comparable light cone.

2.6 Isothermic Surfaces

Now let x be a surface in some 3-dimensional space form $\mathfrak{R}_\kappa$ with constant sectional curvature κ. Again, we restrict our attention to a local coordinate patch $\Sigma \subset \mathcal{M}$ with coordinates (u, v), so assume without loss of generality that $x : \Sigma \to \mathfrak{R}_\kappa$, and let X be its lift. One can, of course, first consider $L : \Sigma \to P(L^4)$ as a map into the conformal 3-sphere, then obtain $X : \Sigma \to M_\kappa$ using $\mathfrak{q}_\kappa$. Viewing L as a null line bundle over Σ, such X is a *section of L*, denoted by $X \in \Gamma L$. In simpler terms, a nowhere zero $X \in \Gamma L$ if and only if $L = \mathrm{span}\{X\}$ over Σ for smooth X.

After a specific value of κ, and hence a *space form projection*, is chosen, let n denote the unit normal vector for x with respect to the metric of $\mathfrak{R}_\kappa$. One could also choose a different section of the same L so that the projection is into $\mathfrak{R}_0 \cong \mathbb{R}^3$, and denote the unit normal with respect to $\mathfrak{R}_0 \cong \mathbb{R}^3$ by n_0.

Now let $X : \Sigma \to M_\kappa$ be a surface, and $x = \phi_\kappa(X) : \Sigma \to \mathfrak{R}_\kappa$. Then by (2.14), we have that at each point in the domain,

$$X_u = \mathcal{T}_{x_u} \in T_X M_\kappa \quad \text{and} \quad X_v = \mathcal{T}_{x_v} \in T_X M_\kappa.$$

Therefore, taking the unit normal n to x in $\mathfrak{R}_\kappa$ (with respect to the correct metric determined by κ), we see that $\mathcal{T}_n \in T_X M_\kappa$ is the unit normal to the surface X in the tangent space of M_κ.

With the notion of normal available to us, we are now able to consider the *first fundamental form* of a surface X,

$$g = \begin{pmatrix} g_{11} & g_{12} \\ g_{21} & g_{22} \end{pmatrix} := \begin{pmatrix} \langle X_u, X_u \rangle & \langle X_u, X_v \rangle \\ \langle X_v, X_u \rangle & \langle X_v, X_v \rangle \end{pmatrix},$$

while the *second fundamental form* is given by

$$b = \begin{pmatrix} b_{11} & b_{12} \\ b_{21} & b_{22} \end{pmatrix} := \begin{pmatrix} \langle X_{uu}, \mathcal{T}_n \rangle & \langle X_{uv}, \mathcal{T}_n \rangle \\ \langle X_{vu}, \mathcal{T}_n \rangle & \langle X_{vv}, \mathcal{T}_n \rangle \end{pmatrix}.$$

As usual, we define the principal curvatures as the eigenvalues of the matrix $g^{-1}b$. Moreover, we can now similarly define the notion of conformal coordinates and curvature line coordinates for surfaces in M_κ, which allows us to define isothermicity:

Definition 2.6.48 Let $X : \Sigma \to M_\kappa$ be a surface. We call (u, v) *isothermic coordinates* if (u, v) are conformal curvature line coordinates for X. An *isothermic surface* is a surface that admits isothermic coordinates away from umbilics.

Since the first and second fundamental forms are defined in terms of the inner products of $\mathbb{R}^{4,1}$, we can easily see the following:

Theorem 2.6.49 *Isothermicity is Möbius invariant. Moreover, an umbilic point of a surface will remain an umbilic point after the application of Möbius transformations.*

Assumption 2.6.50 We will assume from this point on that the surface

- is defined on a simply-connected domain $\Sigma \subset \mathbb{R}^2$, and
- has no umbilics on Σ,

as we are interested in introducing the transformation theory of isothermic surfaces to obtain the notion of discrete isothermic surfaces. The assumption allows us to assume always that isothermic surfaces admit curvature line coordinates on Σ, and introduce transformation theory based on these coordinates.

Exercise 2.6.51 Show that $X : \Sigma \to M_\kappa$ is isothermic if and only if there exist coordinates (u, v) such that

$$\langle X_u, X_u \rangle = \langle X_v, X_v \rangle, \ \langle X_u, X_v \rangle = 0, \ X_{uv} \in \text{span}\{X_u, X_v\}. \tag{2.29}$$

(Hint: after showing that $\langle X_{uv}, X \rangle = 0$, use the fact that $\langle X, \mathfrak{q}_\kappa \rangle = -1$.)

In the next exercises, we show that this definition of isothermicity is compatible with the notion of isothermicity in space forms $\mathbb{S}^3$ and $\mathbb{H}^3$.

Exercise 2.6.52 Let $\tilde{x} : \Sigma \to \mathbb{S}^3 \subset \mathbb{R}^4$ with unit normal $\tilde{n} : \Sigma \to \mathbb{S}^3$. By Exercise 2.2.16, we have that the lift of $\tilde{x}$ is $X = (1, \tilde{x}^t)^t : \Sigma \to M_1$, and let $\phi_1(X) = x : \Sigma \to \mathfrak{R}_1$. Then show that

$$\mathcal{T}_{\mathrm{d}x} = \mathrm{d}X = \begin{pmatrix} 0 \\ \mathrm{d}\tilde{x} \end{pmatrix}, \quad \mathcal{T}_n = \begin{pmatrix} 0 \\ \tilde{n} \end{pmatrix}.$$

Finally, show that $(u, v) \in \Sigma$ are isothermic coordinates for X if and only if they are isothermic coordinates for $\tilde{x}$.

Exercise 2.6.53 Let $\tilde{x} : \Sigma \to \mathbb{H}^3 \subset \mathbb{R}^{3,1}$ with unit normal $\tilde{n} : \Sigma \to \mathbb{S}^{2,1}$. By Exercise 2.2.17, we have that the lift of $\tilde{x}$ is $X = (\tilde{x}^t, 1)^t : \Sigma \to M_{-1}$, and let $\phi_{-1}(X) = x : \Sigma \to \mathfrak{R}_{-1}$. Then show that

$$\mathcal{T}_{\mathrm{d}x} = \mathrm{d}X = \begin{pmatrix} \mathrm{d}\tilde{x} \\ 0 \end{pmatrix}, \quad \mathcal{T}_n = \begin{pmatrix} \tilde{n} \\ 0 \end{pmatrix}.$$

Finally, show that $(u, v) \in \Sigma$ are isothermic coordinates for X if and only if they are isothermic coordinates for $\tilde{x}$.

We can also define the notion of mean curvature H_κ and Gaussian curvature K_κ using the fundamental forms via (2.9). In the following lemma, we explicitly calculate the mean curvature and Gaussian curvature of a surface X in M_κ using its projection x in $\mathfrak{R}_\kappa$.

Lemma 2.6.54 *Let $L : \Sigma \to P(L^4)$ be isothermic with curvature line coordinates $(u, v) \in \Sigma$, and $X_\kappa \in \Gamma L$ so that $X_\kappa : \Sigma \to M_\kappa$. Suppose that x_0 is the space form projection of $X_0 : \Sigma \to M_0 \cong \mathfrak{R}_0$ and let n_0 be the normal of x_0 (so that $|n_0| = \frac{1}{2}$). Denoting x to be the space form projection of X_κ to $\mathfrak{R}_\kappa$, the mean curvature H_κ is then*

$$H_\kappa = (1 + \kappa|x|^2)H_0 + 2\kappa(x \cdot n_0).$$

Then H_κ is constant exactly when $\partial_u H_\kappa = \partial_v H_\kappa = 0$, which is equivalent to

$$\begin{aligned}
(\partial_u H_0)(1 + \kappa|x|^2) &= \kappa \tfrac{k_1 - k_2}{2} \partial_u(|x|^2), \\
(\partial_v H_0)(1 + \kappa|x|^2) &= \kappa \tfrac{k_2 - k_1}{2} \partial_v(|x|^2),
\end{aligned} \tag{2.30}$$

where the $k_1, k_2 : \Sigma \to \mathbb{R}$ are the principal curvatures of X_0. Furthermore, the Gaussian curvature K_κ of X_κ is

$$K_\kappa = (1 + \kappa|x|^2)^2 K_0 + 4(1 + \kappa|x|^2)H_0\kappa(x \cdot n_0) + 4\kappa^2(x \cdot n_0)^2.$$

Proof Since we have curvature line coordinates, $g_{12} = g_{21} = 0$, and

$$\begin{aligned}
g_{11} &= \langle \mathcal{T}_{x_u}, \mathcal{T}_{x_u} \rangle = \frac{4|x_u|^2}{(1 + \kappa|x|^2)^2}, \\
g_{22} &= \langle \mathcal{T}_{x_v}, \mathcal{T}_{x_v} \rangle = \frac{4|x_v|^2}{(1 + \kappa|x|^2)^2}.
\end{aligned} \tag{2.31}$$

Noting that the unit normal vector to the surface is $\mathcal{T}_n$, where $n = (1 + \kappa|x|^2)n_0$, by isothermicity, we have that $b_{12} = b_{21} = 0$. To calculate the diagonal terms of the second fundamental form, we can use (2.16), with the symbol $'$ denoting either ∂_u or ∂_v to obtain that

$$\begin{aligned}
b_{11} &= \langle X_{uu}, \mathcal{T}_n \rangle = \frac{4}{(1 + \kappa|x|^2)^2}(x_{uu} \cdot n) + \frac{8\kappa|x_u|^2}{(1 + \kappa|x|^2)^3}(x \cdot n), \\
b_{22} &= \langle X_{vv}, \mathcal{T}_n \rangle = \frac{4}{(1 + \kappa|x|^2)^2}(x_{vv} \cdot n) + \frac{8\kappa|x_v|^2}{(1 + \kappa|x|^2)^3}(x \cdot n).
\end{aligned} \tag{2.32}$$

Then we have

$$\frac{b_{11}}{g_{11}} = (1 + \kappa|x|^2)k_1 + 2\kappa(x \cdot n_0), \qquad \frac{b_{22}}{g_{22}} = (1 + \kappa|x|^2)k_2 + 2\kappa(x \cdot n_0). \tag{2.33}$$

Finally, use (2.9) and the fact that $H_0 = (k_1 + k_2)/2$ to obtain the expression for H_κ. The rest of the claims can now be checked directly. $\qquad\square$

In the next exercise, we calculate the Codazzi equations:

Exercise 2.6.55 Let $X : \Sigma \to M_\kappa$ parametrized by curvature line coordinates $(u, v) \in \Sigma$ with unit normal $\mathcal{T}_n$. Denoting the first and second fundamental forms by g and b, we then have that the principal curvatures k_1 and k_2 of X_κ are

$$k_1 = \frac{b_{11}}{g_{11}}, \quad k_2 = \frac{b_{22}}{g_{22}}.$$

(1) Show that $\langle (\mathcal{T}_n)_*, X \rangle = 0$, and deduce $(\mathcal{T}_n)_* \in \mathrm{span}\{X, X_u, X_v, \mathcal{T}_n\}$ for $* = u, v$.
(2) Prove the Rodrigues' equation, that is,

$$(\mathcal{T}_n)_u = -k_1 X_u, \quad \text{and} \quad (\mathcal{T}_n)_v = -k_2 X_v. \tag{2.34}$$

(3) Use the compatibility condition of $\mathcal{T}_n$, i.e. $(\mathcal{T}_n)_{uv} = (\mathcal{T}_n)_{vu}$, to show that

$$(k_1)_v X_u + k_1 X_{uv} = (k_2)_u X_v + k_2 X_{uv},$$

and use this relation to obtain the Codazzi equations, that is,

$$2(k_1)_v = \frac{\partial_v g_{11}}{g_{11}}(k_2 - k_1), \quad 2(k_2)_u = \frac{\partial_u g_{22}}{g_{22}}(k_1 - k_2). \tag{2.35}$$

Remark 2.6.56 When a surface $X : \Sigma \to M_\kappa$ has curvature line coordinates, using the metric in (2.15) with $\kappa = 0$, the first and second fundamental forms are

$$g = \begin{pmatrix} g_{11} & 0 \\ 0 & g_{22} \end{pmatrix}, \quad b = \begin{pmatrix} b_{11} & 0 \\ 0 & b_{22} \end{pmatrix} = \begin{pmatrix} k_1 g_{11} & 0 \\ 0 & k_2 g_{22} \end{pmatrix}.$$

Because $(\mathcal{T}_n)_u = -k_1 X_u$ and $(\mathcal{T}_n)_v = -k_2 X_v$, the third fundamental form is

$$\mathrm{III} = \begin{pmatrix} \langle (\mathcal{T}_n)_u, (\mathcal{T}_n)_u \rangle & \langle (\mathcal{T}_n)_u, (\mathcal{T}_n)_v \rangle \\ \langle (\mathcal{T}_n)_v, (\mathcal{T}_n)_u \rangle & \langle (\mathcal{T}_n)_v, (\mathcal{T}_n)_v \rangle \end{pmatrix} = \begin{pmatrix} k_1^2 g_{11} & 0 \\ 0 & k_2^2 g_{22} \end{pmatrix},$$

and then the Cayley-Hamilton identity is immediate:

$$\mathrm{III} - 2Hb + Kg = 0. \tag{2.36}$$

2.7 Sphere Congruences

We start by stating the notion of sphere congruences:

Definition 2.7.57 A *sphere congruence* is a family of spheres depending smoothly on $(u, v) \in \Sigma$. We say that a sphere congruence and a surface x *envelop each other* if the sphere corresponding to any given point $p \in \Sigma$ is tangent to the surface at $x(p)$.

Remark 2.7.58 As we are working in Möbius geometry where the orientations of spheres matter, we will always assume hereafter that the orientations of a sphere congruence and its enveloping surface align.

It is easy to see that any sphere congruence is represented by a smooth map $\mathcal{S} : \Sigma \to \mathbb{S}^{3,1}$. Building on this, we would like to see a Möbius geometric characterization of a sphere congruence enveloping a given surface x in some space form. As the first step, we will characterize the normal to a surface in M_0 in terms of a special sphere congruence.

Lemma 2.7.59 *Let $X_0 : \Sigma \to M_0$ be a surface with unit normal $\mathcal{T}_{n_0} : \Sigma \to \mathbb{S}^{3,1}$, and denote the projection of X_0 to $\mathfrak{R}_0 \cong \mathbb{R}^3$ as x. Then $\mathcal{T}_{n_0}$ projects to a tangent plane congruence (enveloping x).*

Proof Fixing some point in the domain, comparing the form of $\mathcal{T}_{n_0}$ and (2.22) tells us that $\mathcal{T}_{n_0}$ projects to a plane through point x with normal n_0 in $\mathfrak{R}_0 \cong \mathbb{R}^3$. Therefore, at each point in the domain, $\mathcal{T}_{n_0}$ projects to the tangent plane of x. $\square$

In fact, we can show that the above Lemma 2.7.59 is true for any choice of space form (where the tangent planes are now tangent geodesic planes).

Lemma 2.7.60 *Let $L : \Sigma \to P(L^4)$ and $X_\kappa \in \Gamma L$ be a surface in M_κ (so that $\mathrm{span}\{X_\kappa\} = L$) with unit normal $\mathcal{T}_n : \Sigma \to \mathbb{S}^{3,1}$. Denoting the projection to $\mathfrak{R}_\kappa$ as x, we have that $\mathcal{T}_n$ projects to a tangent geodesic plane congruence (enveloping x).*

Proof To show the tangency, we let $X_0 \in \Gamma L$ be a surface in M_0 with unit normal $\mathcal{T}_{n_0} : \Sigma \to \mathbb{S}^{3,1}$. It is not difficult to see that the projection of X_0 to $\mathfrak{R}_0 \cong \mathbb{R}^3$ is also x (albeit with different induced metric). Then through direct calculation, we can see using $n = (1 + \kappa |x|^2)n_0$ that

$$\mathcal{T}_n = \mathcal{T}_{n_0} - \frac{2\kappa x \cdot n_0}{1 + \kappa |x|^2} X_0.$$

Thus, Lemma 2.3.28 tells us that $\mathcal{T}_n$ projects to a sphere tangent to x on each point of the domain. A sphere being tangent to a surface is a Möbius invariant quality; therefore, the conformal equivalence of different space forms gives us that $\mathcal{T}_n$ projects to a sphere congruence enveloping the surface in the space form. Finally, the form of $\mathcal{T}_n$ tells us that $\langle \mathcal{T}_n, \mathfrak{q}_\kappa \rangle = 0$, so Remark 2.3.24 tells us that $\mathcal{T}_n$ projects to geodesic planes. $\square$

We are now ready to state the Möbius geometric characterization of sphere congruences enveloping a surface.

Proposition 2.7.61 *Let $L : \Sigma \to P(L^4)$ be a surface with any $X_\kappa \in \Gamma L$ in M_κ, and let $\mathcal{S} : \Sigma \to \mathbb{S}^{3,1}$. Denote the space form projection of X_κ as x. Then $\mathcal{S}$ projects to a sphere congruence enveloping x if and only if $\langle X, \mathcal{S} \rangle = 0 = \langle \mathrm{d}X, \mathcal{S} \rangle$ for some (and in fact any) $X \in \Gamma L$.*

Proof Suppose that $\langle X, \mathcal{S} \rangle = 0 = \langle \mathrm{d}X, \mathcal{S} \rangle$, then $\mathcal{S} \in \mathrm{span}\{X, X_u, X_v\}^\perp$. Since we have that the signature of $\mathrm{span}\{X, X_u, X_v\}$ is $(+ + 0)$, we see that $\mathrm{span}\{X, X_u, X_v\}^\perp$ has signature $(+0)$. On the other hand, since X and X_κ are both sections of L, we have that $X = \alpha X_\kappa$ for some real function α, implying that $\mathrm{span}\{X, X_u, X_v\} = \mathrm{span}\{X_\kappa, (X_\kappa)_u, (X_\kappa)_v\}$. Therefore, $\mathrm{span}\{X_\kappa, \mathcal{T}_n\} = \mathrm{span}\{X, X_u, X_v\}^\perp$, allowing us to write

$$\mathcal{S} = \mathcal{T}_n + t X_\kappa$$

for some $t \in \mathbb{R}$ since $\mathcal{S}$ is unit length. By Lemmata 2.3.28 and 2.7.60, $\mathcal{S}$ projects to a sphere congruence enveloping x. A simple reversal of the argument then gives the equivalence. $\qquad\qquad\qquad\qquad\qquad\qquad\qquad\qquad\qquad\qquad\qquad\qquad\quad\square$

We now would like to discuss some other special sphere congruences enveloping a surface, but before this, we first calculate the mean curvature of a given sphere via an exercise:

Exercise 2.7.62 Let $\mathcal{S} = (z_0, z, z_4)^t \in \mathbb{R}^{4,1}$ with $\|\mathcal{S}\|^2 = 1$, and let $\tilde{\mathcal{S}}$ be the sphere determined by $\mathcal{S}$ using (2.17). Using Example 2.3.20 and Lemma 2.6.54, show that when projected to a sphere in M_κ, $\tilde{\mathcal{S}}$ has mean curvature

$$H_\kappa = \frac{z_0 + z_4 + \kappa(z_0 - z_4)}{2}.$$

(Hint: If x is the projection of the sphere $\tilde{\mathcal{S}}$ to $\mathfrak{R}_0 \cong \mathbb{R}^3$ with unit normal n_0, then $x = -rn_0 + c$ for center c and radius r, with mean curvature H_0 inverse its radius.)

Exercise 2.7.63 In Exercise 2.3.25, we have seen that for $\kappa = \pm 1$, a sphere represented by $\mathcal{S}$ projects to a geodesic plane in $\mathfrak{R}_\kappa$ if $\langle \mathcal{S}, \mathfrak{q}_\kappa \rangle = 0$. Use Exercise 2.7.62 to show that this is true for any fixed κ.

Exercise 2.7.64 For $X : \Sigma \to M_\kappa$ with unit normal $\mathcal{T}_n : \Sigma \to \mathbb{S}^{3,1}$, let $x : \Sigma \to \mathfrak{R}_\kappa$ be the space form projection, and so n is the unit normal to x. Show that if

$$\mathcal{S}_t := \mathcal{T}_n + t X,$$

then $\mathcal{S}_t$ project to the spheres with mean curvature t. (Hint: Using Lemma 2.3.28, deduce that $\mathcal{S}_t$ represent all of the tangent spheres to x, and then use Exercise 2.7.62 to calculate the mean curvature of the sphere $\mathcal{S}_t$ with respect to the space form M_κ.)

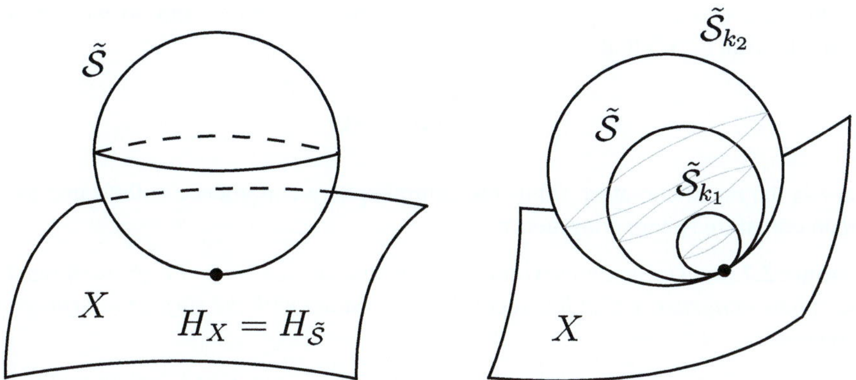

Fig. 2.8 The left figure depicts a mean curvature sphere, while the right figure shows the situation described in Lemma 2.7.65, where the two principal curvature spheres are related by inversion through the sphere $\tilde{\mathcal{S}}$

A *mean curvature sphere congruence* is a sphere congruence enveloping the surface such that each sphere has mean curvature coinciding with that of the corresponding point on the surface, regardless of the choice of space form (i.e. the choice of value κ) (Fig. 2.8). Similarly, a *principal curvature sphere*, also referred to as *curvature sphere*, is a sphere tangent to the surface with the same mean curvature as the principal curvature at that point.

Lemma 2.7.65 (cf. [111, Lemma 3.4.4]) *The mean curvature sphere congruence $\mathcal{S}$ can also be characterized as the* central sphere congruence, *i.e. the sphere congruence whose spheres exchange the principal curvature spheres via inversion (Fig. 2.8).*

Proof Let X a surface in M_κ with space form projection x in $\mathfrak{R}_\kappa$. Take $\mathcal{T}_n : \Sigma \to \mathbb{S}^{3,1}$, the tangent geodesic plane congruence, that is,

$$\langle \mathcal{T}_n, \mathfrak{q}_\kappa \rangle = \langle \mathcal{T}_n, X \rangle = \langle \mathcal{T}_n, dX \rangle = 0,$$

with $\mathfrak{q}_\kappa$ as in (2.12). With $\mathcal{T}_n = (z_0, z, z_4)^t$, in particular we have that

$$(z_0 - z_4)\kappa + z_0 + z_4 = 0.$$

By Exercise 2.7.64, we have that $\mathcal{S}_{k_1}$ and $\mathcal{S}_{k_2}$ are the principal curvature spheres. So by Lemma 2.3.29, when $\mathcal{S}$ is the central sphere congruence, we have

$$- \mathcal{S}_{k_2} = \mathcal{S}_{k_1} - 2\langle \mathcal{S}_{k_1}, \mathcal{S} \rangle \mathcal{S}. \tag{2.37}$$

Note that, as we wish to have an inversion that preserves orientation rather than reversing it, we changed $\mathcal{S}_{k_2}$ to $- \mathcal{S}_{k_2}$ here.

We have that $\mathcal{S} = \mathcal{S}_t$ as in Exercise 2.7.64 for some t, and so we can now compute from (2.37) that

$$t = \frac{1}{2}(k_1 + k_2),$$

i.e. t is the mean curvature. Thus the central sphere congruence is the same as the mean curvature sphere congruence. $\qquad\square$

Lemma 2.7.66 *The mean curvature sphere congruence $\mathcal{S}$ can be characterized as the sphere congruence that has second order contact with the surface in orthogonal directions.*

Proof Principal curvature spheres, second order contact and orthogonality are examples of notions that are invariant under Möbius transformations. Because only Möbius invariant notions appear in this proof, without loss of generality we may assume that the surface $X : \Sigma \to M_0 \cong \mathbb{R}^3$.

Let $\mathcal{S}$ be the mean curvature sphere at a point $X(u_0, v_0)$ of the surface. Then $X(u_0, v_0)$ is one point of the sphere $\mathcal{S}$. Let p be a different point in $\mathcal{S}$ and let $\mathcal{S}_1$ be a sphere with center p that intersects $\mathcal{S}$ transversally. We apply inversion $f_{\mathcal{S}_1}$ of $\mathbb{R}^3$ through the sphere $\mathcal{S}_1$, so that the point p is mapped to infinity and the sphere $\mathcal{S}$ is thus mapped to a flat plane $f_{\mathcal{S}_1}(\mathcal{S})$. The image $f_{\mathcal{S}_1}(X(u, v))$ of $X(u, v)$ under inversion will satisfy $H = 0$ at the point $f_{\mathcal{S}_1}(X(u_0, v_0))$. Thus the asymptotic directions of $f_{\mathcal{S}_1}(X(u, v))$ at that point are perpendicular to each other, and are also the directions of second order contact with $f_{\mathcal{S}_1}(\mathcal{S})$. Since we have been working only with Möbius invariant notions, the lemma follows. $\qquad\square$

We recover another characterization of the mean curvature sphere congruence:

Lemma 2.7.67 *Let $X : \Sigma \to M_\kappa$ be an isothermic surface. Then the mean curvature sphere congruence $\mathcal{S}$ can be characterized as the* conformal Gauss map *of the surface X, i.e. the (unique) sphere congruence with the same induced conformal structure as X.*

Proof Assume that $(u, v) \in \Sigma$ are isothermic coordinates so that

$$\langle dX, dX \rangle = g_{11}(du^2 + dv^2),$$

and denote the principal curvatures of X by k_1 and k_2. Now, let $\mathcal{S}$ be any sphere congruence enveloping X so that $\mathcal{S} = \alpha X + \mathcal{T}_n$. Then since $d\mathcal{S} = d\alpha\, X + \alpha dX + d\mathcal{T}_n$, we can use $\langle X, X \rangle = \langle X, dX \rangle = \langle X, d\mathcal{T}_n \rangle = 0$ to calculate that

$$\langle d\mathcal{S}, d\mathcal{S} \rangle = \langle \alpha dX + d\mathcal{T}_n, \alpha dX + d\mathcal{T}_n \rangle$$

$$= \langle (\alpha - k_1)X_u du + (\alpha - k_2)X_v dv, (\alpha - k_1)X_u du + (\alpha - k_2)X_v dv \rangle$$

$$= (\alpha - k_1)^2 g_{11} du^2 + (\alpha - k_2)^2 g_{11} dv^2.$$

Therefore, $\mathcal{S}$ is conformally equivalent to X if and only if $\alpha = \frac{1}{2}(k_1 + k_2) = H_\kappa$, i.e. $\mathcal{S}$ is the mean curvature sphere congruence. $\square$

2.8   Cross Ratio Characterization of Isothermicity

Let $X : \Sigma \to M_\kappa$ with normal $\mathcal{T}_n : \Sigma \to \mathbb{S}^{3,1}$ parametrized by curvature line coordinates $(u, v) \in \Sigma$. One can always stretch the coordinates, so that $X(u, v) = X(u(\tilde{u})), v(\tilde{v}))$ for any strictly monotonic functions u depending only on $\tilde{u}$, and v depending only on $\tilde{v}$. Note that $\langle X_{\tilde{u}}, X_{\tilde{v}} \rangle = 0$, and $X_{\tilde{u}\tilde{v}} = X_{uv}\frac{du}{d\tilde{u}}\frac{dv}{d\tilde{v}}$ implies $\langle X_{\tilde{u}\tilde{v}}, \mathcal{T}_n \rangle = 0$, so $(\tilde{u}, \tilde{v})$ are also curvature line coordinates. Under such setting we have:

Lemma 2.8.68 ([64]) *The surface is isothermic if and only if*

$$\frac{g_{11}}{g_{22}} = \frac{a(u)}{b(v)},$$

where the function a depends only on u, and b depends only on v.

Exercise 2.8.69 Prove Lemma 2.8.68.

Now consider the cross ratio

$$\mathrm{cr}_\epsilon(u, v) := \mathrm{cr}(X(u, v), X(u + \epsilon, v), X(u + \epsilon, v + \epsilon), X(u, v + \epsilon)).$$

A computation gives

$$\lim_{\epsilon \to 0} \mathrm{cr}_\epsilon = -\frac{g_{11}}{g_{22}}, \tag{2.38}$$

giving us a cross ratio characterization of isothermic surfaces:

Theorem 2.8.70 *Let $X : \Sigma \to M_\kappa$ be a surface with curvature line coordinates (u, v). Then X is isothermic if and only if*

$$\lim_{\epsilon \to 0} \mathrm{cr}_\epsilon = -\frac{a(u)}{b(v)} \tag{2.39}$$

for some functions of one variable $a(u)$ and $b(v)$.

Now consider a map $L : \Sigma \to P(L^4)$ into the projective light cone. For $X : \Sigma \to M_\kappa$ such that $L = \mathrm{span}\{X\}$, or $X \in \Gamma L$, we have that if X is isothermic, then the cross ratio condition (2.39) holds. However, for any other section $X_1 \in \Gamma L$ (so that $X_1 = \alpha X$ for some nowhere zero function α), the cross ratio condition holds as well, allowing us to define a notion of isothermicity for maps into the conformal 3-sphere.

Lemma 2.8.71 *A surface in a space form is isothermic if and only if every lift $\alpha(u, v)X(u, v)$ of the surface is isothermic. In this case, we call $L = \mathrm{span}\{X\}$ isothermic.*

The cross ratio characterization of isothermicity is also central to the theory of discrete isothermic surfaces. This characterization does not involve any stretching of coordinates, which we would not be able to do in the case of discrete surfaces, one of our primary motivations here.

Now that we have the notion of isothermicity available for $L : \Sigma \to P(L^4)$, we have the following:

Exercise 2.8.72 Using Exercise 2.6.51, show that if $L : \Sigma \to P(L^4)$ is isothermic then any non-zero section $X \in \Gamma L$ satisfies

$$\langle X_u, X_u \rangle = \langle X_v, X_v \rangle, \quad \langle X_u, X_v \rangle = 0, \quad X_{uv} \in \mathrm{span}\{X_u, X_v, X\}. \tag{2.40}$$

Exercise 2.8.73 Using Exercise 2.6.51, show that $L : \Sigma \to P(L^4)$ is isothermic if there exists a non-zero section $X \in \Gamma L$ satisfying (2.40). (Hint: define $\tilde{X} \in \Gamma L$ by $\tilde{X} := \beta X$ where $\beta = -\frac{1}{\langle X, \mathfrak{q}_\kappa \rangle}$ so that $\tilde{X} : \Sigma \to M_\kappa$, and show that this $\tilde{X}$ satisfies (2.29).)

2.9   Moutard Lifts

Given an immersion $X : \Sigma \to M_\kappa$ into some space form, let $L = \mathrm{span}\{X\} : \Sigma \to P(L^4)$. If there is some non-zero section $F \in \Gamma L$ such that F satisfies the *Moutard equation* [159], that is,

$$F_{uv} \parallel F, \tag{2.41}$$

then we say that F is a *Moutard lift*.

Theorem 2.9.74 *L is isothermic if and only if L admits Moutard lifts.*

Proof For one direction, let L be isothermic with isothermic coordinates u, v, and $X \in \Gamma L$. Applying Exercise 2.8.72 to X, we have that

$$\langle X_u, X_u \rangle = \langle X_v, X_v \rangle =: \xi.$$

Defining $F \in \Gamma L$ by

$$F := \frac{1}{\sqrt{\xi}} X, \tag{2.42}$$

we have that

$$\langle F_u, F_u \rangle = \frac{1}{\xi} \langle X_u, X_u \rangle = 1 = \frac{1}{\xi} \langle X_v, X_v \rangle = \langle F_v, F_v \rangle.$$

Exercise 2.8.72 also implies that there exist real-valued functions A, B, C so that

$$F_{uv} = A F_u + B F_v + C F,$$

and taking the inner product of this with F_u and with F_v tells us that $A = B = 0$; therefore F is Moutard.

For the other direction, suppose that $F \in \Gamma L$ is Moutard; equivalently, assume that for some real function α, we have that $F_{uv} = \alpha F$. Then we have

$$0 = -\alpha \langle F, F \rangle = -\langle F_{uv}, F \rangle = \langle F_u, F_v \rangle - (\langle F_u, F \rangle)_v = \langle F_u, F_v \rangle.$$

Meanwhile,

$$(\langle F_u, F_u \rangle)_v = 2\langle F_{uv}, F_u \rangle = 0 \quad \text{and} \quad (\langle F_v, F_v \rangle)_u = 2\langle F_{uv}, F_v \rangle = 0.$$

Thus, we can scale each coordinate independently as in the setting of Lemma 2.8.68 so that

$$\langle F_u, F_u \rangle = \langle F_v, F_v \rangle.$$

Finally, Exercise 2.8.73 tells us that L is isothermic. $\qquad \square$

From the above proof, we see that the Moutard equation holds with respect to u, v if and only if they are curvature line coordinates for an isothermic surface.

2.10 Christoffel Transformation

We now explain the Christoffel transformation x^*, or "dual surface" of an isothermic surface x in $\mathbb{R}^3$. Let $x : \Sigma \to \mathbb{R}^3$ be a surface in $\mathbb{R}^3$ with mean curvature H_0 and unit normal n_0 (with respect to the metric in (2.15)). The Christoffel transformation $x^* : \Sigma \to \mathbb{R}^3$ satisfies that (see Fig. 2.9)

- x^* has the same conformal structure as x,
- and x and x^* have parallel tangent planes with opposite orientations at corresponding points.

Automatically, the corresponding principal curvature directions of x and x^* will be parallel.

This description above of the Christoffel transformations turns out to be equivalent to the following definition, and the existence of the integrating factor ρ below is equivalent to the existence of isothermic coordinates.

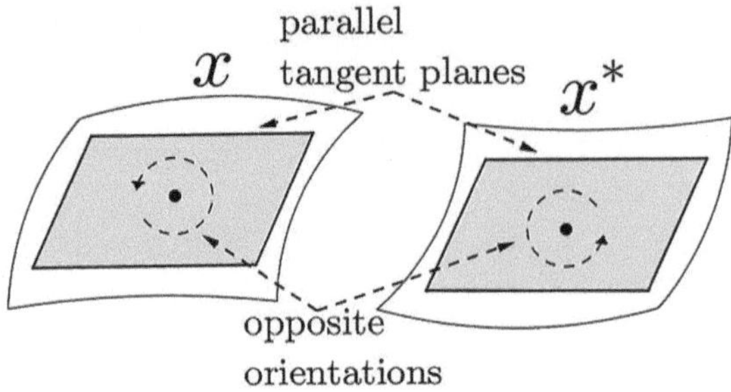

Fig. 2.9 The Christoffel transformation

Definition 2.10.75 ([117, Equation (38)]) A *Christoffel transform* $x^* : \Sigma \to \mathbb{R}^3$ of a surface $x : \Sigma \to \mathbb{R}^3$ is a surface that satisfies

$$\mathrm{d}x^* = \rho(\mathrm{d}n_0 + H_0\mathrm{d}x)$$

for some nonzero $\rho : \Sigma \to \mathbb{R}$ on the surface x.

In the next lemma, we identify the condition for such x^* to exist.

Lemma 2.10.76 *A Christoffel transform x^* exists if and only if x is isothermic.*

Proof First we prove one direction, by assuming x is isothermic and then showing x^* exists. Take x to be isothermic, and take isothermic coordinates u, v for x, so $x_{uv} = Ax_u + Bx_v$ for some A, B. Defining $\mathbb{R}^3$–valued 1-form η via

$$\eta := -\frac{x_u}{|x_u|^2}\mathrm{d}u + \frac{x_v}{|x_v|^2}\mathrm{d}v, \tag{2.43}$$

we can verify using isothermicity that

$$(\eta(\partial_u))_v = -\frac{x_{uv}}{|x_u|^2} + \frac{2x_u \cdot x_{uv}}{|x_u|^4}x_u = \frac{x_{uv}}{|x_v|^2} - \frac{2x_v \cdot x_{uv}}{|x_v|^4}x_v = (\eta(\partial_v))_u.$$

Thus, there exists an x^* such that $\mathrm{d}x^* = \eta$ (since $x_{uv}^* = x_{vu}^*$). Also,

$$\mathrm{d}n_0 + H_0\mathrm{d}x = \tfrac{1}{8}(b_{11} - b_{22})\left(-\frac{x_u}{|x_u|^2}\mathrm{d}u + \frac{x_v}{|x_v|^2}\mathrm{d}v\right),$$

implying that x^* is a Christoffel transform, since $b_{11} - b_{22} \neq 0$ at non-umbilic points.

Now we prove the other direction, by assuming x^* exists and then showing that x is isothermic. Then existence of x^* gives

$$\mathrm{d}(\rho\mathrm{d}n_0 + \rho H_0\mathrm{d}x) = 0,$$

from which it follows via curvature line coordinates (u, v) that

$$\begin{pmatrix} 0 & \frac{b_{11}}{g_{11}} - \frac{b_{22}}{g_{22}} \\ \frac{b_{22}}{g_{22}} - \frac{b_{11}}{g_{11}} & 0 \end{pmatrix} \begin{pmatrix} \rho_u \\ \rho_v \end{pmatrix} = \rho \begin{pmatrix} \left(\frac{b_{11}}{g_{11}} + \frac{b_{22}}{g_{22}} \right)_v \\ \left(\frac{b_{11}}{g_{11}} + \frac{b_{22}}{g_{22}} \right)_u \end{pmatrix}. \tag{2.44}$$

Then because $\rho_{uv} = \rho_{vu}$, we have

$$\left(\frac{(k_2 + k_1)_v}{k_1 - k_2} \right)_u = \left(\frac{(k_1 + k_2)_u}{k_2 - k_1} \right)_v,$$

which implies

$$\frac{2(((k_1)_v)_u + ((k_2)_u)_v)}{k_1 - k_2} + 2(k_2 - k_1)^{-2} \left((k_1)_v(k_2 - k_1)_u + (k_2)_u(k_2 - k_1)_v \right) = 0.$$

Using the Codazzi equations (2.35), we have

$$\left(\log \frac{g_{11}}{g_{22}} \right)_{uv} = 0.$$

In particular, there exist positive functions $a(u)$ and $b(v)$ depending only on u and v, respectively, so that

$$(a(u))^2 g_{11} = (b(v))^2 g_{22}.$$

Therefore, Lemma 2.8.68 tells us that x is isothermic. $\square$

With respect to isothermic coordinates (u, v), (2.44) implies

$$\rho_u = -\frac{g_{11} \partial_u H_0}{g_{11} H_0 - b_{22}} \rho, \qquad \rho_v = -\frac{g_{11} \partial_v H_0}{g_{11} H_0 - b_{11}} \rho. \tag{2.45}$$

The existence of x^* then automatically implies the compatibility condition $(\rho_u)_v = (\rho_v)_u$, with ρ_u and ρ_v as in the right-hand sides of the equations in (2.45). This pair of Eqs. (2.45) tells us that ρ is uniquely determined once its value is chosen at a single point, and thus the solution ρ is unique up to scalar multiplication by a constant factor. Thus the Christoffel transformation x^* is uniquely determined up to homothety and translation via (2.43):

$$\mathrm{d}x^* = -\frac{x_u}{|x_u|^2} \mathrm{d}u + \frac{x_v}{|x_v|^2} \mathrm{d}v.$$

This form tells us that x^* satisfies all the geometric requirements of a Christoffel transform listed at the beginning of this section. Furthermore, we can now easily prove:

Corollary 2.10.77 *Christoffel transformation is involutive, that is, $(x^*)^* = x$ up to homothety and translation.*

Proof Using (2.43), we have that

$$d((x^*)^*) = -\frac{x_u^*}{|x_u^*|^2}du + \frac{x_v^*}{|x_v^*|^2}dv = x_u du + x_v dv = dx,$$

giving us the desired conclusion.   □

Exercise 2.10.78 Let $x : \Sigma \to \mathbb{R}^3$ be an isothermic surface with lift $X_0 : \Sigma \to M_0$, and let x^* be the Christoffel transformation of x. Show that

$$\mathcal{T}_{dx^*} = -\frac{1}{|x_u|^2}X_{0,u}du + \frac{1}{|x_v|^2}X_{0,v}dv.$$

(Hint: $x_u^* = -|x_u|^{-2}x_u$ implies that x_u^* is in the tangent space of x.)

Remark 2.10.79 The function ρ in Definition 2.10.75 is a constant scalar multiple of the multiplicative inverse of the mean curvature of x^*, seen as follows: assuming $(x^*)^* = x$, we have that if the normal of x is n_0, then the normal of x^* is $-n_0$. Thus

$$dx = d((x^*)^*) = \rho^*(dn_0^* + H_0^* dx^*) = \rho^*(-dn_0 + H_0^* \rho(dn_0 + H_0 dx)),$$

and so

$$(1 - \rho\rho^* H_0 H_0^*)dx = (H_0^* \rho\rho^* - \rho^*)dn_0.$$

Since dx and dn_0 are linearly independent away from umbilic points, it follows that

$$\rho H_0^* = \rho^* H_0 = 1.$$

Remark 2.10.80 When H_0 is a nonzero constant and we have isothermic coordinates, (2.45) implies ρ is constant. Then the Christoffel transform x^* and the parallel constant mean curvature (cmc) surface $x^{\|} = x + H_0^{-1}n_0$ differ by only a homothety and translation of $\mathbb{R}^3$. Thus the Christoffel transformation is essentially the same as the parallel cmc surface to x.

Example 2.10.81 The round cylinder gives one simple example of a Christoffel transform's orientation reversing property. For the cylinder $x(u, v) = (\cos u, \sin u, v)$ in $\mathbb{R}^3$ (with metric as in (2.15)), the normal vector is $n_0 = \frac{1}{2}(\cos u, \sin u, 0)$, and the Christoffel transform is $x^*(u, v) = (-\cos u, -\sin u, v)$ with its normal vector $n_0^* = \frac{1}{2}(-\cos u, -\sin u, 0)$. Thus $n_0^* = -n_0$.

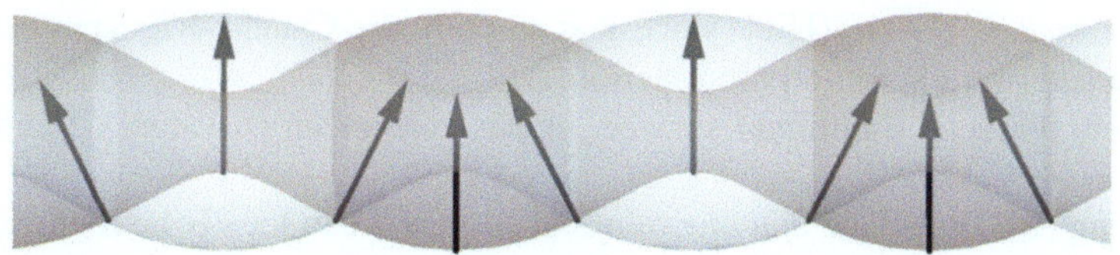

Fig. 2.10 A Delaunay unduloid and its dual surface (Christoffel transform), which is a translated copy of the same surface. The arrows are the unit normals to the dark surface

Example 2.10.82 Delaunay surfaces are cmc surfaces of revolution in $\mathbb{R}^3$. They can be either embedded (unduloids) or nonembedded (nodoids). The Christoffel transforms of Delaunay surfaces are again Delaunay surfaces (see Fig. 2.10).

Using the Christoffel transformations, we can prove some basic facts about cmc surfaces. First we have the following lemma, which is immediate from the proof of Lemma 2.10.76.

Lemma 2.10.83 *We have the relation*

$$\mathrm{d}x^* = \frac{2}{(k_1 - k_2)|x_u|^2}(\mathrm{d}n_0 + H_0\mathrm{d}x).$$

The Hopf differential for a surface in $\mathbb{R}^3$ is defined as (the inner products $\langle \cdot, \cdot \rangle$ and "$\cdot$" are bilinearly extended to apply to complex vectors)

$$\hat{Q}\mathrm{d}z^2, \quad \hat{Q} = \langle \mathcal{T}_{n_0}, (X_0)_{zz} \rangle = 4(n_0 \cdot x_{zz}) \qquad (z = u + \sqrt{-1}v)$$

where (u, v) are conformal coordinates.

Corollary 2.10.84 *If H_0 is constant for the surface $x = x(u, v)$ in $\mathbb{R}^3$ with isothermic coordinates (u, v), then the factor $\hat{Q}$ of the Hopf differential is a real constant.*

Proof The factor

$$\hat{Q} = 4\left(n_0 \cdot \tfrac{1}{4}(x_{uu} - x_{vv})\right) = (k_1 - k_2)|x_u|^2$$

is constant by Lemma 2.10.83 and Remark 2.10.80. It is clearly also real.   □

In the next exercise, we show that the Weierstrass representation of minimal surfaces [218] arises as Christoffel transformations:

Exercise 2.10.85 Let $x : \Sigma \to \mathbb{R}^3$ admit isothermic coordinates (u, v) with unit normal $n_0 : \Sigma \to \mathbb{S}^2$.

(1) Denoting stereographic projection from $\mathbb{S}^2$ as $St : \mathbb{S}^2 \to \mathbb{R}^2 \cong \mathbb{C}$, show that n_0 is isothermic if and only if $g := St \circ n_0 : \Sigma \to \mathbb{C}$ is holomorphic with respect

to $z = u + \sqrt{-1}v$. (Hint: since g is smooth, the Cauchy-Riemann conditions characterize a holomorphic function.)

(2) Show that x is a minimal surface, or has zero mean curvature, if and only if $x^* = n_0$.

(3) Using (2.43), deduce that any minimal surface x can locally be represented as

$$x = \mathrm{Re} \int (1 - g^2, \sqrt{-1}(1 + g^2), 2g)\frac{1}{g_z}\, dz$$

for some holomorphic function $g : \Sigma \to \mathbb{C}$.

Chapter 3
From Smooth to Discrete via Permutability

3.1 Wedge Product

We first explain the wedge product between vectors and differential forms that appears in this text. Note that we use the specific case of $\mathbb{R}^{4,1}$ to explain these concepts, but the definitions can be easily generalized for any $\mathbb{R}^{p,q}$. In this text, we use the fact that the exterior algebra $\wedge^2 \mathbb{R}^{4,1}$ is isomorphic to the Lie algebra $\mathfrak{o}_{4,1}$ without further comment, via

$$\wedge^2 \mathbb{R}^{4,1} \ni A \wedge B \mapsto A \wedge B \in \mathfrak{o}_{4,1}$$

for some $A, B \in \mathbb{R}^{4,1}$, where

$$(A \wedge B)X = \langle A, X \rangle B - \langle B, X \rangle A \tag{3.1}$$

for any vector $X \in \mathbb{R}^{4,1}$. The fact that $A \wedge B$ as in (3.1) is in the Lie algebra $\mathfrak{o}_{4,1}$ can be directly checked:

Lemma 3.1.1 *$A \wedge B$ lies in the Lie algebra $\mathfrak{o}_{4,1}$.*

Proof For

$$D = \begin{pmatrix} -1 & 0 & 0 & 0 & 0 \\ 0 & 1 & 0 & 0 & 0 \\ 0 & 0 & 1 & 0 & 0 \\ 0 & 0 & 0 & 1 & 0 \\ 0 & 0 & 0 & 0 & 1 \end{pmatrix}, \quad A = \begin{pmatrix} A_0 \\ A_1 \\ A_2 \\ A_3 \\ A_4 \end{pmatrix}, \quad B = \begin{pmatrix} B_0 \\ B_1 \\ B_2 \\ B_3 \\ B_4 \end{pmatrix},$$

the matrix $\mathfrak{A}$ that represents $A \wedge B$ is

© The Author(s), under exclusive license to Springer Nature Switzerland AG 2025
J. Cho et al., *Discrete Isothermic Surfaces in Lie Sphere Geometry*, Lecture Notes in Mathematics 2375, https://doi.org/10.1007/978-3-031-95592-1_3

$$\begin{pmatrix} 0 & A_1B_0 - A_0B_1 & A_2B_0 - A_0B_2 & A_3B_0 - A_0B_3 & A_4B_0 - A_0B_4 \\ A_1B_0 - A_0B_1 & 0 & A_2B_1 - A_1B_2 & A_3B_1 - A_1B_3 & A_4B_1 - A_1B_4 \\ A_2B_0 - A_0B_2 & A_1B_2 - A_2B_1 & 0 & A_3B_2 - A_2B_3 & A_4B_2 - A_2B_4 \\ A_3B_0 - A_0B_3 & A_1B_3 - A_3B_1 & A_2B_3 - A_3B_2 & 0 & A_4B_3 - A_3B_4 \\ A_4B_0 - A_0B_4 & A_1B_4 - A_4B_1 & A_2B_4 - A_4B_2 & A_3B_4 - A_4B_3 & 0 \end{pmatrix}$$

which satisfies the condition $\mathfrak{A}\,D + D\,\mathfrak{A}^t = 0$ to lie in $\mathfrak{o}_{4,1}$. $\square$

Remark 3.1.2 Note that the bilinearity and skew-symmetricity of wedge product imply that for any $\alpha_1, \alpha_2 \in \mathbb{R}$ and $X, Y \in \mathbb{R}^{4,1}$,

(1) $B \wedge B = 0$,
(2) $(\alpha_1 A) \wedge (\alpha_2 B) = \alpha_1\alpha_2(A \wedge B)$,
(3) $\langle X, (A \wedge B)Y \rangle = -\langle (A \wedge B)X, Y \rangle$, and
(4) $\|(A \wedge B)X\|^2 = -\langle X, (A \wedge B)^2 X \rangle$.

Lemma 3.1.3 *If* $T \in \mathfrak{o}_{4,1}$ *and* $A, B \in \mathbb{R}^{4,1}$, *then*

$$[T, A \wedge B] = (TA) \wedge B + A \wedge (TB).$$

Proof For $C \in \mathbb{R}^{4,1}$,

$$\langle A, TC \rangle = A^t DTC = -A^t T^t DC = -\langle TA, C \rangle.$$

Then

$$\begin{aligned} [T, A \wedge B]C &= T(A \wedge B)C - (A \wedge B)TC \\ &= T(\langle A, C \rangle B - \langle B, C \rangle A) - (\langle A, TC \rangle B - \langle B, TC \rangle A) \\ &= \langle A, C \rangle TB - \langle B, C \rangle TA - \langle A, TC \rangle B + \langle B, TC \rangle A \\ &= \langle A, C \rangle TB - \langle B, C \rangle TA + \langle TA, C \rangle B - \langle TB, C \rangle A \\ &= ((TA) \wedge B)C + (A \wedge (TB))C, \end{aligned}$$

giving us the desired conclusion. $\square$

Remark 3.1.4 A second proof of Lemma 3.1.3 can be given as follows: Let $g(t)$ be a curve in $O_{4,1}$ such that $g(0) = I$ and $g'(0) = T$. Then

$$\begin{aligned} (g(A \wedge B)g^{-1})C &= g(\langle A, g^{-1}C \rangle B - \langle B, g^{-1}C \rangle A) \\ &= \langle gA, C \rangle (gB) - \langle gB, C \rangle (gA) = ((gA) \wedge (gB))C\,, \end{aligned}$$

so

$$\partial_t (g(A \wedge B)g^{-1})|_{t=0} = \partial_t ((gA) \wedge (gB))|_{t=0},$$

which implies

$$[g'(0), A \wedge B] = (g'(0)A) \wedge B + A \wedge (g'(0)B).$$

The wedge product defined above is different from the wedge product of two differential forms. Here we contemplate two types of wedge products we use in the text for multiplying two differential forms. Suppose $X, A, B, C, D : \Sigma \to V$, and let $\mathcal{B} : V \times V \to W$ be a bilinear map for some vector spaces V and W. We then have that the wedge product of a V-valued 0-form X and a 1-form $\omega = A du + B dv$ with respect to $\mathcal{B}$ is given by the W-valued 1-form

$$X \wedge \omega = \mathcal{B}(X, A) du + \mathcal{B}(X, B) dv,$$

while the wedge product of V-valued 1-forms $\omega = A du + B dv$ and $\tau = C du + D dv$ is given by the W-valued 2-form

$$\omega \wedge \tau = (A du + B dv) \wedge (C du + D dv) = (\mathcal{B}(A, D) - \mathcal{B}(B, C))(du \wedge dv).$$

In other words, $\mathcal{B}$ tells us how to multiply the coefficients. The two types of wedge products of differential forms we encounter in this text are:

- $V = \mathbb{R}^{4,1}$, $W = \wedge^2 \mathbb{R}^{4,1} \cong \mathfrak{o}_{4,1}$, and $\mathcal{B}(A, B) = A \wedge B$. In this case, to avoid confusion between the different type of products, we denote the product between two differential forms ω and τ as $\omega \curlywedge \tau$.
- $V = W = \mathfrak{o}_{4,1}$, and $\mathcal{B}(A, B) = [A, B]$. In this case, we write $[\omega \wedge \tau]$.

Example 3.1.5 Here we show a few examples of wedge product of two differential forms:

- Let $A, B, X : \Sigma \to \mathbb{R}^{4,1}$, and $\omega = A du + B dv$. Then,

$$X \curlywedge \omega = (X \wedge A) du + (X \wedge B) dv.$$

- Let $\omega = A du$ and $\tau = B dv$ for $A, B : \Sigma \to \mathbb{R}^{4,1}$. Then we have

$$\omega \curlywedge \tau = (A du) \curlywedge (B dv) = (A \wedge B)(du \wedge dv).$$

- Now, let $\omega = A du$ and $\tau = B dv$ for $A, B : \Sigma \to \mathfrak{o}_{4,1}$. Then we can calculate that

$$[\omega \wedge \tau] = [A, B](du \wedge dv) = (AB - BA)(du \wedge dv).$$

- Let x be an isothermic surface in $\mathbb{R}^3$ with its Christoffel dual x^*. Then for isothermic coordinates (u, v),

$$\mathrm{d}x \curlywedge \mathrm{d}x^* = (x_u\mathrm{d}u + x_v\mathrm{d}v) \curlywedge \left(-\frac{x_u}{|x_u|^2}\mathrm{d}u + \frac{x_v}{|x_v|^2}\mathrm{d}v\right)$$

$$= \left(x_u \wedge \frac{x_v}{|x_v|^2} + x_v \wedge \frac{x_u}{|x_u|^2}\right)(\mathrm{d}u \wedge \mathrm{d}v)$$

$$= \frac{1}{|x_v|^2}(x_u \wedge x_v + x_v \wedge x_u)(\mathrm{d}u \wedge \mathrm{d}v) = 0,$$

where the $\wedge$ here is similar to (3.1), with the inner product of $\mathbb{R}^3$ replacing that of $\mathbb{R}^{4,1}$ so that we have $\wedge^2 \mathbb{R}^3 \cong \mathfrak{o}_3$ (see [117, Theorem 4], [135, Definition 3.1, Remark 3]).

Exercise 3.1.6 Let $A, B, C, D : \Sigma \to \mathbb{R}^{4,1}$, and let $\omega = (A \wedge B)\mathrm{d}u$, $\tau = (C \wedge D)\mathrm{d}v$ be 1-forms (taking values in $\mathfrak{o}_{4,1}$). Show that

$$[\omega \wedge \tau] = \big((A \wedge B)(C \wedge D) - (C \wedge D)(A \wedge B)\big)(\mathrm{d}u \wedge \mathrm{d}v).$$

3.2 Connections on the Trivial Bundle

In this section we shall give a brief review of connections on vector bundles, focusing only on the special case of trivial vector bundles. Connections generalize the notion of differentiation on Euclidean spaces. A *trivial vector bundle* over a manifold M is of the form $\underline{V} := M \times V$, where V is a vector space. Let $\pi : \underline{V} \to M$ be the projection of $\underline{V}$ onto M, i.e., $\pi(p, v) = p$ for all $p \in M$ and $v \in V$. *Sections* of this bundle are smooth maps $\sigma : M \to \underline{V}$ satisfying $\pi(\sigma(p)) = p$ for all $p \in M$, denoted by $\sigma \in \Gamma\underline{V}$. In particular, we can write $\sigma(p) = (p, v(p))$ for some smooth function $v : M \to V$. In this way, sections of $\underline{V}$ are identified with smooth functions $M \to V$.

A *connection* on $\underline{V}$ is an $\mathbb{R}$-linear map ∇ that takes smooth sections σ of $\underline{V}$ and smooth tangent vector fields Y on M and yields a new smooth section $\nabla_Y\sigma$ of $\underline{V}$, such that ∇ satisfies the Leibniz rule: for any smooth section σ of $\underline{V}$, smooth tangent vector field Y and smooth function $\alpha : M \to \mathbb{R}$,

$$\nabla_Y(\alpha\sigma) = (\mathrm{d}_Y\alpha)\sigma + \alpha\nabla_Y\sigma,$$

where d is the trivial connection induced by the standard differentiation on V. The curvature tensor R^∇ of ∇ is defined by

$$R^\nabla_{Y,Z}\sigma = \nabla_Y\nabla_Z\sigma - \nabla_Z\nabla_Y\sigma - \nabla_{[Y,Z]}\sigma \tag{3.2}$$

for smooth tangent vector fields Y, Z of M and smooth sections σ of $\underline{V}$. A connection is *flat* if $R^\nabla \equiv 0$.

Exercise 3.2.7 Check that R is a tensor, i.e., C^∞-linear in Y, Z and σ.

Using that mixed partial derivatives commute, one can check that this is a flat connection, i.e., $R^d \equiv 0$. Any other connection can be written as

$$\nabla = d + A, \tag{3.3}$$

where A is a 1-form with values in the endomorphism bundle of $\underline{V}$, i.e., $M \times \mathrm{End}(V)$.

Exercise 3.2.8 Show (3.3) by checking that $\nabla - d$ is linear, that is, for any smooth tangent vector field Y of M and scalar function α,

$$(\nabla_Y - d_Y)(\sigma_1 + \sigma_2) = (\nabla_Y - d_Y)\sigma_1 + (\nabla_Y - d_Y)\sigma_2,$$

$$(\nabla_Y - d_Y)(\alpha\sigma_1) = \alpha(\nabla_Y - d_Y)\sigma_1$$

where $\sigma_1, \sigma_2 \in \Gamma\underline{V}$.

The curvature tensor of ∇ is then given by

$$R^\nabla = dA + \frac{1}{2}[A \wedge A]. \tag{3.4}$$

Exercise 3.2.9 Prove the formula (3.4) using (3.2).

Definition 3.2.10 Let ∇ be a connection. We say that σ is a *parallel section* of ∇, or ∇-*parallel*, if $\nabla\sigma = 0$.

We now state the gauging principle for connections. Gauge transformations are maps $g : M \to \mathrm{GL}(V)$, and act on connections as follows: Let ∇ be a connection. Then g acts on ∇ via

$$g \bullet \nabla = g \circ \nabla \circ g^{-1}.$$

Note that we have, for a smooth section σ of $\underline{V}$,

$$\nabla\sigma = 0 \quad \text{if and only if} \quad g \bullet \nabla(g\sigma) = 0. \tag{3.5}$$

Therefore, σ is a parallel section of ∇ if and only if $g\sigma$ is a parallel section of $g \bullet \nabla$. Moreover $R^{g \bullet \nabla} = g \bullet R^\nabla$. Thus ∇ is flat if and only if $g \bullet \nabla$ is flat. We now state the following well-known result about flat connections on trivial bundles:

Proposition 3.2.11 *A connection ∇ is flat if and only if there locally exists a gauge transformation g such that*

$$g \bullet \nabla = d,$$

where d *is the trivial connection on* $\underline{V}$.

Proof First, suppose that a connection $\nabla = d + A$ on $\underline{V}$ is flat. Then $g : M \to$ GL(V) can be defined via

$$dg = gA,$$

since, noting that the term in (3.4) is zero, we would have

$$d(dg) = dg \wedge A + g dA = g(A \wedge A + dA) = g R^{\nabla} = 0.$$

Then for any section $\sigma \in \Gamma\underline{V}$,

$$(g \bullet \nabla)\sigma = g(d(g^{-1}\sigma) + A g^{-1}\sigma)$$
$$= -dg\, g^{-1}\sigma + d\sigma + g A g^{-1}\sigma = d\sigma + (gA - dg)(g^{-1}\sigma) = d\sigma.$$

The other direction can be proved easily from the fact that $R^d = 0$. $\square$

Since parallel sections of d are constant sections of $\underline{V}$, we can deduce the following:

Proposition 3.2.12 *A connection ∇ is flat if and only if for each $p \in M$ and $\sigma_0 \in V$, there locally exists a unique parallel section σ of ∇ such that $\sigma(p) = \sigma_0$.*

Proof First assume that ∇ is flat. Then by Proposition 3.2.11, around p there locally exists a gauge transformation g such that $g \bullet \nabla = d$. By premultiplication by a constant element of GL(V), we may assume that $g_p = id_V$. Now the constant section σ_0 is parallel for d. Thus $\sigma := g^{-1}\sigma_0$ is the unique parallel section of ∇ with $\sigma(p) = \sigma_0$.

To show the converse, at any given p, insert all local parallel sections σ with $\sigma(p) = \sigma_0$ for any $\sigma_0 \in V$ to see that the curvature tensor is zero. $\square$

Connections ∇ give rise to a notion of parallel transport along curves in M. Let $\gamma : [0, 1] \to M$ be a curve in the manifold M. A section σ of $\underline{V}$ is *parallel along γ with respect to ∇* if

$$\nabla_{\gamma'}\sigma = 0,$$

where γ' denotes the tangent vector field of γ.

Let $p, q \in M$ and let $\gamma : [0, 1] \to M$ be a path in M with $\gamma(0) = p$ and $\gamma(1) = q$. Fix $\sigma_0 \in V$. By the existence of solutions of ordinary differential equations, there exists a section σ that is parallel along γ with respect to ∇ such that $\sigma(p) = \sigma_0$. Moreover the restriction of σ to γ is unique. We thus obtain a vector $\sigma(q) = \sigma(\gamma(1)) \in V$. By changing the initial $\sigma_0 \in V$, we obtain an isomorphism

$$\phi_{\gamma} : V \to V, \quad \phi_{\gamma}(\sigma_0) = \sigma(\gamma(1)).$$

ϕ_γ is called the *parallel transport map along γ*.

A pertinent question is: does the parallel transport map depend on the choice of path γ from p to q? By Proposition 3.2.12, the answer is that the parallel transport map depends only on the homotopy class of each curve (with fixed endpoints) if and only if ∇ is a flat connection. We shall see that this point of view will serve as motivation for the definition of discrete flat connections.

Example 3.2.13 For $M_\kappa \subset L^4 \subset \mathbb{R}^{4,1}$, let $\underline{\mathbb{R}}^{4,1} = M_\kappa \times \mathbb{R}^{4,1}$ be the trivial vector bundle, and denote by d the trivial connection. We can consider the tangent bundle TM_κ of M_κ as a subbundle of the trivial vector bundle by restricting to sections σ that are tangent to M_κ, i.e. for $\pi : TM_\kappa \to M_\kappa$ as before, we have

$$\pi^{-1}(X) = (X, T_X M_\kappa).$$

Define the subbundle connection as

$$\tilde{\nabla}_Y \sigma = P_{M_\kappa}(\mathrm{d}_Y \sigma),$$

where P_{M_κ} is orthogonal projection in V to the tangent spaces of M_κ which is well-defined as we are now in the presence of an ambient metric.

Then we have

$$\tilde{\nabla}_{fW+Y} Z = f \tilde{\nabla}_W Z + \tilde{\nabla}_Y Z \tag{3.6a}$$

$$\tilde{\nabla}_W (fY + Z) = W(f)Y + f \tilde{\nabla}_W Y + \tilde{\nabla}_W Z \tag{3.6b}$$

$$\tilde{\nabla}_W Y - \tilde{\nabla}_Y W = [W, Y] \tag{3.6c}$$

$$W \langle Y, Z \rangle = \langle \tilde{\nabla}_W Y, Z \rangle + \langle Y, \tilde{\nabla}_W Z \rangle \tag{3.6d}$$

hold, where W, Y, Z are any smooth tangent vector fields of M_κ, and $f : M_\kappa \to \mathbb{R}$ is any smooth function. Thus $\tilde{\nabla}$ is the (unique) Riemannian connection of M_κ.

The governing equation for geodesics can be interpreted in terms of the parallel transport along a curve with respect to $\tilde{\nabla}$: a curve $c : [a, b] \to M_\kappa$ is a geodesic in M_κ if and only if

$$\tilde{\nabla}_{c'(t)} c'(t) = 0.$$

One can also analogously consider the tangent bundle of a surface $X : \Sigma \to M_\kappa$, thus reducing the dimension of the submanifold further by 1. For this, we use the trivial vector bundle $\underline{\mathbb{R}}^{4,1} := \Sigma \times \mathbb{R}^{4,1}$, and we consider the tangent bundle of the surface X as a subbundle by restricting to sections σ that are tangent to X. Defining the subbundle connection as

$$P_X(\mathrm{d}_Y \sigma),$$

where now P_X is orthogonal projection in V to the tangent planes of X, equations corresponding to (3.6) all hold, where now W, Y, Z are smooth tangent vector fields of X and $f : \Sigma \to \mathbb{R}$, giving the (unique) Riemannian connection of the surface. This connection then determines the first fundamental form for the surface, and noting that for any fixed $p \in \Sigma$, the subspace of $T_{X(p)}M_\kappa$ perpendicular to $T_p X$ provides a unit normal vector to the surface at p, we also obtain the second fundamental form.

3.3   Isothermicity via Flat Connections

For some simply-connected (2-dimensional) domain Σ, consider the trivial vector bundle $\underline{\mathbb{R}}^{4,1} = \Sigma \times \mathbb{R}^{4,1}$, and view a surface into the projective light cone $L : \Sigma \to P(L^4)$ as a null line subbundle of $\underline{\mathbb{R}}^{4,1}$. Choosing $X \in \Gamma L$, we define the *retraction form* of L (cf. [111, Definition and Lemma 5.3.19]) as

$$\tau = \frac{-2}{\langle X_u, X_u \rangle} X \wedge X_u \mathrm{d}u + \frac{2}{\langle X_v, X_v \rangle} X \wedge X_v \mathrm{d}v. \tag{3.7}$$

From the retraction form, we define a connection on the trivial bundle by, for any given $\lambda \in \mathbb{R}$,

$$\Gamma^\lambda := \mathrm{d} + \lambda \tau.$$

Exercise 3.3.14   Show that τ is invariant of the choice of section $X \in \Gamma L$.

We now show the flat connection characterization of isothermic surfaces in the projective light cone.

Theorem 3.3.15 (cf. [38, Theorem 15.4], [41, Proposition 3.6])   *L is isothermic if and only if Γ^λ is flat for all $\lambda \in \mathbb{R}$.*

Proof   We first note that via (3.4),

$$R^{\Gamma^\lambda} = \lambda \, \mathrm{d}\tau + \tfrac{1}{2}\lambda^2 [\tau \wedge \tau].$$

Assume that L is isothermic with isothermic coordinates $(u, v) \in \Sigma$. Taking ∂_u and ∂_v as the basis of the tangent vector fields of Σ, we will show that

$$\partial_v (\tau(\partial_u)) = \partial_u (\tau(\partial_v)) \tag{3.8}$$

and

$$\tau(\partial_u)\tau(\partial_v) = \tau(\partial_v)\tau(\partial_u). \tag{3.9}$$

Expressing τ with any $X \in \Gamma L$, Exercise 2.8.72 tells us that

$$X_{uv} = AX_u + BX_v + CX$$

for some real functions A, B, C. Thus, we have that

$$\partial_v(\|X_u\|^{-2}) = -2A\|X_u\|^{-2}, \quad \partial_u(\|X_v\|^{-2}) = -2B\|X_v\|^{-2}.$$

Using these relations, we can verify the condition (3.8), implying that $d\tau = 0$. On the other hand, we have for $Z \in \mathbb{R}^{4,1}$,

$$\tau(\partial_u)\tau(\partial_v)Z = \tau(\partial_v)\tau(\partial_u)Z = \frac{4}{\|X_u\|^2\|X_v\|^2}\langle X, Z\rangle\langle X_u, X_v\rangle X = 0, \tag{3.10}$$

giving us (3.9), and we conclude that $[\tau \wedge \tau] = 0$.

Now assume that for some $L : \Sigma \to P(L^4)$, the connection $\Gamma^\lambda = d + \lambda\tau$ is flat with τ as in (3.7) with $X_0 \in \Gamma L$. Then arguing similarly as in (3.10), we can check that $[\tau \wedge \tau] = 0$. Hence, we obtain that $d\tau = 0$, allowing us to deduce that

$$\tau\mathfrak{q}_0 = \frac{1}{2|x_u|^2}X_{0,u}du - \frac{1}{2|x_v|^2}X_{0,v}dv$$

is also a closed 1-form (that is, $d(\tau\mathfrak{q}_0) = 0$), where $x : \Sigma \to \mathbb{R}^3$ is the projection of L to $\mathfrak{R}_0 \cong \mathbb{R}^3$. Taking the $\mathfrak{R}_0$-component (the 2nd, 3rd, 4th components) of the above equation then tells us that

$$\eta := \frac{1}{|x_u|^2}x_u du - \frac{1}{|x_v|^2}x_v dv$$

is closed, i.e., there is some $x^* : \Sigma \to \mathbb{R}^3$ such that $dx^* = \eta$. Hence, x admits a Christoffel transformation x^*, and Lemma 2.10.76 tells us that x, and therefore L, is isothermic. $\qquad\square$

Exercise 3.3.16 In this exercise, we will show one direction of the above proof of Lemma 3.3.15 using Moutard lifts.

- Assuming that L is isothermic, the proof of Theorem 2.9.74 (see (2.42)) tells us that

$$F := \frac{\sqrt{2}}{\|X_{0,u}\|}X_0$$

  is a Moutard lift. Using this F, show that

$$\tau = -F \wedge F_u du + F \wedge F_v dv.$$

- Using this expression for τ, show that (3.8) and (3.9) hold.

Remark 3.3.17 Finally, we make a few important remarks about the *retraction form* τ. Noting that the condition $[\tau \wedge \tau] = 0$ is satisfied regardless of the isothermicity of the given surface, we see that the flatness of the connection Γ^λ is equivalent to the closure of τ, i.e. $d\tau = 0$. Now, since $\langle X, X_u \rangle = 0 = \langle X, X_v \rangle$, we see that τ takes values in $L \wedge L^\perp$ for $L = \mathrm{span}\{X\}$. The existence of such closed 1-form is another well-known characterization of isothermic surfaces (see [38, Theorem 15.4] or [41, §3.2.1]).

Furthermore, rewriting τ as

$$\tau = \frac{1}{2} X_0 \wedge \left(-\frac{1}{|x_u|^2} X_{0,u} du + \frac{1}{|x_v|^2} X_{0,v} dv \right) = \frac{1}{2} X_0 \wedge \mathcal{T}_{dx^*}, \tag{3.11}$$

by Exercise 2.10.78, if we let η be the restriction of τ in (3.11) to $\wedge^2 \mathbb{R}^3$ where we use $\mathfrak{R}_0 \cong \mathbb{R}^3$, then we obtain

$$\eta = 2x \wedge \left(-\frac{1}{|x_u|^2} x_u du + \frac{1}{|x_v|^2} x_v dv \right) = 2x \wedge dx^*,$$

which we know to be closed by Example 3.1.5.

Exercise 3.3.18 (cf. [55, Proposition 1.5 (2)]) For an isothermic surface $L : \Sigma \to P(L^4)$, express τ using $X_0 \in \Gamma L$ such that $X_0 : \Sigma \to M_0$. Use Exercise 2.10.78 to show that

$$\tau \mathfrak{q}_0 = -\frac{1}{2} \mathcal{T}_{dx^*}.$$

3.4   Calapso Transformations

The existence of a 1-parameter family of flat connections points to the existence of the spectral deformation via Proposition 3.2.11:

Definition 3.4.19 (cf. [41, §3.3]) Let $L : \Sigma \to P(L^4)$ be isothermic with associated flat connection Γ^λ. A *Calapso transformation* $T^\lambda : \Sigma \to O_{4,1}$ is the orthogonal trivializing gauge transformation of Γ^λ, that is,

$$T^\lambda \bullet \Gamma^\lambda = d.$$

Then $L^\mu := \mathrm{span}\{T^\mu X\} : \Sigma \to P(L^4)$ for $X \in \Gamma L$ is called a *Calapso transform* (with respect to parameter μ), also called a *T-transform* or *conformal deformation* of L.

Exercise 3.4.20 (cf. [116, §3.4]) Show that T^λ is a Calapso transformation if and only if it satisfies

$$(T^\lambda)^{-1}\mathrm{d}T^\lambda = \lambda\tau. \tag{3.12}$$

Remark 3.4.21 If the initial condition for the solution T^λ lies in $\mathrm{O}_{4,1}$, then T^λ will lie in $\mathrm{O}_{4,1}$ for all u and v. In fact, the Calapso transformations are well-defined up to some choice of initial condition, i.e., premultiplication by a constant $\mathrm{O}_{4,1}$ transformation.

Remark 3.4.22 Because of (3.12), $\lambda\tau$ can be thought of as the logarithmic derivative of the Calapso transformation.

The existence of the Calapso transformations (with deformation parameter λ) shows that isothermic surfaces are deformable in Möbius geometry. This deformation preserves second order invariants in Möbius geometry, including the conformal class of the induced metric on surfaces, which is actually a first-order invariant (as the next lemma shows). Note that for surfaces in Euclidean geometry, a nontrivial deformation will never preserve the second order invariants of Euclidean differential geometry, i.e. the first and second fundamental forms.

Lemma 3.4.23 *If $L : \Sigma \to P(L^4)$ is isothermic with isothermic coordinates (u, v), then (u, v) will also be isothermic coordinates for any Calapso transform L^μ.*

Proof The isothermicity does not depend on the choice of the lift; therefore, take any section $X \in \Gamma L$. Exercise 2.8.72 then implies

$$\langle X_u, X_u \rangle = \langle X_v, X_v \rangle, \quad \langle X_u, X_v \rangle = 0, \quad X_{uv} \in \mathrm{span}\{X, X_u, X_v\}.$$

Now, if $X^\mu := T^\mu X \in \Gamma L^\mu$, then (3.12) shows that

$$(T^\mu X)_u = T^\mu X_u, \quad (T^\mu X)_v = T^\mu X_v, \quad \text{and} \quad (T^\mu X)_{uv} = T^\mu X_{uv}.$$

Therefore, one can deduce that

$$(X^\mu)_{uv} \in \mathrm{span}\{X^\mu, X_u^\mu, X_v^\mu\},$$

while the fact that $T^\mu \in \mathrm{O}_{4,1}$ tells us that

$$\langle X_u^\mu, X_u^\mu \rangle = \langle X_v^\mu, X_v^\mu \rangle \quad \text{and} \quad \langle X_u^\mu, X_v^\mu \rangle = 0.$$

In particular, we have $\langle \mathrm{d}X, \mathrm{d}X \rangle = \langle \mathrm{d}X^\mu, \mathrm{d}X^\mu \rangle$. Therefore, Exercise 2.8.73 allows us to conclude that L^μ is isothermic with isothermic coordinates (u, v). $\qquad\square$

Since we have a characterization of isothermic surfaces via the existence of Moutard lifts, via Theorem 2.9.74, we can assume that the Calapso transform also admits Moutard lifts, and prove the following:

Lemma 3.4.24 *If $F \in \Gamma L$ is a Moutard lift, then $T^\mu F \in \Gamma L^\mu$ is a Moutard lift as well.*

Proof As in the proof of Lemma 3.4.23, we have $(T^\mu F)_{uv} = T^\mu F_{uv}$, and so $F_{uv} \parallel F$ implies $(T^\mu F)_{uv} \parallel T^\mu F$. $\square$

The isothermicity of a Calapso transform L^μ implies that it has an associated retraction form τ^μ:

$$\tau^\mu = \frac{-2}{\langle X_u^\mu, X_u^\mu \rangle} X^\mu \wedge X_u^\mu \mathrm{d}u + \frac{2}{\langle X_v^\mu, X_v^\mu \rangle} X^\mu \wedge X_v^\mu \mathrm{d}v \tag{3.13}$$

for any section $X^\mu \in \Gamma L^\mu$. Using (3.13), we can show the following lemma:

Lemma 3.4.25 (cf. [111, Lemma 5.5.8]) *For a Calapso transform L^μ of L, the retraction form τ^μ of L^μ is*

$$\tau^\mu = T^\mu \tau (T^\mu)^{-1}. \tag{3.14}$$

Exercise 3.4.26 Show that $\mathrm{d}\tau^\mu = 0 = [\tau^\mu \wedge \tau^\mu]$.

Exercise 3.4.27 Use Definition 3.4.19 and Lemma 3.4.25 to show that the connection $\Gamma^{\mu,\lambda} := \mathrm{d} + \lambda \tau^\mu$ satisfies

$$\Gamma^{\mu,\lambda} = T^\mu \bullet \Gamma^{\mu+\lambda},$$

where $\Gamma^\lambda = \mathrm{d} + \lambda \tau$ is the flat connection associated with L, and deduce that $\Gamma^{\mu,\lambda}$ is flat.

Remark 3.4.28 If we take the existence of a 1-parameter family of flat connections as in Lemma 3.3.15 to characterize isothermicity, then Exercise 3.4.27 constitutes a short proof of Lemma 3.4.23. Similarly, if we take the existence of a closed 1-form as in Remark 3.3.17 to characterize isothermicity, then Exercise 3.4.26 also efficiently proves the isothermicity of Calapso transforms.

Now for some fixed $\mu \in \mathbb{R}$, let $L^\mu = T^\mu L$ be a Calapso transform of L. Then we can also determine the Calapso transformations of L^μ by solving

$$\mathrm{d}T^{\mu,\lambda} = T^{\mu,\lambda} \lambda \tau^\mu$$

for $T^{\mu,\lambda}$. However, as the next lemma shows, we can find this algebraically:

Lemma 3.4.29 (cf. [116, p. 196]) *For the Calapso transformations $T^{\mu,\lambda}$ and T^λ, we have*

$$T^{\mu+\lambda} = T^{\mu,\lambda} T^\mu$$

for suitable choice of initial conditions.

Proof We have that by Definition 3.4.19 and Exercise 3.4.27,

$$(T^{\mu,\lambda}T^{\mu}) \bullet \Gamma^{\mu+\lambda} = T^{\mu,\lambda} \bullet (T^{\mu} \bullet \Gamma^{\mu+\lambda}) = T^{\mu,\lambda} \bullet \Gamma^{\mu,\lambda} = \mathrm{d} = T^{\mu+\lambda} \bullet \Gamma^{\mu+\lambda},$$

giving us our conclusion. $\square$

3.5 Darboux Transformations

Geometrically, a Darboux transformation of an isothermic surface (see also Fig. 3.1) is one such that

- there exists a sphere congruence enveloped by the original surface and the transform,
- the correspondence, given by the sphere congruence, from the original surface to the other enveloping surface (i.e. the transform), preserves curvature lines (Fig. 3.1),
- this correspondence preserves conformality.

In light of recent developments concerning Darboux transformations [110, p. 38], we define them as follows:

Definition 3.5.30 Let T^{μ} be a Calapso transformation of $L : \Sigma \rightarrow P(L^4)$ for some $\mu \in \mathbb{R}^{\times}$. Then $\hat{L} : \Sigma \rightarrow P(L^4)$ is a *Darboux transformation with parameter* μ of L if $\mathrm{span}\{T^{\mu}\hat{X}\}$ is constant in $P(L^4)$ for non-zero $\hat{X} \in \Gamma\hat{L}$.

The equation that $\mathrm{span}\{T^{\mu}\hat{X}\}$ is constant is called *Darboux's linear system.* Letting $\hat{x}$ be the projection of $\hat{L}$ to $\mathbb{R}^3$, Darboux's linear system implies that

$$\mathrm{d}\left(rT^{\mu} \begin{pmatrix} 1 + |\hat{x}|^2 \\ 2\hat{x}^t \\ 1 - |\hat{x}|^2 \end{pmatrix} \right) = 0 \tag{3.15}$$

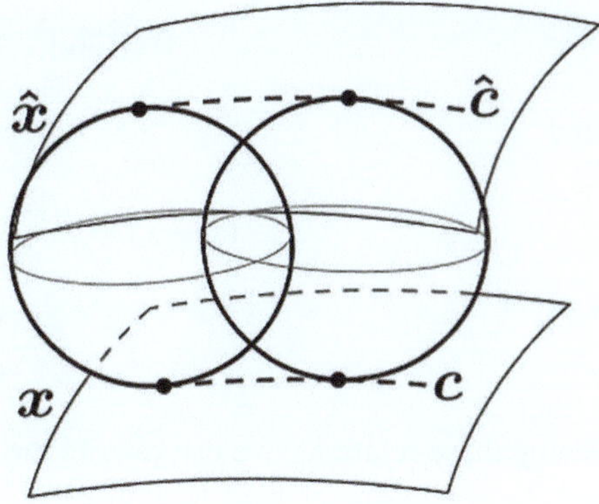

Fig. 3.1 A Darboux transform: when the curve c in the surface x is a curvature line, then the corresponding curve $\hat{c}$ in the surface $\hat{x}$ is also a curvature line

for some nonzero function $r : \Sigma \to \mathbb{R}$. We can reformulate this as in the next lemma:

Lemma 3.5.31 (cf. [117, Theorem 2]) *Darboux's linear system (3.15) is equivalent to the Riccati-type equation*

$$\mathrm{d}\hat{x} = \mu|\hat{x} - x|^2\mathrm{d}x^* - 2\mu((\hat{x} - x) \cdot \mathrm{d}x^*)(\hat{x} - x). \qquad (3.16)$$

From (3.16), it is easy to see that an isothermic surface admits a 4-parameter family of Darboux transforms via the choice of the parameter and an initial condition. An equation of the form $y' = f(y)$, where $f(y)$ is a quadratic polynomial, is called a Riccati equation, so (3.16) is a Riccati-type partial differential equation (where y becomes $\hat{x}$).

From (3.16) we have

$$|\hat{x}_u|^2 = |\hat{x}_v|^2 = \mu^2|x_u|^{-2}|\hat{x} - x|^4, \quad \hat{x}_u \cdot \hat{x}_v = 0,$$

so $\hat{x}$ is conformally parametrized by u and v. Also, we can check that

$$\hat{n} \ \| \ (-|\hat{x} - x|^2 n + 2((\hat{x} - x) \cdot n)(\hat{x} - x))$$

and then that

$$\hat{x}_{uv} \cdot \hat{n} = 0.$$

Therefore, we have:

Theorem 3.5.32 *A Darboux transform of an isothermic surface is isothermic.*

Lemma 3.5.33 *If $\hat{X}$ is a Darboux transform of X, then X is also a Darboux transform of $\hat{X}$, and both are Darboux transforms with respect to the same choice of μ. Hence, we say that X and $\hat{X}$ (or $L = \mathrm{span}\{X\}$ and $\hat{L} = \mathrm{span}\{\hat{X}\}$) are a Darboux pair (with parameter μ).*

Proof From the Riccati-type equation (3.16), we have

$$|x_u|^2|\hat{x}_u|^2 = \mu^2|\hat{x} - x|^4,$$

and

$$x_u \cdot (\hat{x} - x) = \frac{\mu|\hat{x} - x|^2}{|\hat{x}_u|^2}(\hat{x}_u \cdot (\hat{x} - x)),$$

$$x_v \cdot (\hat{x} - x) = -\frac{\mu|\hat{x} - x|^2}{|\hat{x}_u|^2}(\hat{x}_v \cdot (\hat{x} - x)).$$

Using these relations, we can rewrite the Riccati-type equation (3.16) as

$$\hat{x}_u = -\frac{|\hat{x}_u|^2 x_u}{\mu|\hat{x} - x|^2} + 2(\hat{x} - x)|\hat{x} - x|^{-2}(\hat{x}_u \cdot (\hat{x} - x)),$$

$$\hat{x}_v = \frac{|\hat{x}_v|^2 x_v}{\mu|\hat{x} - x|^2} + 2(\hat{x} - x)|\hat{x} - x|^{-2}(\hat{x}_v \cdot (\hat{x} - x)).$$

Isolating the x_u and x_v terms, we find that for the Christoffel transformation $\hat{x}^*$ of $\hat{x}$,

$$x_u = \mu|x - \hat{x}|^2 \hat{x}_u^* - 2\mu(x - \hat{x})(\hat{x}_u^* \cdot (x - \hat{x})),$$

$$x_v = \mu|x - \hat{x}|^2 \hat{x}_v^* - 2\mu(x - \hat{x})(\hat{x}_v^* \cdot (x - \hat{x})),$$

$$\tag{3.17}$$

and these are the same as the Riccati-type equation (3.16), but with the x and $\hat{x}$ switched, proving the lemma. $\qquad\square$

To prove that Darboux transformations as we have defined them have all the required geometric properties mentioned at the beginning of this section, it remains only to find a sphere congruence $\mathcal{S}$ enveloping both $X \in \Gamma L$ and $\hat{X} \in \Gamma\hat{L}$. We assume without loss of generality that both X and $\hat{X}$ takes values in the space form M_0.

Denoting the (geodesic) tangent plane congruence of X in M_0 by $\mathcal{T}_n$, Exercise 2.7.64 tells us that

$$\mathcal{S} = \mathcal{T}_n + tX$$

for some function $t : \Sigma \to \mathbb{R}$. Since $\mathcal{S}$ must also envelop $\hat{X}$, Proposition 2.7.61 implies

$$\langle \mathcal{S}, \hat{X} \rangle = 0 = \langle \mathcal{S}, d\hat{X} \rangle.$$

In particular, $\langle \mathcal{S}, \hat{X} \rangle = 0$ allows us to see

$$t = -\frac{\langle \mathcal{T}_n, \hat{X} \rangle}{\langle X, \hat{X} \rangle}.$$

Now, we have

$$\mathcal{T}_n = 2\begin{pmatrix} x \cdot n \\ n^t \\ -x \cdot n \end{pmatrix}, \quad X = \begin{pmatrix} 1 + |x|^2 \\ 2x^t \\ 1 - |x|^2 \end{pmatrix}, \quad \hat{X} = \begin{pmatrix} 1 + |\hat{x}|^2 \\ 2\hat{x}^t \\ 1 - |\hat{x}|^2 \end{pmatrix},$$

so that we calculate directly and obtain

$$\langle \mathcal{T}_n, \hat{X} \rangle = 4(\hat{x} - x) \cdot n, \quad \langle X, \hat{X} \rangle = -2(\hat{x} - x) \cdot (\hat{x} - x).$$

Thus, we have

$$\mathcal{S} = \mathcal{T}_n + \frac{2(\hat{x} - x) \cdot n}{(\hat{x} - x) \cdot (\hat{x} - x)} X.$$

We leave the checking of the last condition as an exercise:

Exercise 3.5.34 Show that $\langle \mathcal{S}, d\hat{X} \rangle = 0$.

As an application, we recover a characterization of cmc surfaces in terms of Christoffel and Darboux transformations:

Theorem 3.5.35 (cf. [117, Theorem 5]) *A smooth surface x in $\mathfrak{R}_0 \cong \mathbb{R}^3$ has nonzero constant mean curvature if and only if some scaling and translation of the Christoffel transform x^* equals a Darboux transform $\hat{x}$ (for some value of μ).*

Proof Assume x is a cmc surface in $\mathfrak{R}_0 \cong \mathbb{R}^3$. Then $x^* = x + H_0^{-1} n_0$ is the parallel cmc surface, by Remark 2.10.80. To show x^* is a Darboux transform, we can compute that (3.16) holds, for $\mu = H_0^2/|n_0|^2 \in \mathbb{R}$.

Now we show the converse direction. Assume $\hat{x}$ is a Darboux transform of x, and that $\hat{x} = a \cdot x^* + p$ for some constant $a \in \mathbb{R} \setminus \{0\}$ and constant vector $p \in \mathbb{R}^3$. So there exists μ such that (3.16) holds, that is,

$$(\mu|ax^* + p - x|^2 - a)dx^* = 2\mu((ax^* + p - x) \cdot dx^*)(ax^* + p - x).$$

Since we have that x_u^* and x_v^* are linearly independent, the span of the left hand side is 2-dimensional, while that of the right hand side is 1-dimesional. Therefore, the coefficient multiplied to dx^* (on the left hand side) must be zero, that is,

$$|ax^* + p - x|^2 = a\mu^{-1}.$$

The constancy of the relation then allows us to deduce that

$$ax^* + p - x = rn_0$$

for some constant $r \in \mathbb{R}$. So

$$dx^* = a^{-1}dx + ra^{-1}dn_0.$$

Definition 2.10.75 implies x has cmc $H_0 = r^{-1}$. $\square$

Corollary 3.5.36 *Let x be a cmc $H_0 \neq 0$ surface in $\mathfrak{R}_0 \cong \mathbb{R}^3$. Suppose $\hat{x}$ is both a Christoffel and Darboux transform of x, as in Theorem 3.5.35. Then, $|\hat{x} - x|^2 = \frac{1}{4H_0^2}$ is constant, and $\hat{x} - x$ is perpendicular to dx, so $\hat{x}$ is a parallel surface to x.*

3.6 Bianchi Permutability and the Cube Theorem

In this section, we interpret Darboux transformation in terms of the 1-parameter family of flat connections, and use this to prove the permutability of Darboux transformations attributed to Bianchi [10]. Consider an isothermic surface $L : \Sigma \to P(L^4)$, and a section $X \in \Gamma L$ in a space form M_κ, admitting isothermic coordinates u, v with associated flat connection Γ^λ.

The next theorem describes Darboux transformations in terms of the parallel sections of the flat connection:

Theorem 3.6.37 (cf. [111, Definition 5.4.8]) $\hat{L} : \Sigma \to P(L^4)$ *is a Darboux transform of* $L : \Sigma \to P(L^4)$ *with parameter* μ *if and only if there is some* $\hat{X} \in \Gamma\hat{L}$ *such that* $\Gamma^\mu \hat{X} = 0$.

Proof To show one direction, assume that $\hat{L}$ is a Darboux transform of L with parameter μ, and let $\hat{X}$ be some section of $\hat{L}$. Then using Darboux's linear system, we know that $\text{span}\{T^\mu \hat{X}\}$ is constant, that is, $T^\mu \hat{X} = rY_0$ for some function $r : \Sigma \to \mathbb{R}$ and constant vector $Y_0 \in \mathbb{R}^{4,1}$. From this, we see that

$$0 = \mathrm{d}(T^\mu r^{-1}\hat{X}) = \mathrm{d}T^\mu r^{-1}\hat{X} + T^\mu \mathrm{d}(r^{-1}\hat{X})$$

$$= T^\mu \mu\tau r^{-1}\hat{X} + T^\mu \mathrm{d}(r^{-1}\hat{X}) = T^\mu(\mathrm{d} + \mu\tau)(r^{-1}\hat{X}) = T^\mu \Gamma^\mu(r^{-1}\hat{X}),$$

giving us the conclusion.

To show the other direction, suppose that we have $\hat{X} \in \Gamma\hat{L}$ such that $\Gamma^\mu \hat{X} = 0$. Then we have

$$0 = (\mathrm{d} + \mu\tau)\hat{X} = \mathrm{d}\hat{X} + \mu\tau\hat{X} = \mathrm{d}\hat{X} + (T^\mu)^{-1}\mathrm{d}T^\mu \hat{X}$$

$$= (T^\mu)^{-1}(T^\mu \mathrm{d}\hat{X} + \mathrm{d}T^\mu \hat{X}) = (T^\mu)^{-1}\mathrm{d}(T^\mu \hat{X}).$$

Therefore, $T^\mu \hat{X}$ is constant. $\qquad\square$

Remark 3.6.38 Under the assumptions of Theorem 3.6.37, Lemma 3.5.33 tells us that there is an $X \in \Gamma L$ such that $\hat{\Gamma}^\mu X = 0$ as well, where $\hat{\Gamma}^\lambda$ is the flat connection associated with $\hat{L}$.

Lemma 3.6.39 *Let* L *and* $\hat{L}$ *be a Darboux pair with* $\hat{\Gamma}^\mu$-*parallel* $X \in \Gamma L$ *and* Γ^μ-*parallel* $\hat{X} \in \Gamma\hat{L}$. *Then we have that*

$$\langle X, \mathrm{d}\hat{X} \rangle = 0 = \langle \mathrm{d}X, \hat{X} \rangle.$$

Proof The result follows from using the definition of parallel sections, and the definition of τ. $\qquad\square$

Since $\hat{L}$ is also an isothermic surface coming with its own 1-parameter family of flat connections $\hat{\Gamma}^\lambda$, we aim to explain how the new flat connection is derived from the flat connection associated with the given surface L.

To do this, first note that for linearly independent null vectors $V, W \in L^4$, we have that $\langle V, W \rangle \neq 0$; therefore, we can split $\mathbb{R}^{4,1}$ into

$$\mathbb{R}^{4,1} = \mathcal{L}_1 \oplus \mathcal{L}_2 \oplus (\mathcal{L}_1 \oplus \mathcal{L}_2)^{\perp} \tag{3.18}$$

for $\mathcal{L}_1 = \mathrm{span}\{V\}$ and $\mathcal{L}_2 = \mathrm{span}\{W\}$. Here, $\mathcal{L}_1 \oplus \mathcal{L}_2 = \mathrm{span}\{V, W\} \cong \mathbb{R}^{1,1}$ so that $(\mathcal{L}_1 \oplus \mathcal{L}_2)^{\perp} \cong \mathbb{R}^3$. Using this splitting, we define a particular orthogonal transformation:

Definition 3.6.40 ([41, Equation (3.6)]) We define the orthogonal transformation $\Gamma_{\mathcal{L}_2}^{\mathcal{L}_1}(s) \in O_{4,1}$ of $\mathbb{R}^{4,1}$, for $\mathcal{L}_1, \mathcal{L}_2 \in P(L^4)$, by

$$\Gamma_{\mathcal{L}_2}^{\mathcal{L}_1}(s)Y = \begin{cases} sY, & \text{if } Y \in \mathcal{L}_1, \\ s^{-1}Y, & \text{if } Y \in \mathcal{L}_2, \\ Y, & \text{if } Y \in (\mathcal{L}_1 \oplus \mathcal{L}_2)^{\perp}, \end{cases}$$

for some $s \in \mathbb{R}$.

Exercise 3.6.41 Check that $\Gamma_{\mathcal{L}_2}^{\mathcal{L}_1}(s) \in O_{4,1}$.

The explicit form for $\Gamma_{\mathcal{L}_2}^{\mathcal{L}_1}(s)$ as in the next lemma can be easily verified:

Lemma 3.6.42 $\Gamma_{\mathcal{L}_2}^{\mathcal{L}_1}(s)Y$, *for* $Y \in \mathbb{R}^{4,1}$, *can be written as*

$$\Gamma_{\mathcal{L}_2}^{\mathcal{L}_1}(s)Y = Y + \langle V, W \rangle^{-1}\{(s-1)\langle Y, W \rangle V + (s^{-1}-1)\langle Y, V \rangle W\}. \tag{3.19}$$

for any choice of non-zero $V \in \mathcal{L}_1$ *and* $W \in \mathcal{L}_2$.

The form in (3.19) allows us to recover a geometric meaning of the orthogonal transformation $\Gamma_{\mathcal{L}_2}^{\mathcal{L}_1}(s)$. Recall from Remark 2.5.39 that the cross ratio of four points $\mathrm{cr}(X_1, X_2, X_3, X_4) = t$ can be viewed as parametrizing the circle through three of the points via (2.25):

$$X_3 = X_1 + \frac{1}{\langle X_2, X_4 \rangle}\{(t-1)\langle X_1, X_4 \rangle X_2 + (t^{-1}-1)\langle X_1, X_2 \rangle X_4\}.$$

Comparing this with (3.19), we see that if we choose $Y \in L^4$ (so that $\Gamma_{\mathcal{L}_2}^{\mathcal{L}_1}(s)Y \in L^4$), then we have

$$\mathrm{cr}(Y, V, \Gamma_{\mathcal{L}_2}^{\mathcal{L}_1}(s)Y, W) = s. \tag{3.20}$$

Now we show the relationship between the two flat connections of a Darboux pair.

Theorem 3.6.43 (cf. [41, Theorem 3.10]) *Let L and $\hat{L}$ be a Darboux pair with parameter μ, with associated flat connections Γ^λ and $\hat{\Gamma}^\lambda$, respectively. Then*

$$(\mathrm{d} + \lambda\hat{\tau}) =: \hat{\Gamma}^\lambda = \Gamma_L^{\hat{L}}(1 - \tfrac{\lambda}{\mu}) \bullet \Gamma^\lambda,$$

for the retraction form $\hat{\tau}$ of $\hat{L}$.

Proof We show the equality using the splitting

$$\mathbb{R}^{4,1} = L \oplus \hat{L} \oplus (L \oplus \hat{L})^\perp = \mathrm{span}\{X\} \oplus \mathrm{span}\{\hat{X}\} \oplus \mathrm{span}\{X, \hat{X}\}^\perp,$$

with $\hat{\Gamma}^\mu$-parallel $X \in \Gamma L$ and Γ^μ-parallel $\hat{X} \in \Gamma\hat{L}$. Due to Lemma 3.6.39, we may further assume that $\langle X, \hat{X}\rangle \equiv 1$.

First, using the fact that $\hat{\Gamma}^\mu X = 0$ and $\mathrm{d}X \in \mathrm{span}\{X, \hat{X}\}^\perp$, we have

$$\left(\Gamma_L^{\hat{L}}(1 - \tfrac{\lambda}{\mu}) \circ \Gamma^\lambda \circ \Gamma_{\hat{L}}^L(1 - \tfrac{\lambda}{\mu})\right) X = (1 - \tfrac{\lambda}{\mu})\left(\Gamma_L^{\hat{L}}(1 - \tfrac{\lambda}{\mu}) \circ (\mathrm{d} + \lambda\tau)\right) X$$

$$= (1 - \tfrac{\lambda}{\mu})\Gamma_L^{\hat{L}}(1 - \tfrac{\lambda}{\mu})\mathrm{d}X = (1 - \tfrac{\lambda}{\mu})\mathrm{d}X$$

$$= \mathrm{d}X - \tfrac{\lambda}{\mu}(-\mu\hat{\tau}X) = (\mathrm{d} + \lambda\hat{\tau})X.$$

Also, we can easily check, using $\Gamma^\mu\hat{X} = 0$ and $\mathrm{d}\hat{X} \in \mathrm{span}\{X, \hat{X}\}^\perp$, that

$$\left(\Gamma_L^{\hat{L}}(1 - \tfrac{\lambda}{\mu}) \circ \Gamma^\lambda \circ \Gamma_{\hat{L}}^L(1 - \tfrac{\lambda}{\mu})\right) \hat{X} = (1 - \tfrac{\lambda}{\mu})^{-1}\left(\Gamma_L^{\hat{L}}(1 - \tfrac{\lambda}{\mu}) \circ (\mathrm{d} + \lambda\tau)\right) \hat{X}$$

$$= \Gamma_L^{\hat{L}}(1 - \tfrac{\lambda}{\mu})\mathrm{d}\hat{X} = \mathrm{d}\hat{X} = (\mathrm{d} + \lambda\hat{\tau})\hat{X}.$$

Finally, let $Y \perp \mathrm{span}\{X, \hat{X}\}$. Writing $\tau = X \wedge \omega$ for

$$\omega = -\tfrac{2}{\langle X_u, X_u\rangle}X_u\mathrm{d}u + \tfrac{2}{\langle X_v, X_v\rangle}X_v\mathrm{d}v,$$

we can calculate that $-\tfrac{1}{\mu}\mathrm{d}\hat{X} = \tau\hat{X} = \omega - \langle\omega, \hat{X}\rangle X = \omega$ (since $\hat{X} \perp \mathrm{d}X$), implying

$$\lambda\tau Y = -\lambda\langle\omega, Y\rangle X = \tfrac{\lambda}{\mu}\langle\mathrm{d}\hat{X}, Y\rangle X.$$

A symmetric argument gives us

$$\lambda\hat{\tau} Y = \tfrac{\lambda}{\mu}\langle\mathrm{d}X, Y\rangle\hat{X}.$$

Furthermore, since we can write

$$\mathrm{d}Y = -\langle\mathrm{d}\hat{X}, Y\rangle X - \langle\mathrm{d}X, Y\rangle\hat{X} + Z$$

for some 1-form Z taking values in $\mathrm{span}\{X, \hat{X}\}^{\perp}$, we have

$$\Gamma^{\lambda} Y = (\mathrm{d} + \lambda \tau) Y = -(1 - \tfrac{\lambda}{\mu})\langle \mathrm{d}\hat{X}, Y\rangle X - \langle \mathrm{d}X, Y\rangle \hat{X} + Z.$$

Using this, we can calculate that

$$\left(\Gamma_L^{\hat{L}}(1 - \tfrac{\lambda}{\mu}) \circ \Gamma^{\lambda} \circ \Gamma_{\hat{L}}^{L}(1 - \tfrac{\lambda}{\mu})\right) Y = \Gamma_L^{\hat{L}}(1 - \tfrac{\lambda}{\mu})(\Gamma^{\lambda} Y)$$

$$= -\langle \mathrm{d}\hat{X}, Y\rangle X - (1 - \tfrac{\lambda}{\mu})\langle \mathrm{d}X, Y\rangle \hat{X} + Z$$

$$= \mathrm{d}Y + \tfrac{\lambda}{\mu}\langle \mathrm{d}X, Y\rangle \hat{X} = (\mathrm{d} + \lambda\hat{\tau})Y.$$

Due to the splitting, any vector in $\mathbb{R}^{4,1}$ can be written as a linear combination of vectors X, $\hat{X}$, and $Y \perp \mathrm{span}\{X, \hat{X}\}$, giving us the desired conclusion. $\qquad\square$

Now suppose that L_1 and L_2 are both Darboux transforms of L, with parameters μ_1 and μ_2 respectively, such that

$$\mu_1 \neq \mu_2,$$

and let Γ_1^{λ} and Γ_2^{λ} be the respective associated 1-parameter families of flat connections.

Bianchi permutability [10, §5] states that there is a fourth surface L_{12} that is both a Darboux transform of L_1 and L_2 with parameter μ_2 and μ_1, respectively, which we aim to show now. Define L_{12} and L_{21} as

$$L_{12} := \Gamma_L^{L_2}(1 - \tfrac{\mu_1}{\mu_2})L_1, \quad L_{21} := \Gamma_L^{L_1}(1 - \tfrac{\mu_2}{\mu_1})L_2. \tag{3.21}$$

Rewriting (3.21) in terms of cross ratios via (3.20), we have

$$\mathrm{cr}(L_1, L_2, L_{12}, L) = 1 - \frac{\mu_1}{\mu_2}, \quad \mathrm{cr}(L_2, L_1, L_{21}, L) = 1 - \frac{\mu_2}{\mu_1}. \tag{3.22}$$

However, due to Lemma 2.5.41,

$$\mathrm{cr}(L_2, L_1, L_{12}, L) = \frac{1 - \frac{\mu_1}{\mu_2}}{\left(1 - \frac{\mu_1}{\mu_2}\right) - 1} = \mathrm{cr}(L_2, L_1, L_{21}, L),$$

telling us that $L_{12} = L_{21}$.

Using this equality, we now show the statement of Bianchi permutability.

Theorem 3.6.44 *For Darboux transforms L_1 and L_2 of L with parameters μ_1 and μ_2, respectively, there is a fourth surface L_{12} that is both a Darboux transform of L_1 and L_2 with parameter μ_2 and μ_1, respectively (Fig. 3.2).*

Fig. 3.2 Bianchi
permutatibility

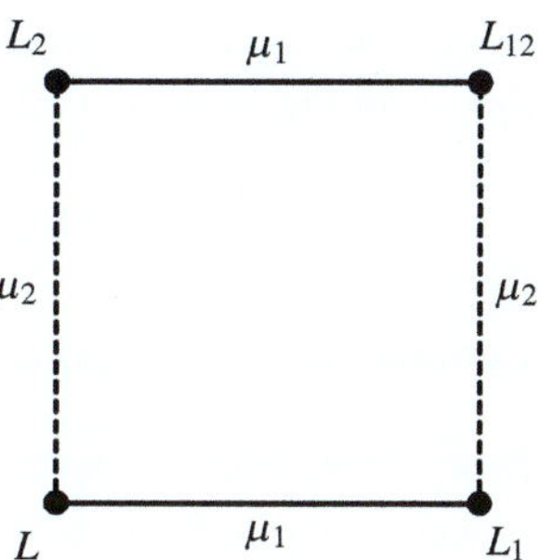

Proof Denote the flat connections of L, L_1, L_2 by $\Gamma^\lambda, \Gamma_1^\lambda, \Gamma_2^\lambda$, respectively. We show that L_{12} is a Darboux transform of L_2 with parameter μ_1. Since L_2 is a Darboux transform of L with parameter μ_2, we have that

$$\Gamma_2^\lambda = \Gamma_L^{L_2}(1 - \tfrac{\lambda}{\mu_2}) \bullet \Gamma^\lambda.$$

Taking L_{12} as in (3.21) and $X_1 \in \Gamma L_1$ so that $\Gamma^{\mu_1} X_1 = 0$, define

$$X_{12} := \Gamma_L^{L_2}(1 - \tfrac{\mu_1}{\mu_2})X_1 \in \Gamma L_{12}.$$

Then, we can calculate that

$$\Gamma_2^{\mu_1} X_{12} = \left(\Gamma_L^{L_2}(1 - \tfrac{\mu_1}{\mu_2}) \bullet \Gamma^{\mu_1}\right)\Gamma_L^{L_2}(1 - \tfrac{\mu_1}{\mu_2})X_1$$

$$= \Gamma_L^{L_2}(1 - \tfrac{\mu_1}{\mu_2})(\Gamma^{\mu_1} X_1) = 0.$$

Hence, L_{12} is a Darboux transform of L_2 with parameter μ_1. The fact that $L_{12} = L_{21}$ is a Darboux transform of L_1 with parameter μ_2 is shown similarly (by switching the indices everywhere). $\qquad\square$

In fact, we can now recover the classical formula of Demoulin that relates the four corresponding points of Bianchi permutability:

Corollary 3.6.45 (cf. [90]) *Let* L, L_1, L_2, L_{12} *be as in Theorem 3.6.44. Then on the corresponding points, we have*

$$\mathrm{cr}(L, L_1, L_{12}, L_2) = \frac{\mu_2}{\mu_1}. \tag{3.23}$$

We call L, L_1, L_2, L_{12} *a* Bianchi quadrilateral.

Proof The proof is a direct application of Lemma 2.5.41 on (3.22). $\qquad\square$

We now state the Bianchi permutability in terms of the gauge transformations of flat connections, with an eye on proving the cube theorem. For this, we need a few more tools regarding orthogonal transformations. In the next few lemmas, we note that from the splitting (3.18), we may write any $Z \in \mathbb{R}^{4,1}$ as

$$Z = \mathrm{Proj}_{\mathcal{L}_1}(Z) + \mathrm{Proj}_{(\mathcal{L}_1 \oplus \mathcal{L}_2)^\perp}(Z) + \mathrm{Proj}_{\mathcal{L}_2}(Z).$$

Lemma 3.6.46 $\Gamma_{\mathcal{L}_2}^{\mathcal{L}_1}(s)$ *is the identity on* $\mathcal{L}_1^\perp/\mathcal{L}_1$.

Proof Note that we have $\mathcal{L}_1^\perp = (\mathcal{L}_1 \oplus \mathcal{L}_2)^\perp \oplus \mathcal{L}_1$. By Definition 3.6.40, $\Gamma_{\mathcal{L}_2}^{\mathcal{L}_1}(s)$ preserves $\mathcal{L}_1$ and acts as the identity on $(\mathcal{L}_1 \oplus \mathcal{L}_2)^\perp$. Hence the result follows. $\square$

Lemma 3.6.47 *Let* $A \in O_{4,1}$ *and let* $\mathcal{L}_1, \mathcal{L}_2 \in P(L^4)$ *be distinct with* $\mathcal{L}_1 = \mathrm{span}\{V\}$. *If*

(1) $AV = sV$ *for some* $s \in \mathbb{R} \setminus \{0\}$,
(2) A *is the identity on* $\mathcal{L}_1^\perp/\mathcal{L}_1$,
(3) $A\mathcal{L}_2 = \mathcal{L}_2$,

then $A = \Gamma_{\mathcal{L}_1}^{\mathcal{L}_2}(s^{-1})$.

Proof As in the proof of Lemma 3.6.46, $\mathcal{L}_1^\perp = (\mathcal{L}_1 \oplus \mathcal{L}_2)^\perp \oplus \mathcal{L}_1$, so the second condition tells us that A is the identity on $(\mathcal{L}_1 \oplus \mathcal{L}_2)^\perp$ (implying that $(\mathcal{L}_1 \oplus \mathcal{L}_2)^\perp$ is an eigenspace of A with eigenvalue 1). Finally, the orthogonality of A tells us that the eigenvalue associated with the eigenspace $\mathcal{L}_2$ must be s^{-1}. $\square$

Lemma 3.6.48 *For any* $\mathcal{Z} \in P(\mathbb{R}^{4,1})$, $\Gamma_{\mathcal{L}_2}^{\mathcal{L}_1}(\infty)\mathcal{Z} = \mathcal{L}_1$. *Similarly,* $\Gamma_{\mathcal{L}_2}^{\mathcal{L}_1}(0)\mathcal{Z} = \mathcal{L}_2$.

Proof We will just prove the first statement. For $Z \in \mathcal{Z}$, we have $t^{-1}\Gamma_{\mathcal{L}_2}^{\mathcal{L}_1}(t)Z = \mathrm{Proj}_{\mathcal{L}_1}(Z) + t^{-1}\mathrm{Proj}_{(\mathcal{L}_1 \oplus \mathcal{L}_2)^\perp}(Z) + t^{-2}\mathrm{Proj}_{\mathcal{L}_1}(Z)$. The statement follows from taking the limit $t \to \infty$. $\square$

With these tools in hand, we now prove:

Proposition 3.6.49 ([41, Lemma 4.7]) *We have*

$$\Gamma_{L_1}^{L_{12}}\left(1-\tfrac{\lambda}{\mu_2}\right)\Gamma_{L}^{L_1}\left(1-\tfrac{\lambda}{\mu_1}\right) = \Gamma_{L_1}^{L_2}\left(\tfrac{1-\lambda\mu_2^{-1}}{1-\lambda\mu_1^{-1}}\right) = \Gamma_{L_2}^{L_{21}}\left(1-\tfrac{\lambda}{\mu_1}\right)\Gamma_{L}^{L_2}\left(1-\tfrac{\lambda}{\mu_2}\right). \tag{3.24}$$

Proof We prove only the first equality, setting $A(\lambda) = \Gamma_{L_1}^{L_{12}}\left(1 - \tfrac{\lambda}{\mu_2}\right)\Gamma_{L}^{L_1}\left(1 - \tfrac{\lambda}{\mu_1}\right)$. We check that $A(\lambda)$ satisfies all three conditions of Lemma 3.6.47. First, note that for any $X_1 \in \Gamma L_1$, we have that $AX_1 = \tfrac{1-\lambda\mu_1^{-1}}{1-\lambda\mu_2^{-1}}X_1$. Then using Lemma 3.6.46 twice tells us that $A(\lambda)$ is identity on $L_1^\perp/L_1$. Now, finally, we wish so show that $AL_2 = L_2$, or, equivalently,

$$\Gamma_{L}^{L_1}\left(1 - \tfrac{\lambda}{\mu_1}\right)L_2 = \Gamma_{L_{12}}^{L_1}\left(1 - \tfrac{\lambda}{\mu_2}\right)L_2.$$

Since (3.23) tells us that both sides of the equation parametrize the same circle with λ as the parameter, we only need to check that they agree on three different points:

(1) at $\lambda = \infty$, by Lemma 3.6.48, both sides of the equations evaluate to L_1,
(2) at $\lambda = 0$, both orthogonal transformations become the identity, so both sides evaluate to L_2, and
(3) at $\lambda = \mu_2$, both sides evaluate to L_{12} by the definition of $L_{21} = L_{12}$ (3.21) and Lemma 3.6.48.

Therefore, we have that $A(\lambda)L_2 = L_2$.

With the conditions of Lemma 3.6.47, we now have that

$$A(\lambda) = \Gamma_{L_1}^{L_2}\left(\frac{1-\lambda\mu_2^{-1}}{1-\lambda\mu_1^{-1}}\right),$$

giving us the desired conclusion. $\square$

Exercise 3.6.50 In this exercise, we will show that (3.24) alone implies the Bianchi permutability, that is, $L_{12} = L_{21}$ and

$$\mathrm{cr}(L, L_1, L_{12}, L_2) = \frac{\mu_2}{\mu_1}.$$

(1) Evaluate (3.24) at L to show that

$$\Gamma_{L_1}^{L_{12}}\left(1 - \tfrac{\lambda}{\mu_2}\right)L = \Gamma_{L_1}^{L_2}\left(\frac{1-\lambda\mu_2^{-1}}{1-\lambda\mu_1^{-1}}\right)L = \Gamma_{L_2}^{L_{21}}\left(1 - \tfrac{\lambda}{\mu_1}\right)L.$$

(2) Let $\lambda \to \infty$ to obtain

$$L_{12} = \Gamma_{L_1}^{L_2}\left(\tfrac{\mu_1}{\mu_2}\right)L = L_{21}. \tag{3.25}$$

(3) Use (3.20) and Exercise 2.5.42 to conclude that

$$\mathrm{cr}(L, L_1, L_{12}, L_2) = \frac{\mu_2}{\mu_1}.$$

The next corollary is an easy consequence of Proposition 3.6.49.

Corollary 3.6.51 *Let Γ_{12}^{λ} be the flat connection of L_{12}. Then we have*

$$\Gamma_{L_1}^{L_{12}}\left(1 - \tfrac{\lambda}{\mu_2}\right) \bullet \left(\Gamma_L^{L_1}\left(1 - \tfrac{\lambda}{\mu_1}\right) \bullet \Gamma^{\lambda}\right) = \Gamma_{12}^{\lambda} = \Gamma_{L_2}^{L_{12}}\left(1 - \tfrac{\lambda}{\mu_1}\right) \bullet \left(\Gamma_L^{L_2}\left(1 - \tfrac{\lambda}{\mu_2}\right) \bullet \Gamma^{\lambda}\right).$$

We now prove the cube theorem (Fig. 3.3).

Theorem 3.6.52 ([41, Theorem 4.9], Cube Theorem) *Given three Darboux transformations L_1, L_2, L_3 of L with distinct parameters μ_1, μ_2, μ_3, respectively, with*

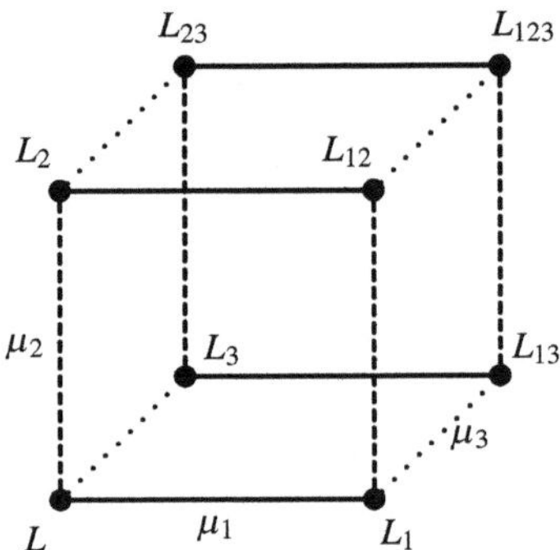

Fig. 3.3 A diagram showing the setting and notation of the Bianchi cube in Theorem 3.6.52

a common Darboux transformation L_{12} of L_1 and L_2, and a common Darboux transformation L_{23} of L_2 and L_3, and a common Darboux transformation L_{31} of L_3 and L_1, there exists a common Darboux transformation L_{123} of L_{12}, L_{23} and L_{31}. We call such a cube a Bianchi cube.

Proof Evaluating the equation in Proposition 3.6.49 on L_3 with $\lambda = \mu_3$ and using (3.21), we have

$$\Gamma_{L_1}^{L_{12}}(1 - \tfrac{\mu_3}{\mu_2})L_{31} = \Gamma_{L_1}^{L_2}\left(\frac{1-\mu_3\mu_2^{-1}}{1-\mu_3\mu_1^{-1}}\right)L_3 = \Gamma_{L_2}^{L_{12}}(1 - \tfrac{\mu_3}{\mu_1})L_{23}.$$

Again, via (3.21), the first term in this equation is a simultaneous Darboux transformation of L_{12} and L_{31}, while the third term in this equation is a simultaneous Darboux transformation of L_{12} and L_{23}. Therefore, we can conclude

$$L_{123} := \Gamma_{L_1}^{L_2}\left(\frac{1-\mu_3\mu_2^{-1}}{1-\mu_3\mu_1^{-1}}\right)L_3$$

is a common Darboux transform of L_{12}, L_{23} and L_{31}, implying the cube theorem.

$\square$

Corollary 3.6.53 *L_1, L_2, L_3 and L_{123} are concircular. Similarly, L, L_{12}, L_{23} and L_{31} are also concircular.*

3.7 Polynomial Conserved Quantities

Now we consider the polynomial conserved quantities of isothermic surfaces.

Definition 3.7.54 ([55, Definition 2.1]) Let $L : \Sigma \to P(L^4)$ be isothermic with associated flat connections Γ^λ. Suppose $P(\lambda)$ is a polynomial whose coefficients are sections of $\mathbb{R}^{4,1}$, that is,

$$P(\lambda) = \sum_{k=0}^{n} \lambda^k P_k,$$

for $P_k : \Sigma \to \mathbb{R}^{4,1}$ for $k = 0, \ldots, n$. If we have

$$\Gamma^\lambda P(\lambda) = 0$$

for all $\lambda \in \mathbb{R}$, then we call $P(\lambda)$ a *polynomial conserved quantity of L of order n*. In particular, when $n = 0$, we call $P(\lambda)$ a constant conserved quantity, while when $n = 1$, we say that $P(\lambda)$ is a linear conserved quantity. We call L a *special isothermic surface of type n* if it admits a polynomial conserved quantity of order n.

For notational simplicity, we sometimes write $Q = P_0$ and $Z = P_n$. Since $\Gamma^\lambda = d + \lambda\tau$, we can calculate that

$$0 = \Gamma^\lambda P(\lambda) = (d + \lambda\tau)\left(\sum_{k=0}^{n} \lambda^k P_k\right) = \sum_{k=0}^{n} \lambda^k dP_k + \sum_{k=0}^{n} \lambda^{k+1}\tau P_k$$

$$= dQ + \sum_{k=1}^{n} \lambda^k(dP_k + \tau P_{k-1}) + \lambda^{n+1}\tau Z.$$

From this we deduce

Lemma 3.7.55 ([55, Proposition 2.2]) $P(\lambda) = \sum_{k=0}^{n} \lambda^k P_k$ *is a polynomial conserved quantity of L if and only if*

(1) Q is constant,
(2) $dP_k = -\tau P_{k-1}$ for $k = 1, \ldots, n$, and
(3) $\tau Z = 0$, equivalently, Z is perpendicular to both X and dX for some non-zero $X \in \Gamma L$.

Corollary 3.7.56 *We have that $\langle Z, Z \rangle$ is constant, and equal to a non-negative number.*

Proof Using Lemma 3.7.55, we have that

$$d(\langle Z, Z \rangle) = 2\langle dZ, Z \rangle = 2\langle -\tau P_{n-1}, Z \rangle = 2\langle P_{n-1}, \tau Z \rangle = 0.$$

Finally, since Z is perpendicular to X, we have that either Z is a scalar multiple of X, or spacelike. $\qquad\square$

Also the third condition immediately tells us:

Corollary 3.7.57 $X \in \Gamma L$ *is perpendicular to both Z and dZ.*

We now wish to see the behavior of polynomial conserved quantities under transformations. First, we treat the Calapso transformations.

Theorem 3.7.58 ([55, Theorem 3.12]) *Suppose the* $L : \Sigma \to P(L^4)$ *is a special isothermic surface of type n. If* L^μ *is a Calapso transform of* L *with parameter* μ, *then it also admits a polynomial conserved quantity of order n.*

Proof Let τ be the retraction form of L, and $L^\mu = T^\mu L$. For any fixed real μ, Lemma 3.4.25 tells us that L^μ has an associated retraction form

$$\tau^\mu = T^\mu \tau (T^\mu)^{-1}.$$

Now suppose $P(\lambda)$ is a polynomial conserved quantity of L. Then using Exercise 3.4.27, we have

$$(\mathrm{d} + \lambda \tau^\mu)(T^\mu P(\lambda + \mu)) = 0,$$

and therefore

$$P^\mu(\lambda) := T^\mu P(\lambda + \mu)$$

is a polynomial conserved quantity for L^μ of the same order as $P(\lambda)$. $\square$

Remark 3.7.59 We note that for $X \in \Gamma L$ and $X^\mu = T^\mu X \in \Gamma L^\mu$,

$$\mathrm{d}X^\mu = \mathrm{d}(T^\mu X) = T^\mu(\mathrm{d}X + (T^\mu)^{-1}\mathrm{d}T^\mu X) = T^\mu(\mathrm{d}X + \mu\tau X) = T^\mu \mathrm{d}X,$$
$$(3.26)$$

and we similarly have, for the top term coefficient $Z^\mu = T^\mu Z$ of $P^\mu(\lambda)$,

$$\mathrm{d}Z^\mu = T^\mu(\mathrm{d}Z + \mu\tau Z) = T^\mu \mathrm{d}Z, \qquad\qquad (3.27)$$

because $\tau Z = 0$, and so

$$\langle \mathrm{d}X^\mu, \mathrm{d}Z^\mu \rangle = \langle T^\mu \mathrm{d}X, T^\mu \mathrm{d}Z \rangle = \langle \mathrm{d}X, \mathrm{d}Z \rangle.$$

Now, we turn our attention to polynomial conserved quantites under Darboux transformations. For this, let Γ^λ and $\hat{\Gamma}^\lambda$ be the associated flat connections for the Darboux pair L and $\hat{L}$ with respect to μ, respectively. Recall that by Theorem 3.6.43, we have that

$$\hat{\Gamma}^\lambda = \Gamma_L^{\hat{L}}(1 - \tfrac{\lambda}{\mu}) \bullet \Gamma^\lambda.$$

Defining $\tilde{P}(\lambda)$ as

$$\tilde{P}(\lambda) := \Gamma_L^{\hat{L}}(1 - \tfrac{\lambda}{\mu})P(\lambda),$$

we have that

$$\hat{\Gamma}^\lambda \tilde{P}(\lambda) = \Gamma_L^{\hat{L}}(1 - \tfrac{\lambda}{\mu})\Gamma^\lambda \left(\Gamma_L^{\hat{L}}(1 - \tfrac{\lambda}{\mu})\right)^{-1} \Gamma_L^{\hat{L}}(1 - \tfrac{\lambda}{\mu})P(\lambda) = 0,$$

so that $\tilde{P}(\lambda)$ is $\hat{\Gamma}^\lambda$-parallel for all $\lambda \in \mathbb{R}$. However, decomposing

$$P(\lambda) = \mathrm{Proj}_L(P(\lambda)) + \mathrm{Proj}_{(L \oplus \hat{L})^\perp}(P(\lambda)) + \mathrm{Proj}_{\hat{L}}(P(\lambda)),$$

we see that $\tilde{P}(\lambda)$ is not polynomial in λ unless $\mathrm{Proj}_L(P(\lambda))$ has a zero at μ, since

$$\tilde{P}(\lambda) = (1 - \tfrac{\lambda}{\mu})^{-1} \mathrm{Proj}_L(P(\lambda)) + \mathrm{Proj}_{(L \oplus \hat{L})^\perp}(P(\lambda)) + (1 - \tfrac{\lambda}{\mu}) \mathrm{Proj}_{\hat{L}}(P(\lambda)).$$

$$(3.28)$$

Therefore, we define

$$\hat{P}(\lambda) := (1 - \tfrac{\lambda}{\mu})\tilde{P}(\lambda), \tag{3.29}$$

which is a polynomial in λ such that $\hat{\Gamma}^\lambda \hat{P}(\lambda) = 0$.

Proposition 3.7.60 ([55, Theorem 3.2 (1)]) *If the initial isothermic surface L admits a polynomial conserved quantity $P(\lambda)$ of order n, then any Darboux transform $\hat{L}$ has a polynomial conserved quantity $\hat{P}(\lambda)$ of order at most $n + 1$.*

Proof Due to the discussions preceding the proposition, we only need to show

$$\hat{P}(\lambda) = \mathrm{Proj}_L(P(\lambda)) + (1 - \tfrac{\lambda}{\mu}) \mathrm{Proj}_{(L \oplus \hat{L})^\perp}(P(\lambda)) + (1 - \tfrac{\lambda}{\mu})^2 \mathrm{Proj}_{\hat{L}}(P(\lambda))$$

has order $n + 1$. This is trivial for the L component and the $(L \oplus \hat{L})^\perp$ component, so we only consider the $\hat{L}$ component. The third condition of Lemma 3.7.55 tells us that $Z \perp X$ for any $X \in \Gamma L$. Therefore, writing $Z = AX + B\hat{X} + Y$ for some $\hat{X} \in \Gamma\hat{L}$ and $Y \in (L \oplus \hat{L})^\perp$, we see

$$0 = \langle Z, X \rangle = B\langle X, \hat{X} \rangle,$$

The fact that $\langle X, \hat{X} \rangle \neq 0$ implies $B = 0$, so $\mathrm{Proj}_{\hat{L}}(Z) = 0$. From this, we can conclude

$$\mathrm{Proj}_{\hat{L}}(P(\lambda)) = \sum_{k=0}^{n} \lambda^k \mathrm{Proj}_{\hat{L}}(P_k)$$

has degree at most $n - 1$. $\qquad\square$

There are special cases when the order of the polynomial conserved quantity is preserved under Darboux transformations; following [52, Definition 4.4], we will refer to these as *Bäcklund transformation for special isothermic surfaces of type n*. To see this, we revisit $\tilde{P}(\lambda)$ as in (3.28), and note that if $\mathrm{Proj}_L(P(\lambda))$ has a zero at μ, then $\tilde{P}(\lambda)$ would be polynomial in λ:

Proposition 3.7.61 ([55, Theorem 3.2 (2)]) *If the initial isothermic surface L admits a polynomial conserved quantity $P(\lambda)$ of order n, and a Darboux transform $\hat{L}$ with parameter μ satisfies $\langle \hat{X}, P(\mu) \rangle \equiv 0$ for any $\hat{X} \in \Gamma\hat{L}$, then $\hat{L}$ has a polynomial conserved quantity $\hat{P}(\lambda)$ of order at most n, given by*

$$\hat{P}(\lambda) := \Gamma_L^{\hat{L}}(1 - \tfrac{\lambda}{\mu})P(\lambda). \tag{3.30}$$

Remark 3.7.62 The condition that $\langle \hat{X}, P(\mu) \rangle = 0$ only needs to be satisfied at one point in the domain. To see this, choose $\hat{X} \in \Gamma\hat{L}$ so that $\Gamma^\mu\hat{X} = 0$. Then we can calculate that

$$d(\langle \hat{X}, P(\mu) \rangle) = \langle d\hat{X}, P(\mu) \rangle + \langle \hat{X}, dP(\mu) \rangle = \langle -\mu\tau\hat{X}, P(\mu) \rangle + \langle \hat{X}, -\mu\tau P(\mu) \rangle = 0.$$

Therefore, a special isothermic surface of type n admits a 3-parameter family of Darboux transforms that are of type n again.

Remark 3.7.63 For a special isothermic surface X of type n with a polynomial conserved quantity $P(\lambda)$, suppose there is some $\mu \in \mathbb{R}$ so that

$$\|P(\mu)\|^2 = \left\| \sum_{k=0}^{n} \mu^k P_k \right\|^2 = 0.$$

Then $P(\mu) : \Sigma \to L^4$, while $\Gamma^\mu P(\mu) = 0$ for the flat connection Γ^λ of X. Therefore, $P(\mu)$ is a Darboux transform of X, and we call such $P(\mu)$ a *complementary surface* of X, which is a special isothermic surface of type at most n by Proposition 3.7.61 (see also [55, Definition 3.3, Proposition 3.4]).

The fact that Darboux transformations come in pairs (Lemma 3.5.33) and Proposition 3.7.60 tells us that there are cases when the order of the polynomial conserved quantity decreases by 1 under Darboux transformations. It is known that if a given special isothermic surface of type n admits a Darboux transform of type $n - 1$, then the transform must be a complementary surface (see [55, Proposition 3.5]).

Bianchi permutability also holds for Bäcklund transformations of special isothermic surfaces of type n:

Theorem 3.7.64 ([55, Theorem 3.6]) *Suppose that $L : \Sigma \to P(L^4)$ is a special isothermic surface of type n, and let L_1 and L_2 be Bäcklund transforms of L with parameter μ_1 and μ_2, respectively. Then there is a fourth surface $L_{12} : \Sigma \to P(L^4)$ that is simultaneously a Bäcklund transform of L_1 and L_2 with parameter μ_2 and μ_1, respectively.*

Proof Letting $P(\lambda)$ and $P_1(\lambda)$ denote the polynomial conserved quantity of L and L_1 respectively, we have that

$$P_1(\mu_2) = \Gamma_L^{L_1}(1 - \tfrac{\mu_2}{\mu_1})P(\mu_2)$$

since L_1 is a Bäcklund transform of L with parameter μ_1. Now, since we have that L_2 is a Bäcklund transform of L with parameter μ_2, Proposition 3.7.61 tells us that $\langle P(\mu_2), X_2 \rangle = 0$ for any $X_2 \in \Gamma L_2$. Let L_{12} be given by the Bianchi permutability of Darboux transformations (Theorem 3.6.44), so that by (3.21), we have

$$L_{12} = \Gamma_L^{L_1}(1 - \tfrac{\mu_2}{\mu_1})L_2.$$

Then,

$$0 = \langle P(\mu_2), X_2 \rangle = \langle \Gamma_L^{L_1}(1 - \tfrac{\mu_2}{\mu_1})P(\mu_2), \Gamma_L^{L_1}(1 - \tfrac{\mu_2}{\mu_1})X_2 \rangle = \langle P_1(\mu_2), X_{12} \rangle$$

for some $X_{12} \in \Gamma L_{12}$. Therefore, Proposition 3.7.61 implies that L_{12} is also a special isothermic surface of type n, that is, L_{12} is a Bäcklund transform of L_1 with parameter μ_2. The fact that L_{12} is a Bäcklund transform of L_2 with parameter μ_1 is shown similarly, by switching the indices everywhere. $\qquad\square$

3.8 Constant Mean Curvature Surfaces and Lawson Correspondence

We will now see how cmc surfaces can be characterized in terms of polynomial conserved quantities.

Theorem 3.8.65 ([55, Proposition 2.4]) *An isothermic surface $L : \Sigma \to P(L^4)$ projects to a part of a sphere (in any space form) if and only if it has a constant conserved quantity $P(\lambda) = Q$.*

Proof Suppose that $X \in \Gamma L$ with $X : \Sigma \to M_\kappa$. For one direction, assume that X has a constant conserved quantity $P(\lambda) = Q$. Then Lemma 3.7.55 tells us that Q is constant and that $\langle X, Q \rangle \equiv 0$, while Corollary 3.7.56 implies that Q is either lightlike or spacelike. If Q is lightlike, then $\mathrm{span}\{Q\} \equiv L$, so the image of X is just a point. Otherwise, let $\tilde{S}$ be the sphere in M_κ defined by Q via (2.17), that is,

$$\tilde{S} = \{Y \in M_\kappa \mid \langle Y, Q \rangle = 0\}.$$

Then it is immediate that $X(\Sigma) \subset \tilde{S}$.

Conversely, suppose that $X(\Sigma) \subset \tilde{S}$ where $\tilde{S}$ is a sphere in M_κ defined by some spacelike $S \in \mathbb{R}^{4,1}$. Then one can use Lemma 3.7.55 to show that $P(\lambda) := S$ is a constant conserved quantity of X. $\qquad\square$

Theorem 3.8.66 (cf. [38, p. 61], [55, Theorem 2.5]) *An isothermic surface L projects to a surface X with constant mean curvature in a space form M_κ (via $\mathfrak{q}_\kappa$) if and only if it admits a linear conserved quantity $P(\lambda) = Q + \lambda Z$ with $Q = \mathfrak{q}_\kappa$.*

Proof Assume that L has a linear conserved quantity with isothermic coordinates $(u, v) \in \Sigma$. Since Q is a constant vector, let M_κ be a space form defined by taking $\mathsf{q}_\kappa = Q$, and let $X \in \Gamma L$ with $\langle X, \mathsf{q}_\kappa \rangle = -1$. Now by Corollary 3.7.56, we have that $\|Z\|^2 \geq 0$, and suppose for contradiction that $\|Z\|^2 = 0$ implying that $Z = \alpha X$ for some real function α. However, we also know that by Lemma 3.7.55

$$\mathrm{d}(\langle Q, Z \rangle) = \langle Q, \mathrm{d}Z \rangle = \langle Q, -\tau Q \rangle = 0,$$

allowing us to deduce that α is a real constant. Then we have that

$$\alpha(X_u \mathrm{d}u + X_v \mathrm{d}v) = \mathrm{d}Z = -\tau Q = -\frac{2}{\|X_u\|^2} X_u \mathrm{d}u + \frac{2}{\|X_v\|^2} X_v \mathrm{d}v,$$

which is a contradiction. Therefore, we may assume without loss of generality that $\|Z\|^2 = 1$.

Again by Lemma 3.7.55, we have $\langle X, Z \rangle = 0 = \langle \mathrm{d}X, Z \rangle$, and therefore Proposition 2.7.61 implies that Z projects to a sphere congruence enveloping X. Therefore, $Z = \beta X + \mathcal{T}_n$ for some real function β where $\mathcal{T}_n$ represents the unit normal of X with respect to M_κ. Then we have

$$\langle Q, Z \rangle = -\beta$$

so that β is a real constant. Denoting the principal curvatures of X with respect to the space form M_κ by k_1 and k_2,

$$(\beta - k_1)X_u \mathrm{d}u + (\beta - k_2)X_v \mathrm{d}v = \mathrm{d}Z = -\tau Q = -\frac{2}{\|X_u\|^2} X_u \mathrm{d}u + \frac{2}{\|X_v\|^2} X_v \mathrm{d}v.$$

Therefore, we see that

$$\beta = k_1 - \frac{2}{\|X_u\|^2} = k_2 + \frac{2}{\|X_v\|^2},$$

and the fact that (u, v) are isothermic coordinates allows us to conclude that $\beta = H_\kappa$. Hence, X is a cmc surface in M_κ.

To prove the converse direction, assume that X is a cmc H_κ surface in M_κ created via q_κ with isothermic coordinate $z = u + \sqrt{-1}v$. The constant mean curvature condition is equivalent to the holomorphicity of the Hopf differential factor (see Corollary 2.10.84 here for the case when the space form is $\mathbb{R}^3$ and [17] for other space forms), and one can always reparametrize so that the Hopf differential factor is constant (see, for example, [18, Lemma 2.3.2]). Thus, assume without loss of generality that

$$b_{11} - b_{22} = 4.$$

Now, let $Q := \mathsf{q}_\kappa$, and

$$Z := H_\kappa X + \mathcal{T}_n. \tag{3.31}$$

Then we have that $\mathrm{d}Q = 0$ while $\tau Z = 0$. Now, if k_1 and k_2 denote the principal curvatures of X in M_κ, we have that

$$\mathrm{d}Z = H_\kappa \mathrm{d}X + \mathrm{d}\mathcal{T}_n = \frac{1}{2}(k_2 - k_1)(X_u \mathrm{d}u - X_v \mathrm{d}v),$$

but we also have

$$\frac{1}{2}(k_2 - k_1) = \frac{1}{2}\left(\frac{b_{22}}{g_{22}} - \frac{b_{11}}{g_{11}}\right) = -\frac{2}{\|X_u\|^2}$$

since (u, v) are isothermic coordinates. Therefore, we have $\mathrm{d}Z = -\tau Q$, and we conclude that $P(\lambda) := Q + \lambda Z$ is a linear conserved quantity of X by Lemma 3.7.55.

$\square$

In the case of $\|Q\|^2 = 0$, and $\langle Q, Z \rangle \neq 0$, Theorem 3.8.66 can be shown using Christoffel duality, as in the following pair of exercises.

Exercise 3.8.67 Let $x : \Sigma \to \mathbb{R}^3$ be a cmc $H_0 \neq 0$ surface, with Christoffel transform x^*. Denote the lifts of x, x^* by $X_0, X_0^* : \Sigma \to M_0$, respectively, and write $\mathsf{q}_0 = Q$.

(1) Check the equation $\langle X_0, X_0^* \rangle = -\frac{1}{2H_0^2}$ using Corollary 3.5.36, and use this to
    show that $\mathrm{d}x^* \cdot x^* = \mathrm{d}x^* \cdot x$. Conclude that $\mathrm{d}X_0^* = \mathcal{T}_{\mathrm{d}x^*}$.
(2) Defining

$$Z := \frac{1}{2}X_0^* - \frac{1}{4H_0^2}Q,$$

show that $\langle Z, X_0 \rangle = 0 = \langle Z, \mathrm{d}X_0 \rangle$, and confirm $\tau Z = 0$ for the retraction form τ of X_0. (Hint: use the Exercise 2.10.78.)
(3) Use Exercise 3.3.18 to verify

$$\mathrm{d}Z = -\tau Q,$$

and deduce that $P(\lambda) = Q + \lambda Z$ is a linear conserved quantity of $L := \mathrm{span}\{X_0\}$ using Lemma 3.7.55.

The converse needs a little more work:

Exercise 3.8.68 Let $L : \Sigma \to \mathbb{R}^3$ be an isothermic surface with linear conserved quantity $P(\lambda) = Q + \lambda Z$ where $\|Q\|^2 = 0$, and normalized so that $\|Z\|^2 = 1$. (Such normalization exists by Corollary 3.7.56.) As in the proof of Theorem 3.8.66, assume without loss of generality that $Q = \mathsf{q}_0$.

(1) Show that $H := -\langle Q, Z \rangle$ is constant, using $\mathrm{d}Z = -\tau Q$ from Lemma 3.7.55.

(2) Assuming $H \neq 0$, define

$$\hat{X}_0 := \frac{1}{2H^2} P(2H) = \frac{1}{2H^2}(Q + 2HZ).$$

Confirm $\hat{X}_0 : \Sigma \to M_0$, and check that $\hat{L} := \mathrm{span}\{\hat{X}_0\}$ is a complementary surface of L (and hence a Darboux transform of L).

(3) Denoting the projection of L to $\mathbb{R}^3$ as x, let x^* be the Christoffel transform of x. Using the definition of parallel sections, verify

$$d\hat{X}_0 = \frac{1}{2H} \mathcal{T}_{dx^*}.$$

(Hint: In addition to the definition of parallel sections, use Exercise 3.3.18.)

(4) Denoting by $\hat{x}$ the projection of $\hat{X}_0$ to $\mathbb{R}^3$, use Lemma 3.6.39 and Exercise 2.10.78 to check that $d\hat{x} = \frac{1}{2H} dx^*$, and conclude that x is a cmc surface using Theorem 3.5.35.

The proof of Theorem 3.8.66 shows that the top term Z of the linear conserved quantity is the mean curvature sphere congruence, which is also the *conformal Gauss map* (see Lemma 2.7.67).

Remark 3.8.69 In light of Proposition 3.7.61 and Remark 3.7.62, we can obtain Darboux transformations of cmc surfaces that are cmc again via:

- take a cmc surface X in some space form M_κ with a linear conserved quantity $P(\lambda) = Q + \lambda Z$,
- pick a value $\lambda = \mu$,
- pick an initial condition $\hat{X}_p$, at some point p in the domain of X, such that $\langle \hat{X}_p, P(\mu)_p \rangle = 0$,
- using the $\mathbb{R}^3$ projection x and $\hat{x}$, solve the Riccati equation (3.16) for $\hat{x}$. (The resulting image will be the stereographic projection of the cmc surface in the same space form M_κ.)

This gives a 3-parameter family of Darboux transformations that keep the cmc condition (Fig. 3.4), and it has been shown that these are equivalent to the Bianchi-Bäcklund transformations, or the simple factor dressings of the associated harmonic maps (see, for example, [36, 71, 117, 138, 157, 175]).

We now treat a simple case of cmc surfaces that admit a complementary surface:

Lemma 3.8.70 *Among cmc surfaces, cmc $\pm \sqrt{-\kappa}$ surfaces in M_κ are the only cases where its Darboux transform can be totally umbilic, i.e. it is of type 0. In particular, if such a Darboux transform exists, then $\kappa \leq 0$.*

Proof When the linear conserved quantity is normalized so that $\|Z\|^2 = 1$, we have

$$\|\lambda Z + Q\|^2 = \lambda^2 - 2H\lambda - \kappa,$$

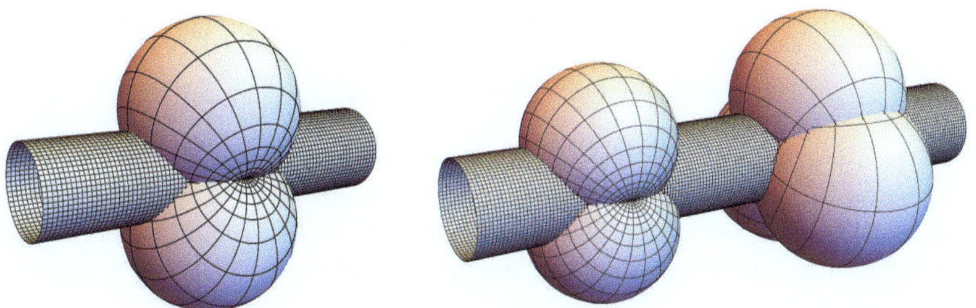

Fig. 3.4 A Darboux transformation of a cylinder on the left, which will have constant mean curvature, called a bubbleton, with curvature lines shown. Then, on the right, a Darboux transformation of that bubbleton, again with constant mean curvature and with curvature lines shown, called a multibubbleton found in [204]

and the discriminant is

$$4(H^2 + \kappa).$$

A condition for the order n of the linear conserved quantity to decrease was identified in [55, Proposition 3.5]: $\langle P(\lambda), P(\lambda) \rangle$ admits a higher order zero $\lambda = \mu$. Therefore, we have

$$H^2 + \kappa = 0,$$

i.e. $H^2 = -\kappa$. $\square$

On the other hand, the construction of the Weierstrass-type representation for cmc-1 surfaces in $\mathbb{H}^3$, implicitly utilizes Lawson correspondence [149] between minimal surfaces in $\mathbb{R}^3$ and cmc-1 surfaces in $\mathbb{H}^3$ (see, for example, [34, 214]). In this section, we explain how this correspondence is related to transformations and linear conserved quantities in Möbius geometry (see also [111, Example 5.5.29]).

Let $L : \Sigma \to P(L^4)$ admit a linear conserved quantity $P(\lambda) = Q + \lambda Z$ (normalized so that $\|Z\|^2 = 1$), implying that it projects to a cmc H surface $X : \Sigma \to M_\kappa$, where Q determines M_κ. Denoting the Calapso transform of L by $L^\mu = T^\mu L$, we have that by Theorem 3.7.58, L^μ also admits a linear conserved quantity

$$P^\mu(\lambda) = T^\mu P(\lambda + \mu).$$

Since

$$Q^\mu = T^\mu(\mu Z + Q),$$

and

$$\langle X^\mu, Q^\mu \rangle = \langle X, Q \rangle = -1,$$

$X^\mu = T^\mu X \in \Gamma L^\mu$ lies in the space form M_{κ^μ} determined by Q^μ, where

$$\kappa^\mu = -\|Q^\mu\|^2 = -\mu^2 + 2\mu H + \kappa.$$

The proof of Lemma 3.4.23 tells us that the metric of the two surfaces in M_κ and M_{κ^μ} coming from X and X^μ, respectively, are the same, while if $\mathcal{T}_n^\mu : \Sigma \to \mathbb{S}^{3,1}$ is the unit normal to X^μ in M_{κ^μ}, then we have $\mathcal{T}_n^\mu = T^\mu \mathcal{T}_n$. Now, to compare the Hopf differentials of the two surfaces, we first note that via (3.12)

$$(T^\mu X)_{uu} = (T^\mu X_u)_u = T_u^\mu X_u + T^\mu X_{uu} = 2\mu T^\mu X + T^\mu X_{uu}$$

$$(T^\mu X)_{vv} = (T^\mu X_v)_v = T_v^\mu X_v + T^\mu X_{vv} = -2\mu T^\mu X + T^\mu X_{vv}.$$

Using in addition that $\langle X, \mathcal{T}_n \rangle = \langle X_{uv}, \mathcal{T}_n \rangle = 0$ and the fact that $T^\mu \in O_{4,1}$, we calculate

$$
\begin{aligned}
4\langle X_{zz}^\mu, \mathcal{T}_n^\mu \rangle &= \langle X_{uu}^\mu - X_{vv}^\mu - 2\sqrt{-1}X_{uv}^\mu, \mathcal{T}_n^\mu \rangle \\
&= \langle 4\mu T^\mu X + T^\mu X_{uu} - T^\mu X_{vv} - 2\sqrt{-1}T^\mu X_{uv}, T^\mu \mathcal{T}_n \rangle \\
&= \langle X_{uu} - X_{vv}, \mathcal{T}_n \rangle = 4\langle X_{zz}, \mathcal{T}_n \rangle.
\end{aligned}
$$

Therefore, the Hopf differentials of the two surfaces are the same. Also, the mean curvatures H and H^μ of X and X^μ, respectively, are related by

$$H^\mu = -\langle T^\mu Z, T^\mu(\mu Z + Q)\rangle = -\mu + H.$$

Thus we conclude that we have the Lawson correspondence between the cmc H surface X in the space form M_κ determined by Q with constant sectional curvature κ, and the cmc H^μ surface X^μ in the space form M_{κ^μ} determined by Q^μ with constant sectional curvature κ^μ.

Example 3.8.71 Minimal surfaces in $\mathbb{R}^3$ and cmc 1 surfaces in $\mathbb{H}^3$ are related to each other by the Lawson correspondence. In this case,

$$\kappa = H = 0 \quad \text{and} \quad \mu = -1,$$

so

$$\kappa^\mu = -1 \quad \text{and} \quad H^\mu = 1.$$

Recommended Further Readings for Chaps. 2 and 3

The classical reference to Möbius geometry is the third book [14] by Blaschke on sphere geometries. In the first four chapters of the book, Möbius geometry is introduced using the Minkowski model, (also called the pentaspherical coordinates) devised by Darboux in [82]. This book stands as one of the most comprehensive monographs on Möbius geometry.

For a more modern introduction to Möbius geometry, using both the Minkowski model and the quaternionic model, we recommend the encyclopedic book [111]. The book generalizes the discussion of Möbius geometry to the conformal n-sphere. Many of our arguments in Chap. 2 are modeled after the arguments in this book, by focusing on the specific case of the conformal 3-sphere. Another remarkable feature of the book is the explanation of the historical context of isothermic surfaces and their transformations. We also mention [42] as a great introduction to understanding surfaces in conformal geometry using the quaternionic vector bundle, while [132] employs a moving frames method to introduce Möbius geometry and isothermic surfaces.

Isothermic surfaces, first defined by Bour in [31], admit Christoffel transfomations in the Euclidean 3-space, found by Christoffel in [76], which can be interpreted in terms of existence of certain closed 1-forms. This is what is precisely done in the paper [135], where isothermic surfaces are explored as an example of Bonnet pairs [28].

A modern treatment of Christoffel transformations and Darboux transformations [85] of isothermic surfaces is given in [117]. Using the quaternionic model, the paper describes Darboux transformations in terms of a Riccati-type equation, and discusses the Darboux transformations that keep the cmc condition, exploring their relationship to the previously known Bianchi-Bäcklund transformation [8]. This relationship is then further clarified in the paper [138].

Isothermic surfaces constitute an integrable class of surfaces, as shown by [77], and hence their transformations can be explained via the dressing methods of [225]. The work [36] takes this point of view, and describes isothermic surfaces in the general conformal n-sphere using Clifford algebra. Furthermore, this work shows that the action of the simple factors in the spirit of [209, 211] in the context of isothermic surfaces amounts to Darboux transformations.

In fact, the integrability of isothermic surfaces can be interpreted by viewing Darboux pairs as a curved flat [103], and this is the approach taken in [46] using the method of moving frames in the Minkowski model, or in [112] using the quaternionic approach.

Building on the latter approach, the work [116] offers a modern interpretation of the Calapso transformation [7, 10, 57, 58] which characterizes the isothermic surfaces as the second-order deformable class of surfaces in conformal geometry as first shown by [60] (see also [163]). As an application, the work [116] interprets the Umehara-Yamada perturbation [214] of minimal surfaces into cmc-c surfaces in hyperbolic spaces with constant sectional curvature $-c^2$ in terms of the Calapso transformation, while the secondary Gauss maps [216] and the duality [213] of those cmc surfaces in hyperbolic spaces are interpreted in terms of Darboux transformations of totally umbilic surfaces.

Further examples of Darboux transformations in the quaternionic setup is given in [69], by considering Darboux transforms of Delaunay surfaces [86] that are cmc surfaces again. The work obtains explicit parametrizations of such surfaces, and shows that these surfaces are non-embedded (see also [136]).

The permutability of Darboux transforms yields non-trivial surfaces only if the parameters of the transformations are distinct. The work [68] considers the case of two-step Darboux transformation with respect to the same parameter in the quaternionic setting, and recovers a Sym-type formula that allows one to calculate every non-trivial two-step Darboux transformations without the need for a second integration.

The unified description for the theory of conformal geometry and isothermic surfaces can be obtained if one utilizes the machinery of the gauge theoretic and vector bundle approach. This is the viewpoint taken in the work [38, 39].

In [41], the concept of isothermicity is generalized by defining isothermicity (and Darboux pairs as curved flats) in symmetric R-spaces, offering conformal geometry as an example (see also [95]). In fact, this work defines isothermicity by the existence of a 1-parameter family of flat connections, a form of zero curvature representation of the integrable structure of isothermic surfaces, and recovers the theory of isothermic surfaces and their transformations in terms of gauge transformations of the flat connections.

For a more elementary introduction to the gauge theoretic approach to surfaces, we recommend the notes [37]. In these notes, the theory of integrable surfaces is presented via the existence of a 1-parameter family of flat connections, and describes the transformations in this context, providing K-surfaces (surfaces with negative constant Gaussian curvature) and isothermic surfaces as examples, including the formulation of the well-known Bäcklund transformations [6] and Darboux transformations, respectively, and their permutability [10, 12].

Another work highly suited for understanding the theory of conformal geometry in the Minkowski picture is the doctoral thesis [194]. The thesis contains numerous explicit calculations that should help anyone interested in the Minkowski model of Möbius geometry.

Within the context of the flat connections approach to isothermic surfaces arises the concept of polynomial conserved quantities. The work [55] develops the theory of polynomial conserved quantities for isothermic surfaces and their behavior under transformations, and characterizes the special isothermic surfaces studied by Darboux and Bianchi [10, 85] in terms of isothermic surfaces with quadratic conserved quantities.

One can also drop the polynomial condition on the conserved quantities and treat conserved quantities which are formal Laurent series. This is the work done in [54], where it is shown that any isothermic surfaces in the conformal 3-sphere admit such conserved quantities globally.

Minimal surfaces in Euclidean space constitute a subclass of isothermic surfaces, and hence they admit Darboux transformations that are minimal again. The work [115] investigates such transformations by interpreting the Weierstrass representation in terms of permutability of transformations. In particular, the Weierstrass data of Darboux pairs of minimal surfaces are shown to be related by a Riccati-type equation.

Chapter 4
Discrete Isothermic Surfaces

4.1 Circular Nets

We first consider discrete nets in the standard Euclidean 3-space $\mathbb{R}^3$ with the standard Euclidean metric. For simplicity, let our domain be $\mathbb{Z}^2$; however, the theory will hold true for subdomains of $\mathbb{Z}^2$. An elementary quadrilateral is a quadrilateral composed of vertices (m, n), $(m + 1, n)$, $(m + 1, n + 1)$, $(m, n + 1)$, for some $m, n \in \mathbb{Z}$, where we often denote the four points by i, j, k, ℓ, respectively, so that they are ordered counterclockwise about the quadrilateral and starting with i at the lower left vertex.

Now, consider $x : \mathbb{Z}^2 \to \mathbb{R}^3$. For some $(m, n) \in \mathbb{Z}^2$, we write $x(m, n)$ as $x_{m,n}$ or x_i. We will always assume for non-degeneracy condition that on every elementary quadrilateral $ijk\ell$ of the surface,

$$\text{the vertices are pairwise distinct, and } \dim(\text{span}\{x_j - x_i, x_k - x_i, x_\ell - x_i\}) \geq 2. \tag{4.1}$$

It would be natural to impose that the points x_i, x_j, x_k and x_ℓ are coplanar, so that they are the vertices of a planar quadrilateral in $\mathbb{R}^3$, and thus the surface is comprised of planar quadrilaterals connecting continuously along edges.

It is even better if the points x_i, x_j, x_k and x_ℓ are concircular (i.e. all lie in one circle). Discrete surfaces with concircular quadrilaterals are called *circular nets*, considered to be discretizations of surfaces with curvature lines [171], since smooth surfaces parametrized by curvature line coordinates have negative infinitesimal cross ratio (Lemma 2.8.68). Furthermore, the circular condition is preserved under Möbius transformations, allowing us to extend the notion of a surface comprised of planar quadrilaterals to the cases that the ambient space is $\mathbb{S}^3$ or $\mathbb{H}^3$. In fact, once the vertices are concircular, there is actually no further need to think about "planar faces", as all the necessary information is encoded in the circle itself.

© The Author(s), under exclusive license to Springer Nature Switzerland AG 2025
J. Cho et al., *Discrete Isothermic Surfaces in Lie Sphere Geometry*, Lecture Notes in Mathematics 2375, https://doi.org/10.1007/978-3-031-95592-1_4

Remark 4.1.1 Discrete surfaces with planar quadrilaterals are often referred to as *Q-nets* (cf. [25, Definition 2.1]). Q-nets are regarded as a discretization of the coordinates of conjugate nets [93, 195, 197]; those surfaces with diagonal second fundamental form.

Thus, we now assume that for every quadrilateral with vertices i, j, k, ℓ, the image points x_i, x_j, x_k, x_ℓ are concircular. The cross ratios $\mathrm{cr}(x_i, x_j, x_k, x_\ell)$ as defined in Sect. 2.5 are then real-valued.

On a circular net x, we have the freedom to arbitrarily define a "normal vector" v_i at one vertex i, and then we define v_* at any adjacent vertex $* = j, \ell$ by the conditions

(1) $v_i, x_* - x_i, v_*$ all lie in one plane,
(2) $|v_i| = |v_*|$,
(3) the angles between $x_* - x_i$ and v_i, and between $x_i - x_*$ and v_*, are equal but oppositely oriented (in other words, $v_i, x_* - x_i$ and v_* form three sides of an isosceles trapezoid).

Continuing to propagate v in this way, we have a well-defined "normal" vector field v over x (see also [199] or [198, Definition 1]):

Lemma 4.1.2 *The above propagation for defining v is consistent.*

Proof We only need to prove this on an elementary quadrilateral $ijk\ell$. By applying isometries and homotheties if necessary, we may assume that

$$x_i, x_j, x_k, x_\ell \in \mathbb{S}^1 \subset \mathbb{R}^2 \cong \{(x_1, x_2, 0) \mid x_1, x_2 \in \mathbb{R}\},$$

where $\mathbb{S}^1$ denotes the unit circle centered at the origin. Then there exist $\theta_* \in \mathbb{R}$ so that, associating $\mathbb{R}^3$ with $\mathbb{C} \times \mathbb{R}$,

$$x_* = (e^{\sqrt{-1}\theta_*}, 0) , \quad * = i, j, k, \ell.$$

We define triplets of vectors (frames)

$$T_* = (\sqrt{-1}e^{\sqrt{-1}\theta_*}, 0), \quad N_* = (-e^{\sqrt{-1}\theta_*}, 0), \quad B_* = (0 + 0\sqrt{-1}, 1)$$

at x_* for $* = i, j, k, \ell$.

If $v_i = \alpha T_i + \beta N_i + \gamma B_i$ for some $\alpha, \beta, \gamma \in \mathbb{R}$, symmetry with respect to reflection across the perpendicular bisecting plane of $x_j - x_i$ shows that $v_j = -\alpha T_j + \beta N_j + \gamma B_j$. Then $v_k = \alpha T_k + \beta N_k + \gamma B_k$, and $v_\ell = -\alpha T_\ell + \beta N_\ell + \gamma B_\ell$. Reflecting this once more to v_i implies consistency of the definition of v. $\square$

Example 4.1.3 In this example (Fig. 4.1), we offer two discretizations of the Clifford torus. For two constant positive integers m_{per}, n_{per}, which represent the lengths of periodicity in the m and n directions, respectively, consider the discrete surface

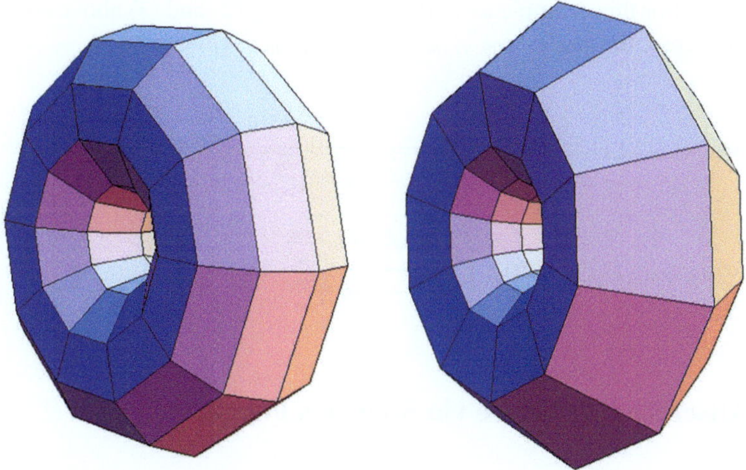

Fig. 4.1 Two different discretizations of the Clifford tori

$$x_{m,n} = \left(\cos\tfrac{2\pi n}{n_{per}}(\sqrt{2}+\cos\tfrac{2\pi m}{m_{per}}),\ \sin\tfrac{2\pi n}{n_{per}}(\sqrt{2}+\cos\tfrac{2\pi m}{m_{per}}),\ \sin\tfrac{2\pi m}{m_{per}}\right)$$

in $\mathbb{R}^3$ with normal vectors at the vertices defined by

$$v_{m,n} = \left(\cos\tfrac{2\pi n}{n_{per}}\cos\tfrac{2\pi m}{m_{per}},\ \sin\tfrac{2\pi n}{n_{per}}\cos\tfrac{2\pi m}{m_{per}},\ \sin\tfrac{2\pi m}{m_{per}}\right)$$

in $\mathbb{S}^2 \subseteq \mathbb{R}^3$. Note that this v and x satisfy conditions (1), (2) and (3) above. In fact, letting $\alpha_{(m,n),(m,n+1)}$ be the common angle between $x_{m,n+1} - x_{m,n}$ and $v_{m,n}$ and also between $x_{m,n} - x_{m,n+1}$ and $v_{m,n+1}$, with similar definitions for $\alpha_{(m,n),(m+1,n)}$, we have

$$\alpha_{(m,n),(m+1,n)} = \frac{\pi(2-m_{per})}{2m_{per}}, \qquad \alpha_{(m,n),(m,n+1)} = \arccos(-\cos\tfrac{2\pi m}{m_{per}}\sin\tfrac{\pi}{n_{per}}).$$

We could also consider the discrete surface

$$x_{m,n} = \frac{1}{\sqrt{2}+\cos\frac{2\pi m}{m_{per}}}\left(\sin\tfrac{2\pi m}{m_{per}},\ \cos\tfrac{2\pi n}{n_{per}},\ \sin\tfrac{2\pi n}{n_{per}}\right)$$

in $\mathbb{R}^3$ with normal vectors at the vertices defined by

$$v_{m,n} = \frac{1}{\sqrt{2}+\cos\frac{2\pi m}{m_{per}}}\left(\sin\tfrac{2\pi m}{m_{per}},\ -(1+\sqrt{2}\cos\tfrac{2\pi m}{m_{per}})\cos\tfrac{2\pi n}{n_{per}},\right.$$

$$\left.-(1+\sqrt{2}\cos\tfrac{2\pi m}{m_{per}})\sin\tfrac{2\pi n}{n_{per}}\right)$$

in $\mathbb{S}^2 \subseteq \mathbb{R}^3$. For this example as well, conditions (1), (2) and (3) above are satisfied, and in this case, the angles $\alpha_{(m,n),(m+1,n)}$, $\alpha_{(m,n),(m,n+1)}$ become

$$\alpha_{(m,n),(m+1,n)} = \arccos\left(\frac{-\sin\frac{\pi}{m_{per}}}{\sqrt{(\sqrt{2}+\cos\frac{2\pi m}{m_{per}})(\sqrt{2}+\cos\frac{2\pi(m+1)}{m_{per}})}}\right),$$

$$\alpha_{(m,n),(m,n+1)} = \arccos\left(\frac{1+\sqrt{2}\cos\frac{2\pi m}{m_{per}}}{\sqrt{2}+\cos\frac{2\pi m}{m_{per}}}\sin\frac{\pi}{n_{per}}\right).$$

4.2 Discrete Curvature via Steiner's Formula

With the notion of normals available, we now introduce the mean and Gaussian curvature of discrete surfaces in $\mathbb{R}^3$.

We first introduce the Steiner's formula for smooth surfaces. Let $x : \Sigma \to \mathbb{R}^3$ be an immersion, parametrized by $(u, v) \in \Sigma$ with unit normal vector field n. The shape (Weingarten) operator is

$$g^{-1}b = \begin{pmatrix} s_{11} & s_{12} \\ s_{21} & s_{22} \end{pmatrix},$$

and so the mean and Gaussian curvatures of the surface are

$$H = \tfrac{1}{2}(s_{11}+s_{22}), \quad K = s_{11}s_{22} - s_{12}s_{21},$$

respectively, and

$$n_u = -s_{11}x_u - s_{21}x_v, \quad n_v = -s_{12}x_u - s_{22}x_v.$$

Considering the parallel surface $x^t = x + tn$ at distance t (with the same normal vector field n), the area A^t of x^t and the area A of x are related by the following Steiner's formula:

$$A^t = \int_\Sigma \det(x_u^t, x_v^t, n)\, du\, dv$$

$$= \int_\Sigma \det\big((1-ts_{11})x_u - ts_{21}x_v, -ts_{12}x_u + (1-ts_{22})x_v, n\big)\, du\, dv$$

$$= \int_\Sigma (1 - 2tH + t^2 K)\, dA,$$

where dA is the area form for x.

We now consider the analogous Steiner's formula for a discrete surface $x(m, n)$ in $\mathbb{R}^3$. For this, we first need to discuss area and mixed area of planar polygons.

Let $\Delta = (p_0, p_1, p_2)$ be a triangle in a plane

$$\mathbb{R}^2 \subset \mathbb{R}^3$$

with vertices p_0, p_1 and p_2, and let $\mathbf{n}$ be the unit normal to that plane. The oriented area of Δ then becomes

$$A(\Delta) = \frac{1}{2} \det(p_1 - p_0, p_2 - p_0, \mathbf{n}).$$

Similarly, for a planar polygon $P = (p_0, ..., p_k)$ with $k + 1$ vertices in the same plane (we define $p_{k+1} := p_0$), its area is

$$A(P) = \frac{1}{2} \sum_{j=1}^{k-1} \det(p_j - p_0, p_{j+1} - p_0, \mathbf{n}) = \frac{1}{2} \sum_{j=0}^{k} \det(p_j, p_{j+1}, \mathbf{n}).$$

Definition 4.2.4 (cf. [201]) For planar polygons $P = (p_0, ..., p_k)$, $Q = (q_0, ..., q_k)$ with parallel corresponding edges all perpendicular to the unit normal $\mathbf{n}$, the *mixed area* is

$$A(P, Q) = \frac{1}{4} \sum_{j=0}^{k} \left(\det(p_j, q_{j+1}, \mathbf{n}) + \det(q_j, p_{j+1}, \mathbf{n}) \right).$$

The following three facts are easily checked:

(1) $A(P, Q)$ is symmetric and bilinear,
(2) $A(P, P) = A(P)$,
(3) for $t \in \mathbb{R}$, $P + tQ$ is defined as having vertices $p_j + tq_j$, and

$$A(P + tQ) = A(P) + 2tA(P, Q) + t^2 A(Q). \tag{4.2}$$

Exercise 4.2.5 ([25, Theorem 4.42]) Let $P = (p_0, p_1, p_2, p_3)$, $Q = (q_0, q_1, q_2, q_3)$ be quadrilaterals with parallel corresponding edges (so that $p_4 = p_0$, $q_4 = q_0$). Writing the edges as $a_j = p_j - p_{j+1}$ and $b_j = q_j - q_{j+1}$ for $j = 0, 1, 2, 3$.

(1) Show that $A(P) = \frac{1}{2}(\det(a_0, a_1, \mathbf{n}) + \det(a_2, a_3, \mathbf{n}))$, and hence deduce that

$$A(P + tQ) = \frac{1}{2}(\det(a_0 + tb_0, a_1 + tb_1, \mathbf{n}) + \det(a_2 + tb_2, a_3 + tb_3, \mathbf{n})).$$

(2) Using the relation (4.2), show that

$$A(P, Q) = \frac{1}{4}(\det(a_0, b_1, \mathbf{n}) - \det(a_1, b_0, \mathbf{n}) + \det(a_2, b_3, \mathbf{n}) - \det(a_3, b_2, \mathbf{n})).$$

(3) Use edge-parallelity to show that

$$A(P, Q) = \frac{1}{2} \det(p_0 - p_2, q_1 - q_3, \mathbf{n}) = \frac{1}{2} \det(q_0 - q_2, p_1 - p_3, \mathbf{n}).$$

Now, for a discrete circular net x with appropriately defined normal vector field v, we have, on each elementary quadrilateral $ijk\ell$,

$$A(x + tv)_{ijk\ell} = A(x)_{ijk\ell} + 2t A(x, v)_{ijk\ell} + t^2 A(v)_{ijk\ell}$$

$$= (1 - 2t H_{ijk\ell} + t^2 K_{ijk\ell}) A(x)_{ijk\ell}$$

where $H_{ijk\ell}$ and $K_{ijk\ell}$ are defined as:

Definition 4.2.6 ([186, Proposition 5]) We call

$$H_{ijk\ell} = -\frac{A(x, v)_{ijk\ell}}{A(x)_{ijk\ell}} \quad \text{and} \quad K_{ijk\ell} = \frac{A(v)_{ijk\ell}}{A(x)_{ijk\ell}}$$

the mean and Gaussian curvatures of the discrete surface x with normal vector field v. Note that they are defined on faces (not vertices, nor edges).

4.3   Discrete Isothermic Surfaces

Recall that in Theorem 2.8.70, isothermic surfaces x parametrized by curvature line coordinates (u, v) are characterized via cross ratios

$$\lim_{\epsilon \to 0} \mathrm{cr}(x(u, v), x(u + \epsilon, v), x(u + \epsilon, v + \epsilon), x(u, v + \epsilon)) = -\frac{a(u)}{b(v)}$$

for some functions of one variable $a(u)$ and $b(v)$. Discretizing this characterization, we first define discrete isothermic surfaces as follows:

Definition 4.3.7 ([22, Definition 6]) A discrete surface $x : \mathbb{Z}^2 \to \mathbb{R}^3$ is called a *discrete isothermic surface* if on every elementary quadrilateral $ijk\ell$, the cross ratios of the four vertices satisfy

$$\mathrm{cr}(x_i, x_j, x_k, x_\ell) = \frac{a_{ij}}{a_{i\ell}} \in \mathbb{R}_{<0} = (-\infty, 0), \tag{4.3}$$

where the function a, defined on (unoriented) edges, is an *edge-labeling*, that is, they satisfy

$$a_{ij} = a_{k\ell} \in \mathbb{R} \quad \text{and} \quad a_{i\ell} = a_{jk} \in \mathbb{R}.$$

We refer to such a function a by *cross ratios factorizing function*.

Remark 4.3.8 We note the following four facts:

- We do not require a priori that x is a circular net as we can deduce this from the cross ratios condition (see Exercise 2.5.38).
- The function a is defined on unoriented edges so that $a_{ij} = a_{ji}$ on any edge ij.
- Because the cross ratios in (4.3) are negative, the associated quadrilaterals are embedded, i.e. the opposite edges of quadrilaterals do not intersect.
- This definition of discrete isothermicity is equivalent to the Toda equation

$$\mathrm{cr}_{(m-1,n-1)}\, \mathrm{cr}_{(m,n)} = \mathrm{cr}_{(m,n-1)}\, \mathrm{cr}_{(m-1,n)}$$

being satisfied [22, Definition 1], where the cross ratios

$$\mathrm{cr}_{(\hat{m},\hat{n})} := \mathrm{cr}\big(x_{(\hat{m},\hat{n})}, x_{(\hat{m}+1,\hat{n})}, x_{(\hat{m}+1,\hat{n}+1)}, x_{(\hat{m},\hat{n}+1)}\big)$$

are all real.

We also have a definition for discrete isothermic surfaces in the narrow sense, by requiring that the cross ratios on elementary quadrilaterals satisfy

$$\mathrm{cr}(x_i, x_j, x_k, x_\ell) = -1,$$

where the motivation for this definition comes from (2.38) in the smooth case, where we use $g_{11} = g_{22}$ [22, Definition 6]. However, with this definition, transformations (such as the Calapso transform) of discrete isothermic surfaces will not remain discrete isothermic. Hence, we take the broader definition given in Definition 4.3.7 as the definition for discrete isothermicity in this text.

One could think of a discrete surface x with cross ratios exactly -1 as being "isothermically parametrized", while a discrete surface x with cross ratios satisfying Definition 4.3.7 is one that could have had its two coordinates independently scaled so that it becomes isothermically parametrized, were it a smooth surface.

Example 4.3.9 Returning to the discrete Clifford tori in Examples 4.1.3, the first example has nonconstant cross ratios

$$\mathrm{cr}(x_i, x_j, x_k, x_\ell) = \frac{-\sin^2 \frac{\pi}{m_{per}}}{\sin^2 \frac{\pi}{n_{per}} \left(\sqrt{2} + \cos \frac{2\pi m}{m_{per}}\right)\left(\sqrt{2} + \cos \frac{2\pi(m+1)}{m_{per}}\right)},$$

which can be readily seen using computational methods as in Sect. 2.5, for example. Thus, the cross ratio factorizing functions can be chosen as

$$a_{ij} = \mathrm{cr}(x_i, x_j, x_k, x_\ell), \quad a_{i\ell} = 1,$$

or any global scalar multiple of these. The second example has constant cross ratios

$$\mathrm{cr}(x_i, x_j, x_k, x_\ell) = \frac{-\sin^2 \frac{\pi}{m_{per}}}{\sin^2 \frac{\pi}{n_{per}}},$$

so we could take

$$a_{ij} = 2\sin^2 \frac{\pi}{m_{per}}, \quad a_{i\ell} = -2\sin^2 \frac{\pi}{n_{per}}.$$

Now consider the lift $X : \mathbb{Z}^2 \to M_0$ of a discrete isothermic surface $x : \mathbb{Z}^2 \to \mathbb{R}^3$ with cross ratios factorizing function a. By the properties of the cross ratio, we then have that

$$\mathrm{cr}(X_i, X_j, X_k, X_\ell) = \mathrm{cr}(x_i, x_j, x_k, x_\ell) = \frac{a_{ij}}{a_{i\ell}}.$$

The additional properties of cross ratios automatically allow us to note that the above cross ratios condition is

(1) Möbius invariant,
(2) extended to include discrete surfaces in any 3-dimensional space form, and
(3) well-defined for maps into the projective light cone $P(L^4)$.

Therefore, for $L : \mathbb{Z}^2 \to P(L^4)$, where $L_i = \mathrm{span}\{X_i\}$ for each vertex i, we have that

$$\mathrm{cr}(L_i, L_j, L_k, L_\ell) = \frac{a_{ij}}{a_{i\ell}} = \mathrm{cr}(x_i, x_j, x_k, x_\ell),$$

for some edge-labeling a. Hence, we can refer to the map L into projective space as being *discrete isothermic* (cf. [111, Definition 5.7.2], [52, Definition 2.1]).

In fact, the cross ratios definition of discrete isothermicity tells us more. Recall from Corollary 3.6.45 that the Bianchi quadrilateral of smooth Darboux transformations enjoys a relationship characterized by cross ratios:

$$\mathrm{cr}(L, L_1, L_{12}, L_2) = \frac{\mu_2}{\mu_1}.$$

Simply switching

$$L, L_1, L_{12}, L_2, \frac{1}{\mu_1}, \frac{1}{\mu_2} \quad \text{to} \quad L_i, L_j, L_k, L_\ell, a_{ij}, a_{i\ell}, \tag{4.4}$$

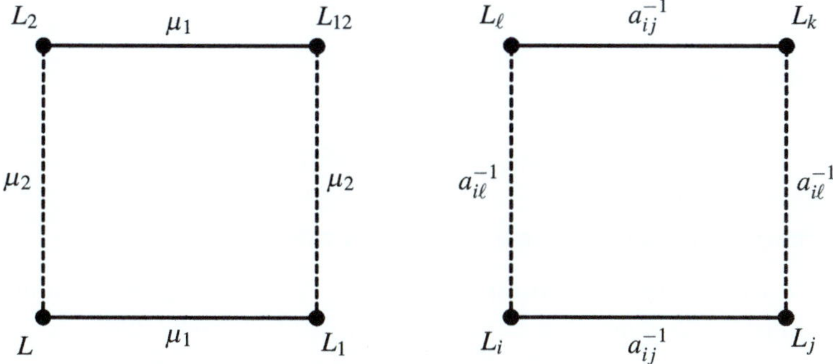

Fig. 4.2 Bianchi permutability of Darboux transformations depicted on the left; a face of a discrete isothermic surface on the right

respectively, we obtain the definition of discrete isothermic surfaces over an elementary quadrilateral. In fact, repeated iterations of Darboux transformations and application of permutability yield the Bianchi lattice where the parameters of Darboux transformations become an edge-labeling (see Fig. 4.2). Therefore, we see that discrete isothermic surfaces and the Bianchi lattice of Darboux transformations of smooth isothermic surfaces are *indistinguishable*, a central theme of *discrete differential geometry with integrability at its heart.*

4.4 Discrete Christoffel Transformations

An important aspect of defining discrete objects is to check which properties of the smooth counterpart hold in the discrete setting. The first of such properties we check is the existence of a discrete Christoffel transformation. Recalling that smooth Christoffel transformations were defined via a 1-form dx^*, we introduce the bare minimum of discrete 1-forms (cf. [91, 121]).

A *discrete 1-form* ω is defined on directed edges, so that $\omega_{ij} = -\omega_{ji}$. We call a discrete 1-form *closed* if we have, on every elementary quadrilateral $ijk\ell$,

$$\omega_{ij} + \omega_{jk} + \omega_{k\ell} + \omega_{\ell i} = 0.$$

For any mapping $\mathcal{F}$ on $\mathbb{Z}^2$, let $d\mathcal{F}_{ij}$ denote the difference between $\mathcal{F}_j$ and $\mathcal{F}_i$ for any adjacent vertices i and j:

$$d\mathcal{F}_{ij} := \mathcal{F}_j - \mathcal{F}_i.$$

Note that we have $d\mathcal{F}_{ij} = -d\mathcal{F}_{ji}$, making $d\mathcal{F}$ a discrete 1-form, and one can easily check that $d\mathcal{F}$ is closed. For the diagonals of an elementary quadrilateral, we also define the notation

$$\mathrm{d}\mathcal{F}_{ik} := \mathcal{F}_k - \mathcal{F}_i.$$

If we also define

$$\mathcal{F}_{ij} := \frac{1}{2}(\mathcal{F}_i + \mathcal{F}_j),$$

then we have the discrete version of the Leibniz rule:

Exercise 4.4.10 (cf. [52, Equation (2.2)]) Let $\mathcal{F}, \mathcal{G} : \mathbb{Z}^2 \to V$ for some vector space equipped with symmetric bilinear product $\odot$. Then show that

$$\mathrm{d}(\mathcal{F} \odot \mathcal{G})_{ij}(= \mathcal{F}_j \odot \mathcal{G}_j - \mathcal{F}_i \odot \mathcal{G}_i) = \mathrm{d}\mathcal{F}_{ij} \odot \mathcal{G}_{ij} + \mathcal{F}_{ij} \odot \mathrm{d}\mathcal{G}_{ij}$$

on any edge ij.

The discrete Christoffel transformations are defined as follows:

Definition 4.4.11 ([22, Theorem 6]) Let $x : \mathbb{Z}^2 \to \mathbb{R}^3$ be a discrete isothermic surface with an edge-labeling a. Then a *Christoffel transform x^** of x is defined by

$$\mathrm{d}x_{ij}^* = \rho \frac{a_{ij}}{|\mathrm{d}x_{ij}|^2} \mathrm{d}x_{ij}, \tag{4.5}$$

for some choice of non-zero constant $\rho \in \mathbb{R}$.

Before checking that such a transformation exists, we first explain how this definition can be viewed as the discrete analogue of the smooth counterpart. Consider the Christoffel transform $x^* : \Sigma \to \mathbb{R}^3$ of a smooth surface $x : \Sigma \to \mathbb{R}^3$ with isothermic coordinates $(u, v) \in \Sigma$. In the smooth case, we have via (2.43) that x^* satisfies

$$\mathrm{d}x^* = -\frac{x_u}{|x_u|^2}\mathrm{d}u + \frac{x_v}{|x_v|^2}\mathrm{d}v,$$

so writing $\mathrm{d}x = x_u \mathrm{d}u + x_v \mathrm{d}v$, we have

$$\mathrm{d}x^*(\partial_u) \cdot \mathrm{d}x(\partial_u) = -1 \quad \text{and} \quad \mathrm{d}x^*(\partial_v) \cdot \mathrm{d}x(\partial_v) = 1.$$

Also, for cr_ϵ as defined in Sect. 2.8, we have

$$\lim_{\epsilon \to 0} \mathrm{cr}_\epsilon = -1 = \frac{\mathrm{d}x^*(\partial_u) \cdot \mathrm{d}x(\partial_u)}{\mathrm{d}x^*(\partial_v) \cdot \mathrm{d}x(\partial_v)}, \quad \mathrm{d}x^*(\partial_u) \parallel \mathrm{d}x(\partial_u), \quad \mathrm{d}x^*(\partial_v) \parallel \mathrm{d}x(\partial_v),$$

by (2.38) in isothermic coordinates.

In discretizing, we generalized the -1 in the right-hand side of (2.38) to the $a_{ij}/a_{i\ell}$ in the right-hand side of (4.3). Because of this, it is natural to consider that

$$\frac{a_{ij}}{a_{i\ell}} = \frac{\mathrm{d}x_{ij}^* \cdot \mathrm{d}x_{ij}}{\mathrm{d}x_{i\ell}^* \cdot \mathrm{d}x_{i\ell}}, \quad \mathrm{d}x_{ij}^* \parallel \mathrm{d}x_{ij}, \quad \mathrm{d}x_{i\ell}^* \parallel \mathrm{d}x_{i\ell},$$

where $\mathrm{d}x_{ij}$, $\mathrm{d}x_{ij}^*$, $\mathrm{d}x_{i\ell}$, $\mathrm{d}x_{i\ell}^*$ now represent discrete analogues of $\mathrm{d}x(\partial_u)$, $\mathrm{d}x^*(\partial_u)$, $\mathrm{d}x(\partial_v)$, $\mathrm{d}x^*(\partial_v)$, respectively.

Now we show the condition for the Christoffel transform to exist.

Theorem 4.4.12 ([22, Theorem 6]) *If x is discrete isothermic, then its Christoffel transform x^* exists.*

Proof Let $x : \mathbb{Z}^2 \to \mathbb{R}^3$ be a discrete isothermic surface with cross ratios factorizing function a. The Christoffel transform x^* exists if and only if the compatibility condition

$$\mathrm{d}x_{ij}^* + \mathrm{d}x_{jk}^* + \mathrm{d}x_{k\ell}^* + \mathrm{d}x_{\ell i}^* = 0 \tag{4.6}$$

holds on each elementary quadrilateral, or $\mathrm{d}x^*$ is closed (so we can apply "discrete integration" to obtain x^*). Therefore, we restrict our attention to an elementary quadrilateral. Without loss of generality, we can apply a rigid motion of $\mathbb{R}^3$ so that x_i, x_j, x_k and x_ℓ all lie in the $x_1 x_2$-plane in $\mathbb{R}^3$, so that we can regard the x, $\mathrm{d}x$ in this quadrilateral as complex numbers, allowing us to multiply and divide with them.

We now prove that (4.6) holds with $\mathrm{d}x^*$ defined as in (4.5). By (4.5), the left hand side of (4.6) can be written as

$$\rho\big(a_{ij}(\mathrm{d}\bar{x}_{ij})^{-1} + a_{jk}(\mathrm{d}\bar{x}_{jk})^{-1} + a_{k\ell}(\mathrm{d}\bar{x}_{k\ell})^{-1} + a_{\ell i}(\mathrm{d}\bar{x}_{\ell i})^{-1}\big), \tag{4.7}$$

where $\bar{x}$ denotes the usual complex conjugation of x. Because x is isothermic, we have that $a_{ij} = a_{\ell k}$, $a_{i\ell} = a_{jk}$, and

$$\frac{a_{ij}}{a_{i\ell}} = \mathrm{cr}(x_i, x_j, x_k, x_\ell) = \frac{\mathrm{d}x_{ij}}{\mathrm{d}x_{jk}}\frac{\mathrm{d}x_{k\ell}}{\mathrm{d}x_{\ell i}} = \frac{\mathrm{d}\bar{x}_{ij}}{\mathrm{d}\bar{x}_{jk}}\frac{\mathrm{d}\bar{x}_{k\ell}}{\mathrm{d}\bar{x}_{\ell i}},$$

since the cross ratios are real. Therefore, we can rewrite (4.7) as

$$a_{ij}\left((\mathrm{d}\bar{x}_{ij})^{-1} + (\mathrm{d}\bar{x}_{k\ell})^{-1} + \frac{\mathrm{d}\bar{x}_{jk}}{\mathrm{d}\bar{x}_{ij}}\frac{\mathrm{d}\bar{x}_{\ell i}}{\mathrm{d}\bar{x}_{k\ell}}((\mathrm{d}\bar{x}_{jk})^{-1} + (\mathrm{d}\bar{x}_{\ell i})^{-1})\right)$$

$$= a_{ij}\frac{\mathrm{d}\bar{x}_{ij} + \mathrm{d}\bar{x}_{jk} + \mathrm{d}\bar{x}_{k\ell} + \mathrm{d}\bar{x}_{\ell i}}{\mathrm{d}\bar{x}_{ij}\mathrm{d}\bar{x}_{k\ell}} = 0. \tag{4.8}$$

since x exists. Therefore, we have that (4.6) holds. $\square$

Remark 4.4.13 As in Definition 4.4.11, the Christoffel transformation of a discrete isothermic surface is well-defined up to translations and homotheties, and we assume without loss of generality that $\rho = 1$, unless noted otherwise.

Lemma 4.4.14 ([22, Equation (40)]) *Let x be a discrete isothermic surface. Then the Christoffel transform x^* is isothermic with the same cross ratios as x.*

Proof By (4.5), we have that $\mathrm{d}x_{ij}$ and $\mathrm{d}x^*_{ij}$ are parallel on any edge ij; therefore as in the proof of Theorem 4.4.12, we can assume that the images of an elementary quadrilateral under x and x^* are in the complex plane. Then we have that

$$\mathrm{cr}(x_i^*, x_j^*, x_k^*, x_\ell^*) = \frac{\mathrm{d}x_{ij}^*}{\mathrm{d}x_{jk}^*} \frac{\mathrm{d}x_{k\ell}^*}{\mathrm{d}x_{\ell i}^*} = \frac{a_{ij}}{a_{jk}} \frac{a_{k\ell}}{a_{\ell i}} \frac{\mathrm{d}\bar{x}_{jk}}{\mathrm{d}\bar{x}_{ij}} \frac{\mathrm{d}\bar{x}_{\ell i}}{\mathrm{d}\bar{x}_{k\ell}} = \mathrm{cr}(x_i, x_j, x_k, x_\ell),$$

giving us the desired result. $\square$

Using Lemma 4.4.14, we can recover another property enjoyed by the Christoffel transformations of smooth isothermic surfaces:

Exercise 4.4.15 Show that discrete Christoffel transformation is involutive, that is, $(x^*)^* = x$, up to translation.

Now, we will show a few lemmata regarding Christoffel transformations.

Lemma 4.4.16 (cf. [25, Theorem 2.25]) *Let x^* be a Christoffel transform of x with cross ratios factorizing function a, and denote*

$$\alpha_{ij} := \frac{a_{ij}}{|\mathrm{d}x_{ij}|^2}.$$

Then

$$\alpha_{ij}\alpha_{k\ell} = \alpha_{i\ell}\alpha_{jk}.$$

Proof Assume that the image of the elementary quadrilateral under x or x^* is in the complex plane. Since we have Lemma 4.4.14,

$$\frac{a_{ij}}{a_{i\ell}} = \frac{\mathrm{d}x_{ij}^*}{\mathrm{d}x_{jk}^*} \frac{\mathrm{d}x_{k\ell}^*}{\mathrm{d}x_{\ell i}^*} = \frac{\alpha_{ij}}{\alpha_{jk}} \frac{\alpha_{k\ell}}{\alpha_{\ell i}} \frac{\mathrm{d}x_{ij}}{\mathrm{d}x_{jk}} \frac{\mathrm{d}x_{k\ell}}{\mathrm{d}x_{\ell i}} = \frac{\alpha_{ij}}{\alpha_{jk}} \frac{\alpha_{k\ell}}{\alpha_{\ell i}} \frac{a_{ij}}{a_{i\ell}}.$$

Therefore, $\frac{\alpha_{ij}}{\alpha_{jk}} \frac{\alpha_{k\ell}}{\alpha_{\ell i}} = 1$. $\square$

Lemma 4.4.17 (cf. [25, Theorem 2.31]) *There exists a function $r : \mathbb{Z}^2 \to \mathbb{R}$ such that*

$$\mathrm{d}x_{ij}^* = r_i r_j \mathrm{d}x_{ij}. \tag{4.9}$$

The Eq. (4.9) is called the Christoffel formula.

Proof Consider an elementary quadrilateral $ijk\ell$, and choose $r_i \in \mathbb{R} \setminus \{0\}$ freely. Define r_j and r_ℓ so that

$$\mathrm{d}x^*_{ij} = r_i r_j \mathrm{d}x_{ij} \quad \text{and} \quad \mathrm{d}x^*_{i\ell} = r_i r_\ell \mathrm{d}x_{i\ell}.$$

We next define r_k and $\hat{r}_k$ by requiring

$$\mathrm{d}x^*_{jk} = r_j r_k \mathrm{d}x_{jk} \quad \text{and} \quad \mathrm{d}x^*_{\ell k} = r_\ell \hat{r}_k \mathrm{d}x_{\ell k}.$$

Our goal is then to show that $r_k = \hat{r}_k$.

Because $r_i r_j = \alpha_{ij}$ and $r_j r_k = \alpha_{jk}$, we have $r_k = r_i \alpha_{ij}^{-1} \alpha_{jk}$. Similarly, $\hat{r}_k = r_i \alpha_{i\ell}^{-1} \alpha_{\ell k}$. Then Lemma 4.4.16 implies $r_k = \hat{r}_k$. $\qquad\square$

Discrete Christoffel transformations enjoy another characterization using mixed areas of the quadrilaterals. For this, we first introduce the following notion.

Definition 4.4.18 ([197], cf. [25, Definition 2.22]) For $x : \mathbb{Z}^2 \to \mathbb{R}^3$, if there is $x^* : \mathbb{Z}^2 \to \mathbb{R}^3$ such that

$$\mathrm{d}x_{ij} \parallel \mathrm{d}x^*_{ij}$$

on every edge ij, and

$$\mathrm{d}x^*_{ik} \parallel \mathrm{d}x_{j\ell} \quad \text{and} \quad \mathrm{d}x^*_{j\ell} \parallel \mathrm{d}x_{ik},$$

on every elementary quadrilateral $ijk\ell$, then we call x a *Koenigs net* and x^* a *Koenigs dual*. In other words, the corresponding edges of x and x^* are parallel, and the non-corresponding diagonals are parallel.

Remark 4.4.19 The Koenigs dual of a Koenigs net is unique up to homotheties and translations (see [25, Lemma 2.20]).

Lemma 4.4.20 *A Koenigs net x has planar quadrilaterals.*

Proof Denoting the Koenigs dual of x by x^*, let γ be a real constant so that

$$\mathrm{d}x_{ik} = \beta \mathrm{d}x^*_{j\ell} = \gamma (\mathrm{d}x^*_{i\ell} - \mathrm{d}x^*_{ij}).$$

Now since x and x^* are also edge-parallel, we have that on every edge ij, $\mathrm{d}x^*_{ij} = \beta_{ij} \mathrm{d}x_{ij}$ for some constant β_{ij} on unoriented edges. Thus,

$$\mathrm{d}x_{ik} = \gamma (\beta_{i\ell} \mathrm{d}x_{i\ell} - \beta_{ij} \mathrm{d}x_{ij}),$$

implying that $\mathrm{d}x_{ij}, \mathrm{d}x_{ik}, \mathrm{d}x_{i\ell}$ are linearly dependent. $\qquad\square$

Since a Koenigs net x has planar quadrilaterals, so does the Koenigs dual x^* by the edge-parallel condition, and we can characterize Koenigs duality via mixed areas.

Theorem 4.4.21 (cf. [186, Equation (8)], [25, Theorem 4.42]) *Two edge-parallel discrete surfaces x and x^* in $\mathbb{R}^3$ are Koenigs dual if and only if $A(x, x^*) = 0$.*

Proof Note that by Exercise 4.2.5,

$$A(x, x^*)_{ijk\ell} = \frac{1}{2}\det(\mathrm{d}x_{ik}, \mathrm{d}x^*_{j\ell}, \mathbf{n}) = \frac{1}{2}\det(\mathrm{d}x^*_{ik}, \mathrm{d}x_{j\ell}, \mathbf{n}),$$

where $\mathbf{n}$ denotes the unit vector normal to the planar quadrilateral. $\square$

We now explain how the Christoffel transformation is related to the Koenigs dual.

Theorem 4.4.22 (cf. [25, Theorem 2.31]) *Let $x : \mathbb{Z}^2 \to \mathbb{R}^3$ be discrete isothermic. Then x^* is a Christoffel transform of x if and only if x is a Koenigs net with Koenigs dual x^*.*

Proof Assuming that x^* is a Christoffel transform of x, the Christoffel formula tells us that x and x^* are edge-parallel. Hence, we only need to show that the opposite diagonals are parallel. Note that we can verify

$$
\begin{aligned}
(r_k - r_i)(r_\ell x_\ell - r_j x_j) &= r_k r_\ell x_\ell + r_i r_j x_j - r_i r_\ell x_\ell - r_k r_j x_j \\
&= (\mathrm{d}x^*_{k\ell} + r_k r_\ell x_k) + (\mathrm{d}x^*_{ij} + r_i r_j x_i) \\
&\quad - (\mathrm{d}x^*_{i\ell} + r_i r_\ell x_i) - (\mathrm{d}x^*_{kj} + r_k r_j x_k) \\
&= (r_\ell - r_j)(r_k x_k - r_i x_i)
\end{aligned}
\tag{4.10}
$$

by the Christoffel formula (4.9) and the compatibility condition of x^*. Therefore, using this relation, we can show

$$
\begin{aligned}
\mathrm{d}x^*_{ik} = \mathrm{d}x^*_{jk} + \mathrm{d}x^*_{ij} &= r_j r_k \mathrm{d}x_{jk} + r_i r_j \mathrm{d}x_{ij} = r_j(r_k x_k - r_i x_i - (r_k - r_i)x_j) \\
&= \frac{r_j(r_k - r_i)}{r_\ell - r_j}\left(\frac{r_\ell - r_j}{r_k - r_i}(r_k x_k - r_i x_i) - (r_\ell - r_j)x_j\right) \\
&= \frac{r_j r_\ell (r_k - r_i)}{r_\ell - r_j}\mathrm{d}x_{j\ell}
\end{aligned}
$$

A symmetric argument shows that $\mathrm{d}x^*_{j\ell} \parallel \mathrm{d}x_{ik}$, and we have that a Christoffel transformation is a Koenigs dual of x.

Finally, since both Koenigs duals and Christoffel transformations are unique up to homotheties and translation, any Koenigs dual of x must be a Christoffel transformation of x. $\square$

We finally arrive at the mixed area characterization of the Christoffel transformation.

Corollary 4.4.23 *Let x be discrete isothermic. Then x^* is a Christoffel transform of x if and only if $A(x, x^*) = 0$.*

Example 4.4.24 For the first discrete Clifford torus in Example 4.1.3, Christoffel transformations of this example can be computed using (4.5) to be

$$
x^*_{m,n} = \left(\frac{-\cos \frac{2\pi n}{n_{per}}}{2\sqrt{2} + 2\cos \frac{2\pi m}{m_{per}}}, \frac{-\sin \frac{2\pi n}{n_{per}}}{2\sqrt{2} + 2\cos \frac{2\pi m}{m_{per}}}, a_m \right),
$$

where a_m satisfies the recursion

$$
a_m = a_{m-1} + \frac{\sin \frac{2\pi m}{m_{per}} - \sin \frac{2\pi (m-1)}{m_{per}}}{2(\sqrt{2} + \cos \frac{2\pi (m-1)}{m_{per}})(\sqrt{2} + \cos \frac{2\pi m}{m_{per}})}.
$$

For the second example of discrete Clifford torus in Example 4.1.3, Christoffel transformations of this example can be computed using (4.5) to be

$$
x^*_{m,n} = \left(y^1_{m,n}, y^2_{m,n}, y^3_{m,n} \right),
$$

where the $y^j_{m,n}$ satisfy the recursions

$$
y^1_{0,n} = y^1_{0,n-1},
$$

$$
y^2_{0,n} = \tfrac{1+\sqrt{2}}{4}(\cos \tfrac{2\pi (n-1)}{n_{per}} - \cos \tfrac{2\pi n}{n_{per}}) + y^2_{0,n-1},
$$

$$
y^3_{0,n} = \tfrac{1+\sqrt{2}}{4}(\sin \tfrac{2\pi (n-1)}{n_{per}} - \sin \tfrac{2\pi n}{n_{per}}) + y^3_{0,n-1},
$$

$$
y^1_{m,n} = \tfrac{1}{2}(\cos \tfrac{\pi}{m_{per}} + \sqrt{2}\cos \tfrac{\pi (2m-1)}{m_{per}}) \sin \tfrac{\pi}{m_{per}} + y^1_{m-1,n},
$$

$$
y^2_{m,n} = \tfrac{1}{4}(\cos \tfrac{2\pi (m-1)}{m_{per}} - \cos \tfrac{2\pi m}{m_{per}}) \cos \tfrac{2\pi n}{n_{per}} + y^2_{m-1,n},
$$

$$
y^3_{m,n} = \tfrac{1}{4}(\cos \tfrac{2\pi (m-1)}{m_{per}} - \cos \tfrac{2\pi m}{m_{per}}) \sin \tfrac{2\pi n}{n_{per}} + y^3_{m-1,n}.
$$

See Fig. 4.3.

4.5   Discrete Isothermicity via Flat Connections

To obtain an analogue of a 1-parameter family of flat connections for an isothermic surface in the discrete case, it is the notion of parallel transport that we discretize. First, we briefly introduce connections, parallel sections, and gauge transformations in the discrete setting [37, §3.9] (see also [52, Defiinition 2.3])]: Let W be a discrete vector bundle, that is, to every vertex i, we assign a vector space V_i.

- A *discrete connection* Γ on W assigns a linear isomorphism $\Gamma_{ij} : V_j \to V_i$ on each oriented edge ij so that

$$
\Gamma_{ij}\Gamma_{ji} = id.
$$

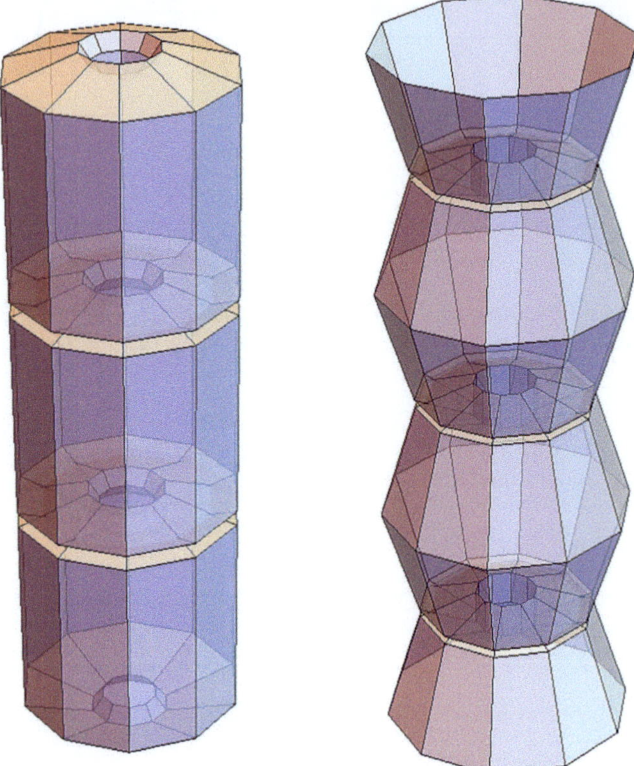

Fig. 4.3 Christoffel transforms of the Example 4.1.3 as given in Example 4.4.24. The domain data for both surfaces is $m_{per} = 5, n_{per} = 10, 0 \leq n \leq n_{per}, 0 \leq m \leq 3m_{per}$

- A map $\sigma : \mathbb{Z}^2 \to \sqcup V_i$ is a *section of* W, denoted $\sigma \in \Gamma W$, if $\sigma_i \in V_i$ for all i.
- A *discrete gauge transformation* g acts on the connection Γ_{ij} via

$$(g \bullet \Gamma)_{ij} = g_i \circ \Gamma_{ij} \circ g_j^{-1},$$

where $g_i : V_i \to V_i$ is a linear isomorphism at each vertex i.
- A discrete connection is a *flat connection* if all holonomies are trivial, i.e.,

$$\Gamma_{ij}\Gamma_{jk}\Gamma_{k\ell}\Gamma_{\ell i} = id$$

on every elementary quadrilateral $ijk\ell$.
- A section σ of V is a *parallel section of* Γ, or Γ-*parallel* if the section is given by parallel transport through Γ, that is,

$$\sigma_i = \Gamma_{ij}\sigma_j$$

on every edge ij.

Example 4.5.25 For some vector space V, the discrete identity map $id_{ij} : V_j \to V_i$ is a discrete connection defined on the trivial bundle $\mathbb{Z}^2 \times V$.

Exercise 4.5.26 Show that discrete gauge transformations act on the connection, i.e., given a linear isomorphism $g_i : V_i \to V_i$ for any vertex i and a discrete connection Γ_{ij} on any edge ij, show that $(g \bullet \Gamma)_{ij}$ is again a discrete connection.

Exercise 4.5.27 Show that discrete gauge transformations preserve the flatness of the connection, that is, show that if Γ is a flat connection, then $(g \bullet \Gamma)$ is also a flat connection.

Exercise 4.5.28 Let Γ be a flat connection, and choose any $v \in V_i$. Show that there is a well-defined parallel section σ of Γ such that $\sigma_i = v$. (Hint: Use the definition of parallel section to propagate σ around an elementary quadrilateral $ijk\ell$, and use the flatness to show that this propagation is consistent as in the proof of Lemma 4.1.2.)

Let $L : \mathbb{Z}^2 \to P(L^4)$ be discrete isothermic, so that we have the cross ratios condition

$$\mathrm{cr}(L_i, L_j, L_k, L_\ell) = \frac{a_{ij}}{a_{i\ell}}$$

for some cross ratios factorizing function a, on every elementary quadrilateral. We can view L as a discrete bundle over $\mathbb{Z}^2$, where the vector space assigned to each vertex i is the null line L_i. Subsequently, a nowhere zero X is a section of L, or $X \in \Gamma L$, if $\mathrm{span}\{X_i\} = L_i$ on every vertex i. Noting that the vertices are pairwise distinct, we have that $L_i \not\perp L_j$ for any adjacent vertices. Therefore, we can define

$$\Gamma_{ij}^{\lambda} := \Gamma_{L_j}^{L_i}(1 - \lambda a_{ij}), \tag{4.11}$$

where the right hand side is the orthogonal transformation defined in Definition 3.6.40. Then we see that such Γ_{ij}^{λ} define a discrete connection on the trivial bundle $\mathbb{Z}^2 \times \mathbb{R}^{4,1}$.

Now, on the Bianchi quadrilateral of Darboux transformations, we have shown through Proposition 3.6.49 and Exercise 3.6.50 that

$$\Gamma_{L_1}^{L_{12}}(1 - \tfrac{\lambda}{\mu_2})\Gamma_{L}^{L_1}(1 - \tfrac{\lambda}{\mu_1}) = \Gamma_{L_2}^{L_{21}}(1 - \tfrac{\lambda}{\mu_1})\Gamma_{L}^{L_2}(1 - \tfrac{\lambda}{\mu_2}),$$

which is equivalent to the conditions $L_{12} = L_{21}$ and

$$\mathrm{cr}(L, L_1, L_{12}, L_2) = \frac{\mu_2}{\mu_1}.$$

The Bianchi quadrilateral and a face of a discrete isothermic surface are indistinguishable; therefore, making the switch as in (4.4), and using the fact that a is an edge-labeling, we see that

$$\Gamma_{L_j}^{L_i}(1 - \lambda a_{ij})\Gamma_{L_k}^{L_j}(1 - \lambda a_{jk})\Gamma_{L_\ell}^{L_k}(1 - \lambda a_{k\ell})\Gamma_{L_i}^{L_\ell}(1 - \lambda a_{\ell i}) = id,$$

giving us the following characterization of discrete isothermicity.

Theorem 4.5.29 ([41, Theorem 4.14], [52, Lemma 2.5]) *Let $L : \mathbb{Z}^2 \to P(L^4)$. L is discrete isothermic with edge-labeling a if and only if the connection Γ on the discrete trivial bundle $\mathbb{Z}^2 \times \mathbb{R}^{4,1}$ defined by*

$$\Gamma_{ij}^\lambda := \Gamma_{L_j}^{L_i}(1 - \lambda a_{ij})$$

is flat. We call such Γ_{ij}^λ the discrete flat connections associated with L.

Proof We only need to show that if Γ^λ is a flat connection, then a is an edge-labeling. First, note that on any edge ij,

$$\Gamma_{ij}^\lambda \Gamma_{ji}^\lambda = id,$$

implies $a_{ij} = a_{ji}$, so a is a function defined on unoriented edges. Next, from Proposition 3.6.49 and the flatness, we know that

$$\Gamma_{L_\ell}^{L_j}\left(\frac{1-\lambda a_{k\ell}}{1-\lambda a_{\ell i}}\right) = \Gamma_{k\ell}^\lambda \Gamma_{\ell i}^\lambda = \Gamma_{kj}^\lambda \Gamma_{ji}^\lambda = \Gamma_{L_j}^{L_\ell}\left(\frac{1-\lambda a_{jk}}{1-\lambda a_{ij}}\right).$$

Therefore, we have that $\frac{1-\lambda a_{k\ell}}{1-\lambda a_{\ell i}}\frac{1-\lambda a_{jk}}{1-\lambda a_{ij}} = 1$. Rewriting this expression, we obtain that

$$(a_{ij}a_{\ell i} - a_{jk}a_{k\ell})\lambda^2 + (a_{jk} + a_{k\ell} - a_{ij} - a_{\ell i})\lambda = 0,$$

for all $\lambda \in \mathbb{R}$, telling us that both coefficients must vanish. Therefore, either

$$a_{ij} = a_{jk} \text{ and } a_{\ell i} = a_{k\ell}, \quad \text{or} \quad a_{ij} = a_{k\ell} \text{ and } a_{\ell i} = a_{jk}.$$

The first of these cases would imply that $\Gamma_{k\ell}\Gamma_{\ell i} = id = \Gamma_{kj}\Gamma_{ji}$, which cannot happen due to the non-degeneracy condition. Therefore, we have that a is an edge-labeling. $\square$

Using (3.19), we can explicitly express Γ_{ij}^λ as

$$\Gamma_{ij}^\lambda Y = Y + \frac{\lambda a_{ij}}{\langle X_i, X_j \rangle}\left\{\left(\frac{1}{1 - \lambda a_{ij}}\right)\langle Y, X_i \rangle X_j - \langle Y, X_j \rangle X_i\right\}.$$

As before, the above expression does not depend on the choice of lift $X \in \Gamma L$ (so that $\mathrm{span}\{X_i\} = L_i$ at each vertex i). The next remark uses this explicit form to compare the discrete flat connections to the smooth flat connections.

Remark 4.5.30 For defining the discrete connection, one might initially naively imitate the smooth case (as in Sect. 3.3), and define parallel transport of a vector Y_i from one vertex i to a vector Y_j at an adjacent vertex j by setting

$$\mathrm{d}Y_{ij} + \lambda \frac{\pm 2}{\langle \mathrm{d}X_{ij}, \mathrm{d}X_{ij} \rangle} (X_i \wedge \mathrm{d}X_{ij}) Y_j = 0.$$

Replacing the "$\pm$" with the cross ratio factorizing function $-a_{ij}$ (in analogy to the discussion in Sect. 4.3), and noting that $X_i \wedge X_i = 0$ and $\langle X_i, X_i \rangle = \langle X_j, X_j \rangle = 0$, we would have

$$Y_i = Y_j + \lambda \frac{a_{ij}}{\langle X_i, X_j \rangle} (X_i \wedge X_j) Y_j.$$

However, this is not an $O_{4,1}$ transformation. To remedy this, we insert an extra factor:

$$Y_j \mapsto Y_i := \Gamma_{ij}^{\lambda} Y_j, \quad \Gamma_{ij}^{\lambda} Y = Y + \frac{\lambda a_{ij}}{\langle X_i, X_j \rangle} \left(\frac{1}{1 - \lambda a_{ij}} \langle X_i, Y \rangle X_j - \langle X_j, Y \rangle X_i \right).$$

Thus the connection is defined by the parallel transport $\Gamma_{ij}^{\lambda} : \mathbb{R}_j^{4,1} \to \mathbb{R}_i^{4,1}$, where $\mathbb{R}_i^{4,1}, \mathbb{R}_j^{4,1} \in \mathbb{Z}^2 \times \mathbb{R}^{4,1}$.

4.6 Discrete Moutard Lifts

Another characterization of smooth isothermic surfaces $L : \Sigma \to P(L^4)$ parametrized by isothermic coordinates $(u, v) \in \Sigma$ is the existence of Moutard lifts $F \in \Gamma L$ satisfying the Moutard equation, that is,

$$F_{uv} \parallel F.$$

To discretize the Moutard equation, consider the quadrilateral with F_i, F_j, F_k and F_ℓ at the vertices. The discrete mixed second derivative of F (analogous to F_{uv} in the smooth case) is

$$(F_k - F_\ell) - (F_j - F_i),$$

defined on faces, so a natural candidate for discrete Moutard equation might naturally be

$$F_k - F_\ell - F_j + F_i = \alpha_1 \tfrac{1}{4}(F_i + F_j + F_k + F_\ell), \quad \alpha_1 \in \mathbb{R},$$

and the $\frac{1}{4}$ can be absorbed into the α_1 as $\alpha_2 = \frac{1}{4}\alpha_1$. Then

$$F_i + F_k = \alpha(F_j + F_\ell),$$

where we have defined α by $\alpha = \frac{1+\alpha_2}{1-\alpha_2}$, giving us the discrete Moutard equation:

$$(F_i + F_k) \parallel (F_j + F_\ell). \tag{4.12}$$

However, in this text, we use the *discrete Moutard equation with a minus sign*,

$$(F_k - F_i) \parallel (F_\ell - F_j), \tag{4.13}$$

and we refer to (4.13) simply as the *discrete Moutard equation*, which appears in the context of the permutability of Moutard transformations (see, for example, [170, Equation (3.13)], [26, Equation (1.5)]).

Now we are ready to define discrete Moutard lifts.

Definition 4.6.31 We say that a section F of a surface $L : \Sigma \to P(L^4)$ is a *discrete Moutard lift* if F satisfies the *discrete Moutard equation*, that is,

$$\mathrm{d}F_{ik} \parallel \mathrm{d}F_{j\ell} \tag{4.14}$$

on every elementary quadrilateral, so that

$$\mathrm{d}F_{ik} = \alpha \mathrm{d}F_{j\ell} \tag{4.15}$$

for some nowhere zero real-valued function α defined on faces.

Remark 4.6.32 The Moutard lift is not unique, and it has more than just the freedom of a constant scalar multiple. For example, if points i correspond to (m, n) in $\mathbb{Z}^2$, we could change a Moutard lift μ_i to $\alpha\mu_i$ when $m + n$ is even and $\beta\mu_i$ when $m + n$ is odd, for any nonzero constants $\alpha, \beta \in \mathbb{R}$, and this gives another Moutard lift.

Using the definition of discrete Moutard lifts, we recover another characterization of discrete isothermic surfaces that the smooth counterparts enjoy.

Theorem 4.6.33 (cf. [26, §3(b)], [52, §2.1]) *For some $L : \mathbb{Z}^2 \to P(L^4)$, L is discrete isothermic if and only if L admits a discrete Moutard lift $F \in \Gamma L$. In fact on every edge ij,*

$$a_{ij} := \langle F_i, F_j \rangle$$

is an edge-labeling of L.

Proof First assume that L is a discrete isothermic surface with cross ratios factorizing function a, and consider an elementary quadrilateral $ijk\ell$. For some choice of $F_i \in L_i$, define $F_j \in L_j$ and $F_\ell \in L_\ell$ so that

$$\langle F_i, F_j \rangle = a_{ij} \quad \text{and} \quad \langle F_i, F_\ell \rangle = a_{i\ell}.$$

If we define

$$F_k := \Gamma_{L_j}^{L_\ell}\left(\tfrac{a_{i\ell}}{a_{ij}}\right) F_i,$$

then we know that $F_k \in L_k$, by (3.25) in Exercise 3.6.50. Using the explicit form (3.19) for the orthogonal transformation above, we then have

$$F_k = F_i + \tfrac{1}{\langle F_j, F_\ell \rangle} \left(\left(\tfrac{a_{i\ell}}{a_{ij}} - 1 \right) \langle F_i, F_j \rangle F_\ell + \left(\tfrac{a_{ij}}{a_{i\ell}} - 1 \right) \langle F_i, F_\ell \rangle F_j \right)$$

$$= F_i + \tfrac{1}{\langle F_j, F_\ell \rangle} \left((a_{i\ell} - a_{ij}) F_\ell + (a_{ij} - a_{i\ell}) F_j \right)$$

$$= F_i + \tfrac{a_{i\ell} - a_{ij}}{\langle F_j, F_\ell \rangle} (F_\ell - F_j).$$

Therefore, F is a Moutard lift.

Taking inner product with F_j on both sides now tells us

$$\langle F_j, F_k \rangle = \langle F_i, F_j \rangle + \tfrac{a_{i\ell} - a_{ij}}{\langle F_j, F_\ell \rangle} \langle F_j, F_\ell \rangle = a_{i\ell} = a_{jk},$$

while taking the inner product with F_ℓ on both sides gives us $\langle F_k, F_\ell \rangle = a_{ij} = a_{k\ell}$, giving us one direction of the proof.

To show the other direction, assume that $F \in \Gamma L$ satisfies the discrete Moutard equation (4.15), that is,

$$F_k - F_i = \alpha(F_\ell - F_j) \tag{4.16}$$

for some nowhere zero real-valued function α defined on faces, and let $a_{ij} := \langle F_i, F_j \rangle$ on any edge ij. Note that we do not yet know that a is an edge-labeling. However, taking the inner product with F_j and then F_ℓ tells us

$$a_{jk} - a_{ij} = \alpha \langle F_j, F_\ell \rangle \qquad a_{k\ell} - a_{i\ell} = -\alpha \langle F_j, F_\ell \rangle$$

from which we can deduce

$$a_{jk} - a_{ij} = a_{i\ell} - a_{k\ell}.$$

Similarly, taking the inner product with F_i and then F_k on both sides of (4.16) further tells us that

$$a_{i\ell} - a_{ij} = a_{jk} - a_{k\ell}.$$

From these two equations involving a, we can then obtain

$$a_{ij} = a_{k\ell} \quad \text{and} \quad a_{i\ell} = a_{jk},$$

that is, a is an edge-labeling. Therefore, (4.16) now reads

$$F_k - F_i = \frac{a_{i\ell} - a_{ij}}{\langle F_j, F_\ell \rangle}(F_\ell - F_j),$$

or equivalently,

$$F_k = F_i + \frac{1}{\langle F_j, F_\ell \rangle}\left((\tfrac{a_{ij}}{a_{i\ell}} - 1)\langle F_i, F_\ell \rangle F_j + (\tfrac{a_{i\ell}}{a_{ij}} - 1)\langle F_i, F_j \rangle F_\ell\right).$$

Therefore, Remark 2.5.39 tells us that

$$\mathrm{cr}(L_i, L_j, L_k, L_\ell) = \frac{a_{ij}}{a_{i\ell}},$$

so that L is isothermic. $\square$

Exercise 4.6.34 Let $L : \mathbb{Z}^2 \to P(L^4)$ be isothermic with cross ratios factorizing function a. Show that if $F \in \Gamma L$ so that $\langle F_i, F_j \rangle = a_{ij}$ on every edge, then F is a Moutard lift.

4.7   Discrete Calapso Transformations

Recall that by Definition 3.4.19, the Calapso transformation $T^\lambda : \Sigma \to O_{4,1}$ of an isothermic surface $L : \Sigma \to P(L^4)$ is given by the gauge transformation that gauges the flat connection Γ^λ associated with an isothermic surface to the trivial connection, that is,

$$T^\lambda \bullet \Gamma^\lambda = T^\lambda \circ \Gamma^\lambda \circ (T^\lambda)^{-1} = \mathrm{d}.$$

The discrete trivial connection was given in Example 4.5.25, serving as a motivation to define the discrete counterpart as follows.

Definition 4.7.35 (cf. [113, Lemma 3.9], [41, Theorem 4.15], [52, Lemma 2.7]) Let $L : \mathbb{Z}^2 \to P(L^4)$ be discrete isothermic with flat connection Γ_{ij}^λ. A *discrete Calapso transformation* is $T^\lambda : \mathbb{Z}^2 \to O_{4,1}$ such that

$$(T^\lambda \bullet \Gamma^\lambda)_{ij} = T_i^\lambda \Gamma_{ij}^\lambda (T_j^\lambda)^{-1} = id_{ij}. \tag{4.17}$$

We call $L^\mu := T^\mu L$ a *discrete Calapso transform of L (with parameter μ).*

Note that once an initial condition $T_i^\lambda \in O_{4,1}$ of (4.17) is chosen, then $T^\lambda \in O_{4,1}$ for all vertices since $\Gamma_{ij}^\lambda \in O_{4,1}$ on every edge.

Exercise 4.7.36 Show that Γ^λ is a flat connection if and only if T^λ defined by (4.17) is well-defined, i.e. after a choice of initial condition T_i^λ, the propagation around an elementary quadrilateral is consistent (as in the proof of Lemma 4.1.2 or Exercise 4.5.28).

Exercise 4.7.37 Show that if $\langle X_i, X_j \rangle \neq 0$ for any edge ij for $X \in \Gamma L$, then $\langle X_i^\mu, X_j^\mu \rangle \neq 0$ where $X^\mu \in \Gamma L^\mu$.

The above Exercise 4.7.36 along with the discrete flat connection characterization of isothermicity in Theorem 4.5.29 shows:

Theorem 4.7.38 *L is isothermic if and only if discrete Calapso transformations exist.*

We are now ready to show the following:

Theorem 4.7.39 (cf. [113, Lemma 3.12], [52, Equation (2.16)]) *Let $L : \mathbb{Z}^2 \to P(L^4)$ be a discrete isothermic surface. Then its Calapso transform L^μ is also discrete isothermic. Furthermore, the flat connection $\Gamma^{\mu,\lambda}$ of L^μ satisfies*

$$\Gamma_{ij}^{\mu,\lambda} := \Gamma_{L_j^\mu}^{L_i^\mu}(1 - \lambda a_{ij}^\mu) = (T^\mu \bullet \Gamma^{\lambda+\mu})_{ij} = T_i^\mu \Gamma_{ij}^{\lambda+\mu}(T_j^\mu)^{-1},$$

where a_{ij}^μ is a cross ratios factorizing function of L^μ such that

$$a_{ij}^\mu = \frac{a_{ij}}{1 - \mu a_{ij}}.$$

Proof Let $A^\lambda := T^\mu \bullet \Gamma^{\lambda+\mu}$. First, for any section $X^\mu \in \Gamma L^\mu$, let $X \in \Gamma L$ so that $X^\mu = T^\mu X$. Then we have

$$A_{ij}^\lambda X_i^\mu = T_i^\mu \Gamma_{ij}^{\lambda+\mu}(T_j^\mu)^{-1} T_i^\mu X_i = T_i^\mu \Gamma_{ij}^{\lambda+\mu} \Gamma_{ji}^\mu X_i = (1 - \mu a_{ij})^{-1} T_i^\mu \Gamma_{ij}^{\lambda+\mu} X_i$$

$$= (1 - \mu a_{ij})^{-1}(1 - (\lambda + \mu)a_{ij})T_i^\mu X_i = \left(1 - \lambda \frac{a_{ij}}{1 - \mu a_{ij}}\right) X_i^\mu.$$

Similarly, we can also verify

$$A_{ij}^\lambda X_j^\mu = T_i^\mu \Gamma_{ij}^{\lambda+\mu}(T_j^\mu)^{-1} T_j^\mu X_j = T_i^\mu \Gamma_{ij}^{\lambda+\mu} X_j$$

$$= (1 - (\lambda + \mu)a_{ij})^{-1} T_i^\mu X_j = (1 - (\lambda + \mu)a_{ij})^{-1} T_j^\mu \Gamma_{ji}^\mu X_j$$

$$= (1 - (\lambda + \mu)a_{ij})^{-1}(1 - \mu a_{ij})T_j^\mu X_j = \left(1 - \lambda \frac{a_{ij}}{1 - \mu a_{ij}}\right)^{-1} X_j^\mu.$$

Finally, let $Y \in (L_i^\mu \oplus L_j^\mu)^\perp$. Since

$$T_i^\mu(\text{span}\{X_j, X_i\}) = \text{span}\{T_j^\mu X_j, T_i^\mu X_i\} = T_j^\mu(\text{span}\{X_j, X_i\}),$$

we deduce that $(T_i^\mu)^{-1}Y, (T_j^\mu)^{-1}Y \in (L_i \oplus L_j)^\perp$. Hence, we calculate

$$A_{ij}^\lambda Y = T_i^\mu \Gamma_{ij}^{\lambda+\mu}((T_j^\mu)^{-1}Y) = T_i^\mu(T_j^\mu)^{-1}Y = T_i^\mu \Gamma_{ji}^\mu(T_i^\mu)^{-1}Y = Y.$$

Therefore,

$$A^\lambda = \Gamma^{\mu,\lambda} \quad \text{with } a_{ij}^\mu = \frac{a_{ij}}{1-\mu a_{ij}}.$$

The fact that a^μ is an edge-labeling follows from the above relationship and the fact that a is an edge-labeling, and since $A^\lambda = \Gamma^{\mu,\lambda}$ is a discrete gauge transformation of a flat connection, $\Gamma^{\mu,\lambda}$ is again a flat connection by Exercise 4.5.27. $\square$

Exercise 4.7.40 Let L be a discrete isothermic surface where its cross ratios factorizing function satisfies $a_{ij} = -a_{i\ell}$. For the cross ratios factorizing function a^μ of the Calapso transform L^μ, show that $a_{ij}^\mu \neq -a_{i\ell}^\mu$ unless $\mu = 0$.

Remark 4.7.41 Above Exercise 4.7.40 shows the reason for taking the "broad definition" of discrete isothermicity, as the Calapso transform of a discrete isothermic surface in the narrow sense would not be discrete isothermic in the narrow sense.

Exercise 4.7.42 Let $F : \Sigma \to L^4$ be a Moutard lift of an isothermic surface L. Show that $F^\mu := T^\mu F$ is a Moutard lift of the Calapso transform L^μ of L. (Hint: use Exercise 4.6.34.)

4.8 Discrete Darboux Transformations

Recall via Theorem 3.6.37 that in the smooth case, Darboux transformations of an isothermic surface are given by finding the parallel sections of the flat connection with respect to some parameter μ. Since we have parallel sections in the discrete case as well, we define discrete Darboux transformations as follows:

Definition 4.8.43 (cf. [52, Definition 4.1]) Let $L : \mathbb{Z}^2 \to P(L^4)$ be discrete isothermic, with associated flat connection Γ^λ. We call $\hat{L} : \mathbb{Z}^2 \to P(L^4)$ a *Darboux transform of L with parameter* μ if there is some $\hat{X} \in \Gamma\hat{L}$ such that $\hat{X}$ is Γ^μ-parallel, i.e.

$$\hat{X}_i = \Gamma_{ij}^\mu \hat{X}_j. \tag{4.18}$$

Note that the existence of $\hat{X} \in \Gamma\hat{L}$ satisfying (4.18) implies that

$$\hat{L}_i = \Gamma_{ij}^\mu \hat{L}_j = \Gamma_{L_j}^{L_i}(1 - \mu a_{ij})\hat{L}_j,$$

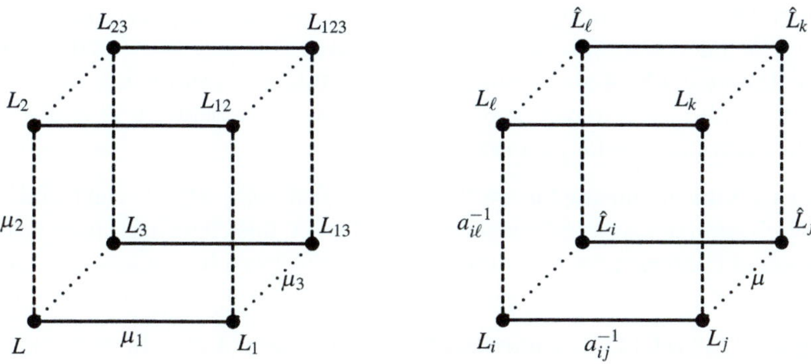

Fig. 4.4 Comparison of the Bianchi cube to Darboux transformation of discrete isothermic surfaces

and interpreting this in terms of cross ratios, we have

$$\mathrm{cr}(\hat{L}_j, L_i, \hat{L}_i, L_j) = 1 - \mu a_{ij},$$

equivalently,

$$\mathrm{cr}(L_i, L_j, \hat{L}_j, \hat{L}_i) = \mu a_{ij}. \tag{4.19}$$

We therefore see that the quadrilateral $L_i, L_j, \hat{L}_j, \hat{L}_i$ is again a Bianchi quadrilateral where

$$L_i, L_j, \hat{L}_j, \hat{L}_i, a_{ij}, \mu \quad \text{replaces} \quad L, L_1, L_{12}, L_2, \frac{1}{\mu_1}, \mu_2.$$

Therefore, we have by (3.24)

$$\Gamma_{L_j}^{\hat{L}_j}(1 - \tfrac{\lambda}{\mu})\Gamma_{L_i}^{L_j}(1 - \lambda a_{ij}) = \Gamma_{\hat{L}_i}^{\hat{L}_j}(1 - \lambda a_{ij})\Gamma_{L_i}^{\hat{L}_i}(1 - \tfrac{\lambda}{\mu}).$$

Furthermore, we have that the image of the elementary quadrilateral under L and $\hat{L}$ (for a total of eight points) forms a Bianchi cube as in Theorem 3.6.52 (see Fig. 4.4), allowing us to conclude as follows:

Theorem 4.8.44 (cf. [52, Lemma 4.2]) *Let $\hat{L}$ be a Darboux transform of a discrete isothermic L with parameter μ. Then $\hat{L}$ is discrete isothermic with the same cross ratios factorizing function as L. Moreover, the flat connection $\hat{\Gamma}^\lambda$ associated with $\hat{L}$ satisfies*

$$\hat{\Gamma}^\lambda = \Gamma_L^{\hat{L}}(1 - \tfrac{\lambda}{\mu}) \bullet \Gamma^\lambda.$$

Remark 4.8.45 The orthogonal transformation used in the gauge transformation between the flat connections for the smooth case in Theorem 3.6.43 and the discrete case in Theorem 4.8.44 are identical. However, this is not surprising, since in both cases, the orthogonal transformation signifies an edge between L and $\hat{L}$ in the Bianchi quadrilateral with parameter μ.

Remark 4.8.46 Requiring that (4.19) holds true on every edge ij can be taken as a characterization of Darboux transformations. In fact, this characterization is the first definition of Darboux transformations of discrete isothermic surfaces given in [114, §4].

Exercise 4.8.47 (cf. [113, Lemma 3.17], [52, Equation (4.2)]) In the smooth case, Darboux transformations are also characterized by *Darboux's linear system* (see Definition 3.5.30). Show that this also holds true in the discrete case, i.e. $\hat{L}$ is a Darboux transform of L with parameter μ if and only if $\mathrm{span}\{T^{\mu}\hat{X}\}$ is constant for $\hat{X} \in \Gamma\hat{L}$.

4.9 Polynomial Conserved Quantities of Discrete Isothermic Surfaces

The characterization of isothermicity via flat connections also allows us to define polynomial conserved quantities in the discrete setting. Using the notion of parallel sections in the discrete case, polynomial conserved quantities of discrete flat connections are defined as follows.

Definition 4.9.48 ([52, Definition 3.1]) Let $L : \mathbb{Z}^2 \to P(L^4)$ be isothermic with flat connection Γ^{λ}. We call

$$P^{\lambda} = \sum_{r=0}^{n} \lambda^r P^{(r)}$$

a *polynomial conserved quantity of order n* of L if

$$P_i^{\lambda} = \Gamma_{ij}^{\lambda} P_j^{\lambda}, \tag{4.20}$$

where $P^{(k)} : \mathbb{Z}^2 \to \mathbb{R}^{4,1}$ are non-zero functions. If L admits a polynomial conserved quantity of order n, we call L a *discrete special isothermic surface of type n*.

For notational simplicity, we sometimes write $P^{(0)} = Q$ and $P^{(n)} = Z$.

Remark 4.9.49 The above definition of discrete special isothermic surfaces of type n is, in fact, consistent with the ethos of discretization with integrability at its heart, as demonstrated in Sect. 4.3. This can be seen as follows: we first note that the condition (4.20) on the edge (ij) is equivalent to the condition (3.30) as seen in Bäcklund transformation for special isothermic surfaces of type n (Proposition 3.7.61). Thus, the elementary quadrilaterals of discrete special isothermic surfaces of type n are the Bianchi quadrilaterals coming from Bäcklund transformations for special isothermic surfaces of type n seen in Theorem 3.7.64.

Exercise 4.9.50 Let T^λ be the Calapso transformation of a discrete isothermic surface L. Show that P^λ is a polynomial conserved quantity of L if and only if

$$T_j^\lambda P_j^\lambda = T_i^\lambda P_i^\lambda.$$

Lemma 4.9.51 ([52, Lemma 3.7]) *P^λ is a polynomial conserved quantity of L if and only if*

$$dP_{ij}^\lambda = \frac{\lambda a_{ij}}{\langle X_i, X_j \rangle} \left(\langle P_j^\lambda, X_j \rangle X_i - \langle P_i^\lambda, X_i \rangle X_j \right), \qquad (4.21)$$

for any section $X \in \Gamma L$.

Proof Suppose that P^λ is a polynomial conserved quantity of L. Then, on any edge ij, writing out the explicit form for Γ_{ij}^λ using (3.19), we see that (4.20) reads

$$dP_{ij}^\lambda = \frac{\lambda a_{ij}}{\langle X_i, X_j \rangle} \left(\langle P_j^\lambda, X_j \rangle X_i - \frac{1}{1 - \lambda a_{ij}} \langle P_j^\lambda, X_i \rangle X_j \right).$$

Taking the inner product with X_i on both sides, we obtain

$$\langle P_i^\lambda, X_i \rangle = \left(\frac{1}{1 - \lambda a_{ij}} \right) \langle P_j^\lambda, X_i \rangle.$$

Now suppose (4.21) holds. Taking the inner product with X_i on both sides, we obtain $\langle P_j^\lambda, X_i \rangle = (1 - \lambda a_{ij}) \langle P_i^\lambda, X_i \rangle$. Substituting this into (4.21), we see that P^λ is a polynomial conserved quantity of L. $\square$

Similar to the smooth case, we can show the following concerning polynomial conserved quantities.

Lemma 4.9.52 ([52, Corollaries 3.8, 3.9]) *Q is constant (i.e. independent of choice of vertex in the domain $\mathbb{Z}^2$) and $\langle Z, X \rangle = 0$ for $X \in \Gamma L$.*

Proof Comparing the λ^0 term of (4.21) gives us that $dQ_{ij} = 0$ on every edge ij, while comparing the λ^{n+1} term tells us that

$$\langle X_j, Z_j \rangle X_i - \langle X_i, Z_i \rangle X_j = 0.$$

Finally, the linear independence of X_i and X_j implies $\langle X, Z \rangle = 0$ on both i and j. $\square$

Lemma 4.9.53 *The inner products* $\langle P^\lambda, P^\lambda \rangle$, $\langle Z, Z \rangle$ *are constant.*

Proof Noting that $T^\lambda \in O_{4,1}$, we have using Exercise 4.9.50,

$$\langle P_j^\lambda, P_j^\lambda \rangle = \langle T_j^\lambda P_j^\lambda, T_j^\lambda P_j^\lambda \rangle = \langle T_i^\lambda P_i^\lambda, T_i^\lambda P_i^\lambda \rangle = \langle P_i^\lambda, P_i^\lambda \rangle$$

for all edges ij and all choices of λ. Evaluating the top term this equality now gives $\langle Z_j, Z_j \rangle = \langle Z_i, Z_i \rangle$. $\qquad\square$

We would like to now see the behavior of the polynomial conserved quantities under transformations. The gauge transformations between flat connections in both the smooth and the discrete case being identical allows one to easily guess the form of the new polynomial conserved quantities (see Theorem 4.7.39 for Calapso transformations and Remark 4.8.45 for Darboux transformations).

Theorem 4.9.54 ([52, Theorem 3.13]) *Let L be a discrete special isothermic surface of type n. Then the Calapso transform L^μ of L with parameter μ is of type n.*

Proof Let P^λ be the polynomial conserved quantity of L, and define

$$\tilde{P}^\mu := T^\mu P^{\lambda+\mu}.$$

A straightforward calculation using the form of the flat connection of L^μ in Theorem 4.7.39 shows that this is a polynomial conserved quantity of L^μ. $\qquad\square$

Theorem 4.9.55 ([52, Theorem 4.3]) *Let L be a discrete special isothermic surface of type n. Then the Darboux transform $\hat{L}$ of L (with parameter μ) is at most of type $n+1$.*

Proof Let P^λ be the polynomial conserved quantity of order n of L. Defining

$$\hat{P}^\lambda := (1 - \tfrac{\lambda}{\mu})\Gamma_L^{\hat{L}}(1 - \tfrac{\lambda}{\mu})P^\lambda,$$

it is easy to check that $\hat{P}^\lambda$ is a polynomial conserved quantity of $\hat{L}$, using Theorem 4.8.44. Then arguing identically as in Proposition 3.7.60, we have that $\hat{P}^\lambda$ is of order at most $n+1$ via Lemma 4.9.52. $\qquad\square$

Again, an identical argument to the paragraph preceding Proposition 3.7.61 gives us:

Theorem 4.9.56 ([52, Theorem 4.3]) *Let L be a discrete special isothermic surface of type n, and $\hat{L}$ be a Darboux transform of L with parameter μ such that $\langle P^\mu, \hat{X} \rangle = 0$ for any $\hat{X} \in \Gamma \hat{L}$. Then $\hat{L}$ has a polynomial conserved quantity $\tilde{P}^\lambda$ of order at most n.*

Exercise 4.9.57 Prove that $\langle P_i^\mu, \hat{X}_i \rangle = 0$ on one vertex i implies $\langle P^\mu, \hat{X} \rangle = 0$ everywhere. (Hint: Show that $\mathrm{d}(\langle P^\mu, \hat{X} \rangle)_{ij} = 0$ for any edge ij.)

4.10   Discrete Minimal and cmc Surfaces in $\mathbb{R}^3$

As an application of the theory of discrete isothermic surfaces, we study discrete minimal and cmc surfaces in Euclidean 3-space. In this section, we extensively use the Möbius geometry; however, to keep the metric of $M \cong \mathbb{R}^3$ the standard Euclidean metric, we use the space form vector as in Exercise 2.2.14.

Recalling that discrete isothermic surfaces are circular nets, we have the notion of mean curvature available on each quadrilateral of the surface via Definition 4.2.6. Therefore, discrete isothermic minimal and cmc surfaces are defined as follows.

Definition 4.10.58 Let $x : \mathbb{Z}^2 \to \mathbb{R}^3$ be isothermic. Then x is called a *discrete (isothermic) minimal surface* if the mean curvature H vanishes on every elementary quadrilateral in the domain. Similarly, x is called a *discrete (isothermic) cmc surface* if the mean curvature $H = c \in \mathbb{R} \setminus \{0\}$ on every elementary quadrilateral in the domain.

We first consider discrete minimal surfaces. The following characterization of discrete minimal surfaces is easy to check using Corollary 4.4.23.

Corollary 4.10.59 (cf. [22, Theorem 8]) $x : \mathbb{Z}^2 \to \mathbb{R}^3$ *is a discrete (isothermic) minimal surface for some choice of unit normal v if and only if v is a Christoffel transform of x.*

Therefore, we see that the image of the unit normal to a discrete minimal surface is also isothermic.

Now, we have seen in Exercise 2.10.85 that $g : \Sigma \to \mathbb{C}$ is a holomorphic function with respect to $z = u + \sqrt{-1}v$ if and only if $g : \Sigma \to \mathbb{R}^2$ is isothermic with isothermic coordinates (u, v). Such characterization of holomorphicity can be discretized as follows (Fig. 4.5).

Definition 4.10.60 ([22, Definition 8]) Let $g : \mathbb{Z}^2 \to \mathbb{C} \cong \mathbb{R}^2 \subset \mathbb{R}^3$. We call g a *discrete holomorphic function* if g is discrete isothermic (as a surface in $\mathbb{R}^2 \subset \mathbb{R}^3$).

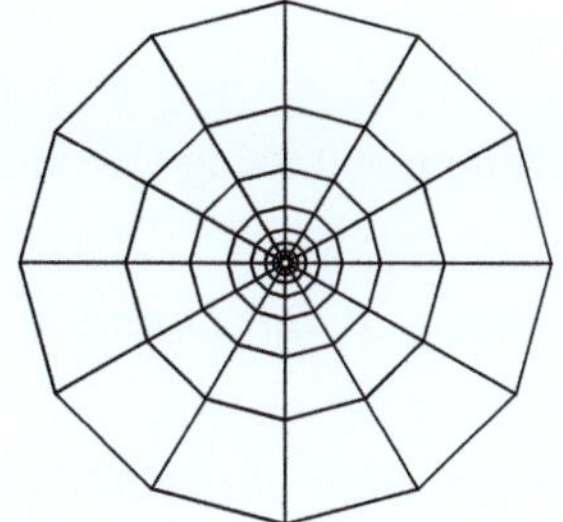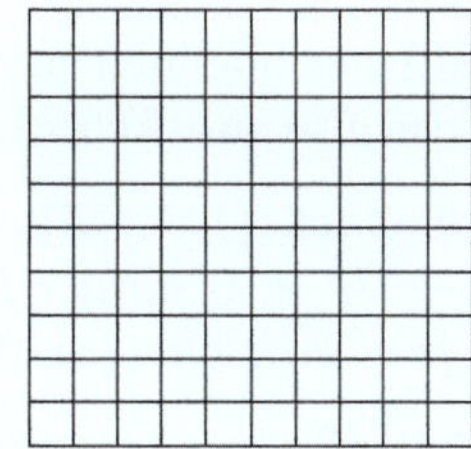

Fig. 4.5   Some examples of discrete holomorphic functions

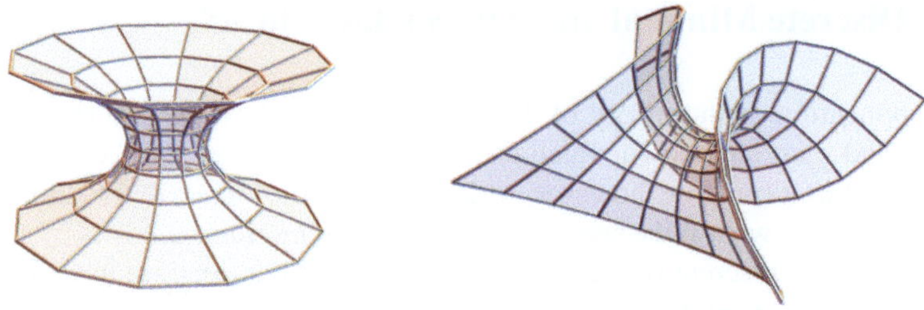

Fig. 4.6 Discrete isothermic minimal surfaces in $\mathbb{R}^3$, the catenoid and Enneper surface, corresponding to the discrete holomorphic functions in Fig. 4.5

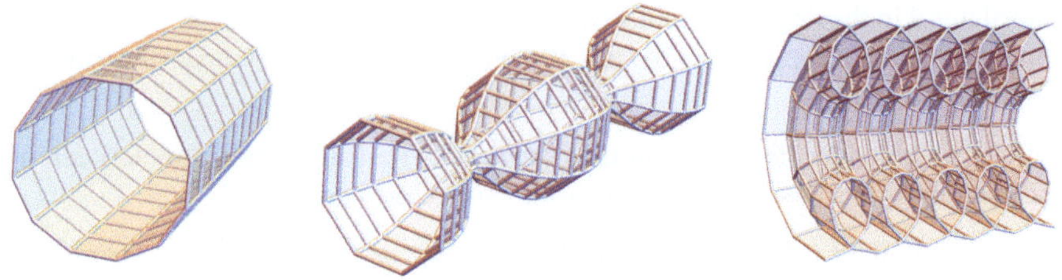

Fig. 4.7 Discrete cmc surfaces in $\mathbb{R}^3$, the cylinder, Delaunay unduloid and Delaunay nodoid

Combining these results, we can obtain the *Weierstrass representation* for discrete minimal surfaces, as in Exercise 2.10.85 for the smooth case (Fig. 4.6).

Exercise 4.10.61 ([22, Theorem 9]) Let $g : \mathbb{Z}^2 \to \mathbb{C}$ be discrete holomorphic with cross ratios factorizing function a. Defining $v = St^{-1} \circ g : \mathbb{Z}^2 \to \mathbb{S}^2$, where St is the stereographic projection from $\mathbb{S}^2$, show that

$$dx_{ij} := dv_{ij}^* = \frac{a_{ij}}{2} \operatorname{Re}\left((1 - g_i g_j, \sqrt{-1}(1 + g_i g_j), g_i + g_j) \frac{1}{dg_{ij}} \right).$$

We now shift our attention to discrete cmc surfaces (Fig. 4.7). Using Definition 4.2.6, we can also easily check the following result using Corollary 4.4.23.

Corollary 4.10.62 (cf. [114, §5]) *We have that* $x : \mathbb{Z}^2 \to \mathbb{R}^3$ *is a discrete (isothermic) cmc* $H \neq 0$ *surface for some choice of unit normal* v *if and only if* $x^* = x + \frac{1}{H} v$ *is a Christoffel transform of* x.

Corollary 4.10.62 gives us another characterization of discrete cmc surface as follows:

Corollary 4.10.63 (cf. [114, §5]) x *is a discrete (isothermic) cmc* $H \neq 0$ *surface (for some choice of unit normal* v*) if and only if* $|x^* - x|^2$ *is a nonzero constant for a Christoffel transform* x^* *of* x.

Proof For a discrete cmc surface x with appropriate normal v, we can write $x^* = x + \frac{1}{H}v$ so that

$$|x^* - x|^2 = \frac{1}{H^2}. \tag{4.22}$$

giving us one direction.

On the other hand, suppose that for a discrete isothermic x and its Christoffel dual x^*, we have $|x^* - x|^2 = c^2$ for some nonzero constant $c \in \mathbb{R}$. Since we have that $dx^*_{ij} \parallel dx_{ij}$, the quadrilateral comprised of x_i, x_j, x^*_j, x^*_i is either a parallelogram or an isosceles trapezoid. However, since x and x^* are Koenigs dual as well, and the opposite diagonals must be parallel, the quadrilateral must be an isosceles trapezoid. Therefore, if we define $v = \frac{1}{c}(x^* - x)$ so that it is unit length, then v becomes a well-defined normal for x. Hence, we can write $x^* = x + cv$, and Corollary 4.10.62 tells us that x is a discrete cmc $H = \frac{1}{c}$ surface. $\qquad\square$

We would now like to recover the result for smooth surfaces that a surface is a cmc surface if and only if its Christoffel transform is a Darboux transform (see Theorem 3.5.35).

Lemma 4.10.64 (cf. [114, §5]) *Let x be a discrete isothermic surface with Christoffel transformation x^*. Then x^* is a Darboux transformation of x if and only if $|x^* - x|^2$ is constant.*

Proof Suppose that x^* is a Darboux transformation of x, and take any edge ij. Since $dx_{ij} \parallel dx^*_{ij}$ and the quadrilateral x_i, x_j, x^*_j, x^*_i is concircular by (4.19) (and Remark 4.8.46), we see that the quadrilateral is symmetric with respect to the perpendicular bisector of dx_{ij}. Therefore, the quadrilateral must be an isosceles trapezoid, and we have that $|x^* - x|^2$ is constant.

Now assume that $|x^* - x|^2$ is constant. On any edge ij, by an identical argument to that in the proof of Corollary 4.10.63, we see that the quadrilateral x_i, x_j, x^*_j, x^*_i is an isosceles trapezoid. Now, to calculate the cross ratio of these four points, we assume without loss of generality that $x_i, x_j, x^*_j, x^*_i \in \mathbb{C}$ and that $dx_{ij} \in \sqrt{-1}\mathbb{R}$. Then for $x_j - x^*_j = c \in \mathbb{C}$, we have $x^*_i - x_i = -\bar{c}$, so

$$\mathrm{cr}(x_i, x_j, x^*_j, x^*_i) = \frac{dx_{ij}}{c}\frac{dx^*_{ij}}{\bar{c}} = -\frac{a_{ij}}{|c|^2}.$$

Therefore, by Remark 4.8.46, x^* is a Darboux transform of x with parameter $-\frac{1}{|c|^2}$. $\qquad\square$

With the above lemma, Corollary 4.10.63 now gives us the following characterization of discrete cmc surfaces.

Theorem 4.10.65 (cf. [114, §5]) *Let x be a discrete isothermic surface in $\mathbb{R}^3$. Then x is a discrete cmc surface (for some unit normal v) if and only if there is a Christoffel transformation that is also a Darboux transformation of x.*

Finally, we will show the characterization of discrete cmc surfaces in $\mathbb{R}^3$ via the existence of a linear conserved quantity. First, we prove a simple lemma.

Lemma 4.10.66 ([52, p. 169]) *Let x be a discrete cmc surface with Christoffel dual x^*, and let X and X^* denote the lifts of x and x^* into M_0, respectively. Then we have*

$$\mathrm{d}X_{ij}^* = \frac{a_{ij}}{|\mathrm{d}x_{ij}|^2}\mathrm{d}X_{ij}.$$

Proof Take any edge ij. Since we have that $|x^* - x|^2$ is constant, we can calculate using the discrete Leibniz rule in Exercise 4.4.10,

$$0 = \mathrm{d}(|x^* - x|^2)_{ij} = 2(\mathrm{d}x_{ij}^* - \mathrm{d}x_{ij})\cdot(x_{ij}^* - x_{ij}) = 2\left(1 - \frac{|\mathrm{d}x_{ij}|^2}{a_{ij}}\right)\mathrm{d}x_{ij}^*\cdot(x_{ij}^* - x_{ij}),$$

so that $x_{ij}^* \cdot \mathrm{d}x_{ij}^* = x_{ij} \cdot \mathrm{d}x_{ij}^*$. Now, using this relation, we can verify

$$\mathrm{d}(|x^*|^2)_{ij} = 2\mathrm{d}x_{ij}^* \cdot x_{ij}^* = 2\frac{a_{ij}}{|\mathrm{d}x_{ij}|^2}\mathrm{d}x_{ij} \cdot x_{ij} = \frac{a_{ij}}{|\mathrm{d}x_{ij}|^2}\mathrm{d}(|x|^2)_{ij}.$$

Finally,

$$\mathrm{d}X_{ij}^* = \begin{pmatrix} \frac{1}{2}\mathrm{d}(|x^*|^2)_{ij} \\ (\mathrm{d}x_{ij}^*)^t \\ \frac{1}{2}\mathrm{d}(|x^*|^2)_{ij} \end{pmatrix} = \frac{a_{ij}}{|\mathrm{d}x_{ij}|^2}\begin{pmatrix} \frac{1}{2}\mathrm{d}(|x|^2)_{ij} \\ (\mathrm{d}x_{ij})^t \\ \frac{1}{2}\mathrm{d}(|x|^2)_{ij} \end{pmatrix} = \frac{a_{ij}}{|\mathrm{d}x_{ij}|^2}\mathrm{d}X_{ij},$$

giving us the desired conclusion. $\square$

We now prove the main theorem of this section.

Theorem 4.10.67 ([52, Theorem 5.5]) *Suppose that L is an isothermic surface. Then L admits a linear conserved quantity $P^\lambda = Q + \lambda Z$ with $\|Q\|^2 = 0$, $\langle Q, Z \rangle \neq 0$, and $\|Z\|^2 \neq 0$ if and only if L projects to a discrete cmc surface in the Euclidean space determined by the constant term Q.*

Proof Suppose $x : \mathbb{Z}^2 \to \mathbb{R}^3$ is a discrete cmc H surface with lift $X : \mathbb{Z}^2 \to M_0$ and $L = \mathrm{span}\{X\}$. Let $Q = \mathfrak{q}_0$ that determines M_0, and define

$$Z := -\frac{1}{2}X^* + \frac{1}{4H^2}Q.$$

Since $\langle X, X^* \rangle = -\frac{1}{2}|x^* - x|^2 = -\frac{1}{2H^2}$ by (4.22), we can check that $\langle Z, X \rangle \equiv 0$ while Lemma 4.10.66 tells us

$$\mathrm{d}Z_{ij} = -\frac{1}{2}\mathrm{d}X_{ij}^* = -\frac{a_{ij}}{2|\mathrm{d}x_{ij}|^2}\mathrm{d}X_{ij}.$$

Now if we let $P^\lambda := Q + \lambda Z$, then $\langle P^\lambda, X \rangle \equiv -1$. Using this, we can calculate

$$\mathrm{d}P^\lambda_{ij} = -\frac{\lambda a_{ij}}{2|\mathrm{d}x_{ij}|^2}\mathrm{d}X_{ij} = \frac{\lambda a_{ij}}{\langle X_i, X_j \rangle}(X_j - X_i)$$

$$= \frac{\lambda a_{ij}}{\langle X_i, X_j \rangle}(\langle P^\lambda_j, X_j \rangle X_i - \langle P^\lambda_i, X_i \rangle X_j).$$

Therefore, by Lemma 4.9.51, we have that P^λ is a linear conserved quantity of L.

To show the other direction, let $P^\lambda = Q + \lambda Z$ be a linear conserved quantity of L and assume without loss of generality that $Q = \mathfrak{q}_0$ and $\|Z\|^2 = 1$. Then we note that Lemma 4.9.53 allows us to define $H := -\langle Q, Z \rangle$, which is constant. If we let

$$\hat{X} := \frac{1}{2H^2}P^{2H} = \frac{1}{2H^2}(Q + 2HZ),$$

then we see that $\langle \hat{X}, \hat{X} \rangle = 0$, $\langle \hat{X}, Q \rangle = -1$, and

$$\hat{X}_i = \frac{1}{2H^2}P_i^{2H} = \frac{1}{2H^2}\Gamma_{ij}^{2H}P_j^{2H} = \Gamma_{ij}^{2H}\hat{X}_j.$$

Therefore, $\hat{L} := \mathrm{span}\{\hat{X}\}$ is a Darboux transform of L with parameter $2H$. Choosing $X \in \Gamma L$ so that $X : \mathbb{Z}^2 \to M_0$, we have $\langle P^{2H}, X \rangle = -1$ allowing us to verify

$$\mathrm{d}\hat{X}_{ij} = \frac{1}{2H^2}\mathrm{d}P_{ij}^{2H} = \frac{a_{ij}}{H\langle X_i, X_j \rangle}\mathrm{d}X_{ij},$$

where the last equality comes from (4.21). Denoting the projection of L and $\hat{L}$ to $\mathbb{R}^3$ as x and $\hat{x}$, respectively, we therefore have

$$\mathrm{d}\hat{x}_{ij} = -\frac{a_{ij}}{2H|\mathrm{d}x_{ij}|^2}\mathrm{d}x_{ij} = -\frac{1}{2H}\mathrm{d}x^*_{ij}.$$

Thus, by Theorem 4.10.65, we have that L projects to a discrete cmc surface in the space form determined by Q. $\qquad\square$

Thus, we conclude that discrete cmc surfaces in space forms come from the Bianchi quadrilaterals of Darboux transformations keeping the cmc condition, another application of *discretization with integrability at its heart*.

Recommended Further Readings for Chap. 4

Historically, the various works of Sauer and Wunderlich including those on discrete K-surfaces in $\mathbb{R}^3$ [196, 197, 220] mark the starting point of discrete differential geometry, with many classical results on discrete surfaces included in the book by Sauer [195].

For a modern approach to discrete differential geometry with integrability at its heart, the standard text is the broadly emcompassing book [25]. The book is methodical in explaining how discrete objects can be defined via the permutability of transformations of the corresponding smooth objects, the central theme to the theory we adopt here. Furthermore, the transformations of discrete objects are explained via the multidimensional consistency of the permutability, thus allowing the discrete objects to be defined on a general n-dimensional lattice. The material in the book is a culmination of many defining results that formulated the theory of discrete differential geometry, a few of which we name here.

Following on from the classical approach, the work [23] revisits discrete K-surfaces. Exploiting the well-known relationship between smooth K-surfaces and the (hyperbolic) sine-Gordon equation, the work interprets discrete K-surfaces as manifestations of the solutions to the discrete sine-Gordon equation defined in [122].

Discrete curvatures are defined in [199], en route to discretizing the concept of O-surfaces [200], where many different integrable classes of surfaces (including isothermic surfaces) are characterized via the existence of certain Combescure transformations. The mixed area characterization of discrete curvatures is given in [186], on the way to exploring the application of the theory of discrete surfaces to architecture.

Now focusing on the works related to discrete isothermicity, discrete isothermic surfaces are defined in [22] via the cross ratios condition. Furthermore, the work recovers the discrete analogue of Christoffel duality, paving the way for defining discrete minimal surfaces as those discrete isothermic surfaces with the Christoffel dual lying on a sphere. In doing so, discrete holomorphic functions are defined, and the Weierstrass representation for discrete minimal surfaces is derived.

The Möbius geometric characterization of isothermic surfaces is also discretized in [26]. In the work, isothermic surfaces are viewed in the light cone model of Möbius geometry, and focuses on the characterization via the existence of Moutard lifts, and interprets discrete isothermic surfaces as the Bianchi quadrilateral of Moutard transformations of Moutard lifts.

Discrete Darboux transformations coming from the Bianchi cube of smooth Darboux transformations is defined in [114]. Using quaternions, the work characterizes the Bianchi cube in terms of the cross ratios condition, arriving at the cross ratios definition of Darboux transformations. Furthermore, the work defines discrete cmc surfaces as those surfaces with parallel nets as Christoffel transformations.

Building on the quaternionic approach to discrete isothermic surfaces, the work [113] recovers the Calapso transformations for discrete isothermic surfaces. Furthermore, the work uses both Calapso transformations and Darboux transformations to define and explore discrete cmc-1 surfaces in hyperbolic 3-space.

The flat connections approach is again well-suited to studying the theory of discrete isothermic surfaces and their transformations. This approach is explored in [52] where the work characterizes discrete isothermic surfaces in Möbius geometry via the existence of a 1-parameter family of discrete flat connections, and defines discrete polynomial conserved quantities (see also [51]). Then the polynomial conserved quantities are used to identify discrete cmc surfaces among discrete isothermic surfaces, making contact with the previous definition of discrete cmc surfaces. The work [19] then characterizes discrete cmc surfaces in different space forms via the existence of certain Koenigs duals.

In fact, the flat connections approach to discrete isothermicity admits a natural generalization to symmetric R-spaces just as for the smooth counterpart. Continuing on the discussions of transformations of smooth isothermic surfaces in symmetric R-spaces, the work [41] offers discrete isothermicity in symmetric R-spaces as Bianchi quadrilaterals of Darboux transformations of smooth isothermic surfaces.

The work [128] recovers a characterization of discrete isothermic surfaces with spherical curvature lines in terms of certain discrete holomorphic functions, and constructs examples of discrete isothermic tori using this method.

An elementary introduction to discrete gauge theory and discrete isothermicity is given in [37]. The work quickly shows the flat connections characterization of smooth isothermic surfaces and the permutability of Darboux transformations via gauge transformations, and introduces discrete isothermic surfaces via discrete flat connections.

Discrete isothermicity can also be similarly defined in the context of space forms with constant sectional curvature given an indefinite signature. For spacelike surfaces in Minkowski 3-space, discrete isothermicity is defined in [221] en route to defining discrete maximal surfaces. Smooth maximal surfaces are known to admit certain types of non-degenerate singularities, defined as *maxfaces* in [215] (see also [104]), and this work further explores the concept of singular faces of discrete surfaces.

Independently of discrete isothermicity, a discrete version of smooth conformal mappings has been rapidly developed by relating the geometry of circle packings. A comprehensive introduction can be found in [203]. In particular, [202] investigated circle patterns with the combinatorics of the square grid, and the equivalence of such circle patterns and discrete holomorphic functions in the narrow sense has been since shown (see [25] for example).

Connecting the theory of discrete conformal mapping and discrete isothermicity is the study of discrete minimal surfaces. However, an obstacle to creating numerous global examples of discrete minimal surfaces lies in the difficulty of obtaining globally defined discrete holomorphic functions. Several examples of discrete holomorphic functions have been defined in addition to the discrete identity function and discrete exponential function: The discrete power functions are defined in [16] with their embeddedness proven in [1] (see also [3]) using an integrable systems approach. On the other hand, [160] uses a more geometric approach to construct the discrete counterpart of the family of exponential functions, while the work [75] creates numerical examples of discrete holomorphic functions using certain boundary conditions, which results in the discrete counterpart of the minimal k-noids found in [133].

Smooth cmc surfaces have their own Weierstrass-type representation known as the DPW method [96], and the work [124] recovers a discrete analogue for discrete cmc surfaces in Euclidean space, while the work [174] generalizes this to other Riemannian space forms.

The discrete minimal and cmc surfaces introduced above are restricted to the case of discrete isothermic minimal surfaces. To allow for a more flexible "parametrization" of discrete surfaces in the Euclidean space, [127] offers the concept of edge-constraint nets, and gives a generalized Weierstrass representation of discrete minimal surfaces, recovering a discretization of smooth minimal surfaces with non-real constant Hopf differential. Another approach to such flexible parametrization is the concept of face-edge-constraint nets, explored in [125, 129] with their applications to discrete minimal and cmc surface theory.

A shortcoming of defining discrete minimal and cmc surfaces via transformation theory is the apparent lack of the variational characterization enjoyed by their smooth counterparts. Therefore, one can also define discrete minimal and cmc surfaces by requiring an appropriate variational condition, which is the work done in [183] and [185].

The work [146] offers a different approach to discrete minimality using trivalent nets. The discrete minimality defined in this work recovers both characterizations of smooth minimal surfaces: via transformation theory and the variational approach. An analogous work for discrete maximal surfaces is then done in [147], which investigates a broader class of discrete maximal surfaces. In fact, this setting allows for the analysis of the "singularities" appearing on these surfaces, including an investigation of duality of singularities between the original discrete trivalent maximal surface and its conjugate, where the analogous problem for the smooth cases was considered in [104, 173].

On a different note, one can also consider surfaces that are parametrized by a smooth parameter and a discrete parameter, which are called "semi-discrete" surfaces. Semi-discrete isothermic surfaces are first defined in [162] via the tangential cross ratios condition, and Christoffel duality for semi-discrete isothermic surfaces is given.

Building on this definition, [191] recovers the Weierstrass representation of semi-discrete minimal surfaces, while the maximal case was investigated in [222]. The work [161] discusses semi-discrete minimal and cmc surfaces, and explores their characteristics in a very geometrical fashion. The associated families of semi-discrete cmc surfaces are given in [59] via the Lax pair representation.

Semi-discrete isothermic surfaces also enjoy a characterization via transformation theory, as shown in [45]. The work shows that every semi-discrete isothermic surface in the conformal n-sphere can be obtained as an iteration of Darboux transformations of a polarized curve, arriving at the flat connections characterization of semi-discrete isothermic surfaces. The flat connections characterization is then utilized to define Calapso transformations of semi-discrete isothermic surfaces via the trivializing gauge, while the permutability of Darboux transformations of curves is utilized to define the Darboux transformations of semi-discrete isothermic surfaces.

Darboux transformations of discrete polarized curves are considered in [74] and [73], and by considering their integrable reductions to bicycle transformations

[30, 206, 207], the semi-discrete and discrete mKdV equations are obtained as integrability conditions (see also [131, 134, 158]). Then the monodromy problem of Darboux transformations of discrete polarized curves are considered in [70], allowing for the explicit construction of discrete analogues of cmc bubbletons and isothermic tori [67].

Many works mentioned here aimed to recover the discrete analogue of the smooth theory; recently, results in the field of discrete differential geometry started to be used to obtain new results in the smooth theory. One example is the Björling problem [13] for isothermic surfaces, as the work [35] solves the problem of finding isothermic surfaces containing a given curve and a unit normal vector field along the curve, using discrete isothermic surface theory and its convergence under continuum limit. Another example is the global Bonnet conjecture: to find a compact isometric pair of surfaces with the same mean curvature. The problem is solved in [20] via finding all isothermic tori with planar curvature lines, an ansatz found via structure preserving discretization.

Chapter 5
Ω-surfaces in Lie Sphere Geometry

5.1 Lie Sphere Geometry

We now explain Lie sphere geometry, since it is the natural setting for considering Ω-surfaces. The first part of this chapter reiterates material found in [65], so we will keep the descriptions here brief.

Consider $\mathbb{R}^{4,2}$ with the metric $(\cdot, \cdot)$ of signature $(-++++-)$. For some $\mathfrak{p} \in \mathbb{R}^{4,2}$ such that $(\mathfrak{p}, \mathfrak{p}) = -1$, we can split $\mathbb{R}^{4,2}$ as

$$\mathbb{R}^{4,2} = \text{span}\{\mathfrak{p}\}^{\perp} \oplus \text{span}\{\mathfrak{p}\},$$

where $\text{span}\{\mathfrak{p}\}^{\perp} \cong \mathbb{R}^{4,1}$. Without loss of generality, we assume that

$$\mathfrak{p} = (0, 0, 0, 0, 0, 1)^t, \tag{5.1}$$

and we will refer to $\text{span}\{\mathfrak{p}\}^{\perp}$ as $\mathbb{R}^{4,1}$. We denote the light cone of $\mathbb{R}^{4,2}$ by

$$L^5 := \{S \in \mathbb{R}^{4,2} \mid (S, S) = 0\}.$$

Recall that in Möbius geometry, points are represented as lightlike vectors $X \in \mathbb{R}^{4,1}$, while spheres (and planes) with orientations are represented as unit length spacelike vectors $\mathcal{S} \in \mathbb{R}^{4,1}$.

Lie sphere geometry makes no distinction between points (called point spheres now) and spheres (including planes). To do this, let point spheres $X \in L^4$ be considered as vectors in $L^5 \subset \mathbb{R}^{4,2}$, while

$$S = \mathcal{S} + \mathfrak{p}$$

for any spacelike unit $\mathcal{S} \in \mathbb{R}^{4,1}$. Then it is easy to check that $S \in L^5$, since $(\mathcal{S}, \mathfrak{p}) = 0$ when $\mathcal{S}$ is considered as a vector in $\mathbb{R}^{4,2}$. Therefore, in Lie sphere

© The Author(s), under exclusive license to Springer Nature Switzerland AG 2025
J. Cho et al., *Discrete Isothermic Surfaces in Lie Sphere Geometry*, Lecture Notes in Mathematics 2375, https://doi.org/10.1007/978-3-031-95592-1_5

geometry $\mathbb{R}^{4,2}$, we have that the lightlike vectors represent point spheres, spheres, and planes, and we refer to all three simply as *spheres*. We point out here that a sphere with two opposite orientations is represented by $\pm\,\mathcal{S}$ in Möbius geometry, and thus is represented by different lightlike vectors in Lie sphere geometry.

Now that all spheres are represented as lightlike vectors, their scaling does not matter, analogous to points in Möbius geometry. Therefore, we consider points in the projectivized light cone

$$P(L^5) := \{\mathrm{span}\{S\} \in P(\mathbb{R}^{4,2}) \mid S \in L^5 \setminus \{0\}\}$$

$$= \{\text{null lines through origin in } \mathbb{R}^{4,2}\}$$

as the set of all spheres, called the *Lie quadric*. When we make contact with Möbius geometry, we take the representative $(\mathcal{S}^t, 1)^t \in \mathrm{span}\{S\}$ for $\mathcal{S} \in \mathbb{S}^{3,1}$, if $(S, \mathfrak{p}) \neq 0$.

On the other hand, if $(S, \mathfrak{p}) = 0$ for any lightlike vector $S \in L^5$, we see that S can be considered as a lightlike vector $X \in L^4 \subset \mathbb{R}^{4,1} \cong \mathrm{span}\{\mathfrak{p}\}^\perp$. Therefore, one can see that the vector $\mathfrak{p}$ determines the point spheres in Möbius geometry $\mathbb{R}^{4,1} \cong \mathrm{span}\{\mathfrak{p}\}^\perp$. In other words, we cannot distinguish the point spheres from other spheres and planes without a choice of $\mathfrak{p}$. For this reason, we call $\mathfrak{p}$ a *point sphere complex*.

Once we drop down into Möbius geometry from Lie sphere geometry using $\mathfrak{p}$ via $\mathbb{R}^{4,1} \cong \mathrm{span}\{\mathfrak{p}\}^\perp$, then we can obtain 3-dimensional space forms with constant sectional curvature κ using $\mathfrak{q} \in \mathbb{R}^{4,1} \subset \mathbb{R}^{4,2}$ such that $(\mathfrak{q}, \mathfrak{q}) = -\kappa$, as we have in Sect. 2.2. Explicitly, for some choice of $\mathfrak{q}$ with $(\mathfrak{p}, \mathfrak{q}) = 0$ (so that $\mathfrak{q} \in \mathbb{R}^{4,1} \cong \mathrm{span}\{\mathfrak{p}\}^\perp$), we obtain the 3-dimensional space form M_κ via

$$M_\kappa := \{X \in L^5 \subset \mathbb{R}^{4,2} \mid (X, \mathfrak{p}) = 0, (X, \mathfrak{q}) = -1\}$$

$$= \{X \in L^4 \subset \mathbb{R}^{4,1} \mid \langle X, \mathfrak{q} \rangle = -1\}.$$

Thus, as in Möbius geometry, we call $\mathfrak{q}$ a *space form vector*.

For example, with $\mathfrak{p}$ as in (5.1), let $\mathfrak{q}_1 = (1, 0, 0, 0, 0, 0)^t$ to obtain $M_1 \cong \mathbb{S}^3$, while to obtain $M_{-1} \cong \mathbb{H}^3$, let $\mathfrak{q}_{-1} = (0, 0, 0, 0, -1, 0)^t$. To obtain $M_0 \cong \mathbb{R}^3$, we use a different $\mathfrak{q}_0 = (1, 0, 0, 0, -1, 0)$, so that the metric defined on M_0 matches that of $\mathbb{R}^3$.

We will illustrate lifting points and spheres in Möbius geometry to Lie sphere geometry with examples from $\mathbb{R}^3 \cong M_0$. Note that in the following examples, the representatives in Möbius geometry have changed, since the space form vector is also changed.

Example 5.1.1 Recall that a point $x \in \mathbb{R}^3$ is represented by $X \in M_0 \subset L^4$ in Möbius geometry via

$$X = \left(\tfrac{1}{2}(1 + |x|^2), x, \tfrac{1}{2}(1 - |x|^2)\right)^t.$$

Now the points are lifted into L^5 by viewing X as a vector in $\mathbb{R}^{4,2}$, i.e.,

$$X = \begin{pmatrix} \frac{1}{2}(1 + |x|^2) \\ x^t \\ \frac{1}{2}(1 - |x|^2) \\ 0 \end{pmatrix}.$$

On the other hand, in Möbius geometry, a sphere with center $c \in \mathbb{R}^3$ and (signed) radius $r \in \mathbb{R}$ is represented by vectors in $\mathbb{S}^{3,1}$ as

$$\mathcal{S} = \frac{1}{r}\left(\tfrac{1}{2}(1 + |c|^2 - r^2), c, \tfrac{1}{2}(1 - |c|^2 + r^2) \right)^t.$$

Now, such a sphere is represented by

$$S = \mathcal{S} + \mathfrak{p} = \frac{1}{r} \begin{pmatrix} \frac{1}{2}(1 + |c|^2 - r^2) \\ c^t \\ \frac{1}{2}(1 - |c|^2 + r^2) \\ r \end{pmatrix} \tag{5.2}$$

in Lie sphere geometry. (Note that the sign of r changes as the orientation of the sphere changes, since r is the signed radius. Also note that projectively, if $r = 0$ so that we have a point sphere with center c, then this is the representative in Lie sphere geometry for the point c.)

Finally, a plane through the point x and with unit normal n (so that $|n|^2 = 1$) is represented by

$$\mathcal{S} = (x \cdot n, n, -x \cdot n)^t \in \mathbb{S}^{3,1}$$

in Möbius geometry, which is represented in Lie sphere geometry by

$$S = \mathcal{S} + \mathfrak{p} = \begin{pmatrix} x \cdot n \\ n^t \\ -x \cdot n \\ 1 \end{pmatrix}.$$

Remark 5.1.2 If we choose a point sphere complex $\mathfrak{p} \in \mathbb{R}^{4,2}$ such that $(\mathfrak{p}, \mathfrak{p}) = 1$, then we have

$$\operatorname{span}\{\mathfrak{p}\}^{\perp} \cong \mathbb{R}^{3,2}.$$

This is the Minkowski model of *pseudoconformal geometry*, and the Lorentzian space forms with constant sectional curvature such as Minkowski 3-space $\mathbb{R}^{2,1}$, de

Sitter 3-space $\mathbb{S}^{2,1}$, and anti-de Sitter 3-space $\mathbb{H}^{2,1}$ appear as subgeometries. For more information, we refer the readers to [2] or [66], for example.

Example 5.1.3 (Spheres in $\mathbb{H}^3$) Let $S = (m^t, q, p) \in L^5$ for some $m \in \mathbb{R}^{3,1}$ and real p, q. Assuming that p is non-zero, we see that the Möbius geometry representative corresponding to S is $\mathcal{S} = \frac{1}{p}(m^t, q) \in \mathbb{S}^{3,1}$. Therefore, choosing the space form vector $\mathfrak{q} = \mathfrak{q}_{-1}$, one obtains that S represents a sphere $S_{\mathbb{H}^3}[m, q]$. Then we have that

$$\langle m, m \rangle_{3,1} = p^2 - q^2 = p^2 \left(1 - \frac{q^2}{p^2}\right) = p^2 \left(1 - \frac{(S, \mathfrak{q})^2}{(S, \mathfrak{p})^2}\right).$$

Together with Remark 2.3.23, we can deduce that $S \in L^5$ represents

- unbounded spheres when

$$\left(\frac{(S, \mathfrak{q})}{(S, \mathfrak{p})}\right)^2 < 1,$$

- horospheres or the ideal boundary case (with one orientation or the other) when

$$\left(\frac{(S, \mathfrak{q})}{(S, \mathfrak{p})}\right)^2 = 1,$$

- or bounded spheres when

$$\left(\frac{(S, \mathfrak{q})}{(S, \mathfrak{p})}\right)^2 > 1.$$

Exercise 5.1.4 Using Remark 2.3.23, show that for $\mathfrak{q} = \mathfrak{q}_{-1}$, we have $\mathfrak{p} \pm \mathfrak{q} \in L^5$ projects to the ideal boundary in $M_{-1} \cong \mathfrak{R}_{-1} \cong \mathbb{H}^3$.

We now show the advantage of expressing spheres as lightlike vectors in $\mathbb{R}^{4,2}$:

Proposition 5.1.5 *Let two linearly independent lightlike vectors $S_1, S_2 \in L^5$ represent spheres. Then the spheres represented by S_1 and S_2 are in oriented (tangential) contact if and only if $(S_1, S_2) = 0$.*

Proof Let $\mathfrak{p}$ be as in (5.1). If both $S_1, S_2 \in \mathrm{span}\{\mathfrak{p}\}^{\perp}$, then we can consider $X_1, X_2 \in L^4$ as natural projections of S_1, S_2, respectively, and see that

$$(S_1, S_2) = \langle X_1, X_2 \rangle.$$

Therefore, $(S_1, S_2) = 0$ if and only if they project to the same point in Möbius geometry.

If only $S_1 \in \mathrm{span}\{\mathfrak{p}\}^\perp$, then write

$$S_1 = (X_1^t, 0)^t \quad \text{and} \quad S_2 = (\mathcal{S}_2^t, 1)^t$$

for some $X_1 \in L^4$ and $\mathcal{S}_2 \in \mathbb{S}^{3,1}$. Then we see that

$$(S_1, S_2) = \langle X_1, \mathcal{S}_2 \rangle,$$

so $(S_1, S_2) = 0$ if and only if S_1 projects to a point in the sphere defined by S_2 in Möbius geometry.

Finally, assume that $S_1, S_2 \notin \mathrm{span}\{\mathfrak{p}\}^\perp$. Writing

$$S_1 = (\mathcal{S}_1^t, 1)^t \quad \text{and} \quad S_2 = (\mathcal{S}_2^t, 1)^t,$$

we have that $(S_1, S_2) = 0$ if and only if $\langle \mathcal{S}_1, \mathcal{S}_2 \rangle = 1$, telling us that the two spheres are in oriented contact by Lemma 2.3.27. $\qquad\square$

We need the condition "oriented" in the statement of Proposition 5.1.5. For example, a sphere with two different orientations is represented as $\mathcal{S}_1 = -\mathcal{S}$ and $\mathcal{S}_2 = \mathcal{S}$ for some $\mathcal{S} \in \mathbb{S}^{3,1}$ in Möbius geometry. Lifting these into Lie sphere geometry as S_1 and S_2, respectively, we see that

$$(S_1, S_2) = ((\mathcal{S}_1^t, 1)^t, (\mathcal{S}_2^t, 1)^t) = \langle -\mathcal{S}, \mathcal{S} \rangle - 1 = -2.$$

Corollary 5.1.6 *Let $S_1, S_2 \in L^5$ such that $(S_1, S_2) = 0$, with $\Lambda := \mathrm{span}\{S_1, S_2\}$. For any $S \in \Lambda$, S projects to a sphere that is in oriented contact with S_1 and S_2.*

The above Corollary 5.1.6 is immediate from Theorem 5.1.5, but we can also interpret the result in terms of Möbius geometry as in the next exercise.

Exercise 5.1.7 Recall from Lemma 2.3.28 that if $\mathcal{S} \in \mathbb{S}^{3,1}$ is a sphere that includes the point $Y \in L^4$, then $\mathcal{S}_t = \mathcal{S} + tY \in \mathbb{S}^{3,1}$ for any t represents a sphere tangent to $\mathcal{S}$ at point Y, sharing the same orientation. Denoting the lifts of $\mathcal{S}$ and Y into Lie sphere geometry as S_1 and S_2, respectively, show that for the lift S_t of $\mathcal{S}_t$, we have $S_t \in \mathrm{span}\{S_1, S_2\}$.

Hence, in Lie sphere geometry, the null (or isotropic) 2-planes represent the *(parabolic) pencil of spheres* that are all tangent at one point, sharing the same orientation. Let $\mathcal{Z}$ be the Grassmannian of null 2-planes, so that for any $\Lambda \in \mathcal{Z}$, one can find linearly independent $S_1, S_2 \in L^5$ such that $\Lambda = \mathrm{span}\{S_1, S_2\}$.

Definition 5.1.8 Let $\Lambda \in \mathcal{Z}$ be a 2-dimensional null subspace, which projectivizes to a line in $P(L^5)$. Then such Λ determines a pencil of spheres all tangent at one point, called a *contact element*.

Remark 5.1.9 We note here that $\mathcal{Z}$ is a contact manifold. For more information, we refer the readers to [65, § 4.1].

5.2   Lie Sphere Transformations

As opposed to Möbius transformations, which map point spheres to point spheres, Lie sphere transformations map any types of spheres to any types of spheres, so that they preserve the oriented contact between the spheres. In the current set up, the group of Lie sphere transformations can be represented by the orthogonal transformations $O_{4,2}$ acting on the light cone L^5 via left matrix multiplication. In fact, the group of Lie sphere transformations is isomorphic to $O_{4,2}/\{\pm id\}$. Here, we give some examples of Lie sphere transformations.

Example 5.2.10   Let $A_1 \in O_{4,1}$. Then we have

$$A = \begin{pmatrix} A_1 & 0 \\ 0 & 1 \end{pmatrix} \in O_{4,2}$$

is a Lie sphere transformation. Thus any Möbius transformation is a Lie sphere transformation. Furthermore, any Lie sphere transformations that preserve the point sphere complex $\mathfrak{p}$ (so that $A\mathfrak{p} = \mathfrak{p}$) are Möbius transformations of Möbius geometry $\mathrm{span}\{\mathfrak{p}\}^{\perp}$. For this reason, we see that Möbius geometry is a subgeometry of Lie sphere geometry.

Exercise 5.2.11   Show that X projects to a point sphere in $\mathrm{span}\{\mathfrak{p}\}^{\perp}$ if and only if AX projects to a point sphere in $\mathrm{span}\{\mathfrak{p}\}^{\perp}$ for A as in Example 5.2.10.

The next example is a Lie sphere transformation that does not reduce to a Möbius transformation, because it takes point spheres to true spheres, and takes some true spheres to point spheres as well.

Example 5.2.12   Set

$$A = \begin{pmatrix} 1 - \frac{1}{2}s^2 & 0 & 0 & 0 & -\frac{1}{2}s^2 & -s \\ 0 & 1 & 0 & 0 & 0 & 0 \\ 0 & 0 & 1 & 0 & 0 & 0 \\ 0 & 0 & 0 & 1 & 0 & 0 \\ \frac{1}{2}s^2 & 0 & 0 & 0 & 1 + \frac{1}{2}s^2 & s \\ s & 0 & 0 & 0 & s & 1 \end{pmatrix} \in O_{4,2} \ . \tag{5.3}$$

For a sphere S that projects to $\mathbb{R}^3$ as a sphere with center c and radius r, we have seen that

$$S = \left(\tfrac{1}{2}(1 + |c|^2 - r^2), c, \tfrac{1}{2}(1 - |c|^2 + r^2), r \right)^t .$$

Then, one can calculate directly that

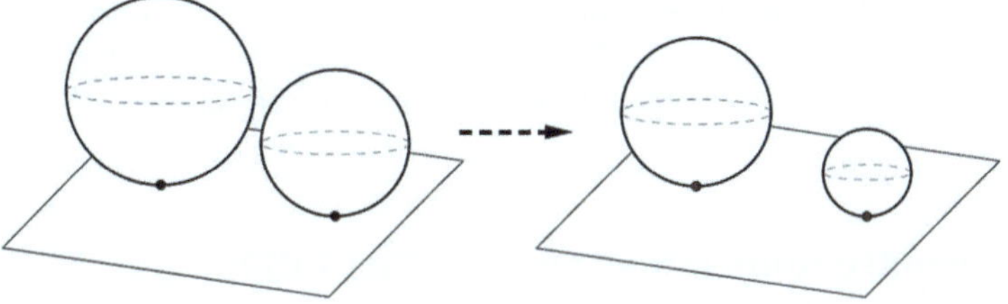

Fig. 5.1 A Laguerre transformation

$$AS = \left(\tfrac{1}{2}(1 + |c|^2 - (r + s)^2), c, \tfrac{1}{2}(1 - |c|^2 + (r + s)^2), r + s\right)^t,$$

and A is the transformation mapping spheres of radius r to spheres of radius $r + s$, while fixing the center. Thus spheres of radius $r = -s$ are mapped to point spheres by the Lie sphere transformation A.

In fact, the Lie sphere transformation introduced in Example 5.2.12 satisfies $A\mathfrak{q}_0 = \mathfrak{q}_0$ for $\mathfrak{q}_0 = (1, 0, 0, 0, -1, 0)^t$, which is the space form vector for $M_0 \cong \mathbb{R}^3$ telling us where the point at infinity is. This transformation is an example of a Lie sphere transformation having $\mathfrak{q}_0$ as an eigenvector; such transformations are called *Laguerre transformations*. Since Laguerre transformations preserve the point at infinity for $\mathbb{R}^3$, point spheres and true spheres are mapped to point spheres and true spheres in $\mathbb{R}^3$, while planes are mapped to planes.[1] Just as Möbius transformations can be treated with a choice of the point sphere complex $\mathfrak{p}$, Laguerre transformations can be obtained after a choice of a null space form vector $\mathfrak{q}$.

A configuration consisting of two spheres in oriented contact with one plane will be mapped by a Laguerre transformation to a configuration of the same type in that $\mathbb{R}^3$ (see Fig. 5.1). If the two spheres are in oriented contact with the plane at two points p and q, then the tangential distance between the two spheres is the distance between p and q inside the plane. One can show that Laguerre transformation preserves the ratio of the tangential distance between the spheres (in oriented contact with a plane). For more information on Laguerre transformations, see [65, Section 3.4].

In geometric terms, Lie sphere transformations preserve oriented tangential contact between spheres; Möbius transformations are Lie sphere transformations that further preserve the angle of intersection between spheres (where the angle 0 and π are viewed separately), while Laguerre transformations Lie sphere transformations that further preserve the ratio of tangential distances between spheres. In fact, we have the following theorem:

---

[1] Recall that planes are spheres including the point at infinity.

Fact (cf. [65, Theorem 3.16]) The group of Lie sphere transformations is equal to the union of the group of Möbius transformations and group of Laguerre transformations, and subsequent compositions of transformations.

5.3   Legendre Immersions and Contact Lifts

In this section, we examine the relationship between surfaces in space forms and maps into the set of contact elements in detail.

Definition 5.3.13 Let $\Lambda : \Sigma \rightarrow \mathcal{Z}$ be a smooth map into the set of contact elements. We call Λ *Legendre* if it satisfies the contact condition:

$$\mathrm{d}S_1 \perp S_2$$

for any two sections $S_1, S_2 \in \Gamma\Lambda$. A Legendre map further satisfying the immersion condition on any $p \in \Sigma$,

$$\mathrm{d}S(\mathbf{v}) \in \Lambda(p) \text{ for all sections } S \in \Gamma\Lambda \text{ implies } \mathbf{v} = \mathbf{0},$$

for any choice of $\mathbf{v} \in T_p\Sigma$, is called a *Legendre immersion*.

Remark 5.3.14 The immersion condition in Definition 5.3.13 can be restated in terms of a basis $S_1, S_2 \in \Gamma\Lambda$, that is, $\Lambda = \mathrm{span}\{S_1, S_2\}$, as

$$\mathrm{d}S_1(\mathbf{v}), \mathrm{d}S_2(\mathbf{v}) \in \Lambda(p) \text{ implies } \mathbf{v} = \mathbf{0}$$

for all $\mathbf{v} \in T_p\Sigma$, and one can then check that this condition is independent of the choice of basis S_1, S_2.

Now, we have seen that in Lie sphere geometry, we do not distinguish point spheres from any other types of spheres, and only after the choice of a point sphere complex and the subsequent drop into Möbius geometry do we separate point spheres from other types of spheres. Therefore, we use the concept of Legendre immersions to lift surfaces into Lie sphere geometry.

Let us explain this process in more detail. For a surface x over Σ in some space form of constant sectional curvature $\mathfrak{R}_\kappa$ denote by n the unit normal of the surface in the space form. Then recall from Sect. 2.6 that we can lift x and n as $X : \Sigma \rightarrow M_\kappa$ and $\mathcal{T}_n : \Sigma \rightarrow \mathbb{S}^{3,1}$, respectively, into Möbius geometry where $\mathcal{T}_n$ represents the tangent geodesic plane congruence of x. Note that we have $\langle \mathrm{d}X, \mathcal{T}_n \rangle = 0$.

Therefore, we can lift these maps further into Lie sphere geometry $\mathbb{R}^{4,2}$ as $X, N : \Sigma \rightarrow L^5$, respectively, where $N = \mathcal{T}_n + \mathfrak{p}$, and we see that $(X, N) = 0$ since they are in (oriented) contact. Hence, $\mathrm{span}\{X, N\}$ at each point of the domain is a null 2-plane through the origin, allowing us to define a map into the set of contact elements, $\Lambda : \Sigma \rightarrow \mathcal{Z}$, via

$$\Lambda := \operatorname{span}\{X, N\}.$$

We treat Λ as a null 2-dimensional subbundle of the trivial bundle $\Sigma \times \mathbb{R}^{4,2}$, or as lines in the Lie quadric $P(L^5)$.

Then we know that $(dX, N) = 0$, implying $(X, dN) = 0$. Now for any sections $S_1, S_2 \in \Gamma\Lambda$, let $a_1, b_1, a_2, b_2 : \Sigma \to \mathbb{R}$ such that

$$S_1 = a_1 X + b_1 N, \quad S_2 = a_2 X + b_2 N.$$

Then, it is straightforward to check that $(dS_1, S_2) = 0$, so that Λ satisfies the contact condition. For this reason, we call such Λ the *Legendre lift* or *contact lift* of x.

On the other hand, one can also reverse the process, and recover the surface (in Möbius geometry) from a given Legendre map $\Lambda : \Sigma \to \mathcal{Z}$ after a choice of a (timelike) point sphere complex $\mathfrak{p}$. Explicitly, we find a 1-dimensional subbundle $L : \Sigma \to P(L^5)$ of Λ (so that $L \subset \Lambda$ in $P(L^5)$) such that $(X, \mathfrak{p}) = 0$ for any section $X \in \Gamma L$. For such L, we then have that

$$L : \Sigma \to P(L^5 \cap \operatorname{span}\{\mathfrak{p}\}^\perp) \cong P(L^4),$$

allowing us to view L as a map into the conformal 3-sphere. One can then use Möbius geometry to further project L into a space form M_κ using a space form vector $\mathfrak{q} \in \operatorname{span}\{\mathfrak{p}\}^\perp$, that is, obtain a section $X \in \Gamma L$ such that $(X, \mathfrak{q}) = -1$. In short, we recover the surface $X : \Sigma \to M_\kappa$ from a given Legendre map $\Lambda : \Sigma \to \mathcal{Z}$ by requiring that $X \in \Gamma\Lambda$ satisfies $(X, \mathfrak{p}) = 0$ and $(X, \mathfrak{q}) = -1$ for a point sphere complex $\mathfrak{p}$ and space form vector $\mathfrak{q}$ with $(\mathfrak{q}, \mathfrak{q}) = -\kappa$, $(\mathfrak{p}, \mathfrak{q}) = 0$. Similarly, one can recover the normal $\mathcal{T}_n : \Sigma \to T_X M_\kappa$ by projecting $N \in \Gamma\Lambda$ satisfying $(N, \mathfrak{p}) = -1$ and $(N, \mathfrak{q}) = 0$ to $\operatorname{span}\{\mathfrak{p}\}^\perp$, that is, $\mathcal{T}_n = N - \mathfrak{p}$. Such sections X and N are called the *point sphere map* and *tangent (geodesic) plane congruence* with respect to the point sphere complex $\mathfrak{p}$ and space form vector $\mathfrak{q}$, respectively.

To examine when the Legendre lifts are immersed, and hence becomes Legendre immersions, we first note the following result, left as an exercise.

Proposition 5.3.15 *Let $x : \Sigma \to \mathbb{R}^3$ be an immersion with unit normal $n : \Sigma \to \mathbb{S}^2$. Show that the Legendre lift $\Lambda : \Sigma \to \mathcal{Z}$ of x is a Legendre immersion, i.e. Λ satisfies the immersion condition.*

Exercise 5.3.16 Prove Proposition 5.3.15.

Proposition 5.3.15 shows that the surface in a space form being immersed is sufficient for its contact lift to be a Legendre immersion; however, it is not necessary, as seen in the next examples which have non-immersed points:

Exercise 5.3.17 Let $x : \Sigma = \mathbb{R}^2 \to \mathbb{R}^3$ be a surface defined by

$$x(u, v) = (u, v^2, v^3).$$

(1) Show that x fails to immerse along $v = 0$, that is, there is some non-zero $\mathbf{v} \in T_{(u,v)}\Sigma$ such that $dx(\mathbf{v}) = 0$ when $v = 0$.
(2) Show that $n : \Sigma \to \mathbb{S}^2$ given by

$$n(u, v) = \frac{1}{\sqrt{4 + 9v^2}}(0, -3v, 2)$$

satisfies $n \perp dx$.
(3) Denoting the Legendre lift of x by Λ, show that Λ is a Legendre immersion. (Hint: only check when $v = 0$ since x immerses everywhere else.)

On the other hand, there are examples whose contact lifts fail the immersion condition, as the next pair of exercises will show:

Exercise 5.3.18 Let $x : \Sigma = \mathbb{R}^2 \to \mathbb{R}^3$ be a surface defined by

$$x(u, v) = (u, v^2, 0).$$

(1) Show that x fails to immerse along $v = 0$.
(2) Show that $n : \Sigma \to \mathbb{S}^2$ given by

$$n(u, v) = (0, 0, 1)$$

satisfies $n \perp dx$.
(3) Show that the Legendre lift Λ of x fails the immersion condition along $v = 0$.

Exercise 5.3.19 Let $x : \Sigma = \mathbb{R}^2 \to \mathbb{R}^3$ be a surface defined by

$$x(u, v) = (u, v^2, v^5).$$

(1) Show that x fails to immerse along $v = 0$.
(2) Show that $n : \Sigma \to \mathbb{S}^2$ given by

$$n(u, v) = \frac{1}{\sqrt{4 + 25v^6}}(0, -5v^3, 2)$$

satisfies $n \perp dx$.
(3) For the Legendre lift Λ of x, show that Λ fails the immersion condition along $v = 0$.

To see what the Legendre maps and Legendre immersions project to in space forms, we recall the following definitions (see, for example, [4, 5, 104, 212]):

Definition 5.3.20 Let $x : \Sigma \to \mathfrak{R}_\kappa$ be a mapping, and n be unit length with respect to its ambient space. If we have that

$$n \perp \mathrm{d}x,$$

then such x is called a *frontal* and n its *unit normal vector field*. Furthermore, if x and n considered together form an immersion, that is, at each point in the domain,

$$\mathrm{d}x(\mathbf{v}) = \mathrm{d}n(\mathbf{v}) = \mathbf{0} \text{ implies } \mathbf{v} = \mathbf{0},$$

then such x is called a *front*, or *wave front*.

It is then easy to check that the Legendre maps correspond to frontals, while the Legendre immersions correspond to fronts under space form projections:

Exercise 5.3.21 Let $x : \Sigma \to \mathbb{R}^3$ be a front and $n : \Sigma \to \mathbb{S}^2$ its unit normal vector field. Show that the contact lift Λ of x is a Legendre immersion.

Remark 5.3.22 The Examples 5.3.17, 5.3.18, 5.3.19 all contain points at which the mapping fails to immerse; such points are called *singularities*. The singularities appearing in these examples are respectively called *cuspidal edges*, *fold singularities*, and *5/2-cuspidal edges*. Cuspidal edges are well-known examples of *front singularities* while fold singularities and *5/2*-cuspidal edges are well-known examples of *frontal singularities*. For more information, we refer the readers to works such as [4, 193, 210, 212, 219].

We finish this section by treating Lie sphere transformations of a surface. Starting with some surface x and its unit normal n in some space form $\mathfrak{R}_\kappa$, let $\Lambda : \Sigma \to \mathcal{Z}$ be its Legendre lift. Since any Lie sphere transformation $A \in O_{4,2}$ is an isometry of $\mathbb{R}^{4,2}$, an application of a Lie sphere transformation A to such Λ will result in another Legendre map $A\Lambda : \Sigma \to \mathcal{Z}$. Hence, one can project $A\Lambda$ to the same space form $\mathfrak{R}_\kappa$, obtaining another surface $\tilde{x}$. Such $\tilde{x}$ is called a *Lie sphere transformation* of x.

In the next lemma, we show that the parallel surfaces in $\mathbb{H}^3$ amount to Lie sphere transformations.

Lemma 5.3.23 *Let* $x : \Sigma \to \mathbb{H}^3$ *be a surface with unit normal n. A parallel surface* $\tilde{x} : \Sigma \to \mathbb{H}^3$ *of x is a Lie sphere transformation of x.*

Proof Noting that the parallel surface $\tilde{x}$ can be written as

$$\tilde{x} = x \cosh \phi + n \sinh \phi$$

with its unit normal

$$\tilde{n} = x \sinh \phi + n \cosh \phi,$$

denote the lifts of $\tilde{x}$ and $\tilde{n}$ by $\tilde{X}$ and $\tilde{N}$, respectively, with respect to the specific choices of $\mathfrak{p}$ and $\mathfrak{q} = \mathfrak{q}_{-1}$, and let $\tilde{\Lambda} = \mathrm{span}\{\tilde{X}, \tilde{N}\}$. For a Lie sphere transformation,

$$A = \begin{pmatrix} I_4 & 0 & 0 \\ 0 & \cosh\phi & \sinh\phi \\ 0 & \sinh\phi & \cosh\phi \end{pmatrix} \in O_{4,2},$$

consider $\Lambda := A\tilde{\Lambda} : \Sigma \to \mathcal{Z}$. Then, writing $c := \cosh\phi$ and $s := \sinh\phi$ for brevity, we have at any $(u, v) \in \Sigma$,

$$\Lambda(u, v) = A\left(\left\{ a \begin{pmatrix} (xc + ns)^t \\ 1 \\ 0 \end{pmatrix} + b \begin{pmatrix} (xs + nc)^t \\ 0 \\ 1 \end{pmatrix} \,\middle|\, a, b \in \mathbb{R} \right\} \right)$$

$$= \left\{ (ac + bs) \begin{pmatrix} x^t \\ 1 \\ 0 \end{pmatrix} + (as + bc) \begin{pmatrix} n^t \\ 0 \\ 1 \end{pmatrix} \,\middle|\, a, b \in \mathbb{R} \right\}.$$

Therefore, we see that $\Lambda = \mathrm{span}\{X, N\}$ where X and N are the lifts of x and n, respectively. $\qquad\square$

Remark 5.3.24 The Lie sphere transformations A giving parallel surfaces as in Lemma 5.3.23 fix the 2-plane $\mathrm{span}\{\mathfrak{p}, \mathfrak{q}\}$, while act as the identity on $\mathrm{span}\{\mathfrak{p}, \mathfrak{q}\}^\perp$.

In the next pair of exercises, we will obtain the analogous result to Lemma 5.3.23 for $\mathbb{R}^3$ and $\mathbb{S}^3$.

Exercise 5.3.25 Let $x : \Sigma \to \mathbb{R}^3$ be a surface with unit normal $n : \Sigma \to \mathbb{S}^2$, and denote the lifts of x and n as X and N, respectively, with respect to the specific choices of $\mathfrak{p} = (0, 0, 0, 0, 0, 1)^t$ and $\mathfrak{q} = \mathfrak{q}_0 = (1, 0, 0, 0, -1, 0)^t$. For $\Lambda = \mathrm{span}\{X, N\}$, let $\tilde{\Lambda} = A\Lambda$ for the Lie sphere transformation $A \in O_{4,2}$ given in (5.3) so that $A\mathfrak{q} = \mathfrak{q}$.

(1) Show that $A\mathfrak{p} = \mathfrak{p} - s\mathfrak{q}$, and check that Remark 5.3.24 also applies to this case, i.e., A fixes the 2-plane $\mathrm{span}\{\mathfrak{p}, \mathfrak{q}\}$.
(2) Calculate that $(AX, \mathfrak{q}) = -1$, $(AX, \mathfrak{p}) = -s$, $(AN, \mathfrak{q}) = 0$, and $(AN, \mathfrak{p}) = -1$.
(3) Show that the point sphere map $\tilde{X}$ of $\tilde{\Lambda}$ is given by $\tilde{X} = AX - sAN$, and the tangent plane congruence $\tilde{N}$ of $\tilde{\Lambda}$ by $\tilde{N} = AN$.
(4) Denoting by $\tilde{x}$ the projection of $\tilde{X}$ to $\mathbb{R}^3$, conclude that $\tilde{x} = x - sn$ with unit normal $\tilde{n} = n$, i.e. $\tilde{x}$ is a parallel surface of x.

Exercise 5.3.26 Find the Lie sphere transformations $A \in O_{4,2}$ that amount to generating parallel surfaces in $\mathbb{S}^3$.

In fact, one can show that taking the parallel surfaces, along with Möbius transformations, generate all Lie sphere transformations:

Fact ([65, Theorem 3.18]) All Lie sphere transformations are generated (by composition) from Möbius transformations and parallel surfaces in $\mathbb{R}^3$, $\mathbb{S}^3$ and $\mathbb{H}^3$.

5.4 Sphere Congruences Enveloping a Surface

Under the Lie sphere geometric setup, sphere congruences are smooth maps $s : \Sigma \to P(L^5)$ into the Lie quadric. Sphere congruences enveloping a surface (with aligning orientations) admit a simple characterization in Lie sphere geometry:

Proposition 5.4.27 *Let $\Lambda : \Sigma \to \mathcal{Z}$ be Legendre with any point sphere map X with respect to any choice of point sphere complex and space form vector. Then $s : \Sigma \to P(L^5)$ projects to a sphere congruence enveloping X if and only if s is a subbundle of Λ.*

Proposition 5.4.27 implies that the notion of enveloped sphere congruence is purely a Lie sphere geometric notion; thus, we will often refer to s as a *sphere congruence enveloping Λ*.

Corollary 5.4.28 *Let $s : \Sigma \to P(L^5)$ be a sphere congruence enveloping $\Lambda : \Sigma \to \mathcal{Z}$. Suppose $X, N \in \Gamma\Lambda$ are the point sphere map and tangent plane congruence of Λ, respectively, so that $\Lambda = \mathrm{span}\{X, N\}$. If $S \in \Gamma s$, then $S = \alpha X + \beta N$ for some real functions $\alpha, \beta : \Sigma \to \mathbb{R}$.*

Now we examine the curvature spheres of a given surface in the realm of Lie sphere geometry. Recall that from (2.34) of Exercise 2.6.55, $X : \Sigma \to M_\kappa \subset L^4 \subset L^5$ with tangent geodesic plane congruence $\mathcal{T}_n : \Sigma \to \mathbb{S}^{3,1}$ satisfies the Rodrigues' equation

$$(\mathcal{T}_n)_i = -k_i X_i,$$

where $\partial_1 = \partial_u$ and $\partial_2 = \partial_v$ for curvature line coordinates $(u, v) \in \Sigma$. Letting $N = \mathcal{T}_n + \mathfrak{p} : \Sigma \to L^5$, it is easy to check that the Rodrigues' equation still holds in Lie sphere geometry:

$$N_i = (\mathcal{T}_n)_i = -k_i X_i. \tag{5.4}$$

Furthermore, from Exercise 2.7.64, we have that curvature sphere congruences $s_i = \mathrm{span}\{K_i\}$ are given by

$$K_i := k_i X + N.$$

However, the next lemma shows that the notion of curvature spheres is independent of the choice of space forms, i.e. the notion of curvature spheres is purely Lie sphere geometric.

Lemma 5.4.29 *Denote by* $X, N \in \Gamma\Lambda$ *any point sphere map and tangent plane congruence of a Legendre immersion* Λ. *Then* $s = \mathrm{span}\{kX + N\}$ *is a curvature sphere congruence in some principal curvature direction if and only if there exists a direction* $\mathbf{v}$ *such that for any* $K \in \Gamma s$, *we have* $\partial_{\mathbf{v}} K \in \Gamma\Lambda$, *and then* $\mathbf{v}$ *will be the principal curvature direction for the principal curvature* k.

Proof For some function k and some directional derivative $\partial_{\mathbf{v}}$, assume that

$$\partial_{\mathbf{v}}(kX + N) = (\partial_{\mathbf{v}}k)X + k\partial_{\mathbf{v}}X + \partial_{\mathbf{v}}N \in \Gamma\Lambda.$$

Then one can deduce that

$$k\partial_{\mathbf{v}}X + \partial_{\mathbf{v}}N = 0,$$

so that the Rodrigues' equation holds, i.e. k is the principal curvature with principal curvature direction $\partial_{\mathbf{v}}$. Conversely, we have $\partial_i K_i = \partial_i k_i \ X \in \Gamma\Lambda$. $\qquad\square$

Remark 5.4.30 As curvature spheres are a Lie sphere geometric notion, they are preserved under Lie sphere transformations.

Exercise 5.4.31 For curvature line coordinates $(u, v) \in \Sigma$, use the Rodrigues' equations (5.4) and the Codazzi equations (2.35) to show that

$$X_{uv} = \frac{k_{1,v}}{k_2 - k_1}X_u + \frac{k_{2,u}}{k_1 - k_2}X_v, \quad N_{uv} = \frac{k_{1,v}k_2}{k_1 - k_2}X_u + \frac{k_{2,u}k_1}{k_2 - k_1}X_v.$$

Exercise 5.4.32 Given curvature sphere congruences $s_i := \mathrm{span}\{K_i\} : \Sigma \to P(L^5)$ for $i = 1, 2$, show that for any section $S_i \in \Gamma s_i$, the principal curvatures of X in the space form defined by the point sphere complex $\mathfrak{p}$ and space form vector $\mathfrak{q}$ can be expressed as

$$k_i = \frac{(S_i, \mathfrak{q})}{(S_i, \mathfrak{p})}.$$

Remark 5.4.33 A Legendre immersion has an umbilic point at a domain point (u_0, v_0) (umbilic points on Legendre immersions are the points where the two curvature spheres coincide) if and only if its projection to a space form has an umbilic at (u_0, v_0). We assume throughout the text that Legendre immersions are umbilic-free on Σ, unless otherwise noted.

5.5 Dupin Cyclides

Dupin cyclides [97] are the simplest Legendre immersions in Lie sphere geometry, after spheres, and can be constructed as follows:

1. Split $\mathbb{R}^{4,2}$ into two fixed orthogonal $V_1 \cong \mathbb{R}^{2,1}$ and $V_2 \cong \mathbb{R}^{2,1}$ so that each V_j has a light cone for $j = 1, 2$.
2. Denoting the light cone of V_j by L_j^2, we can view each projectivized light cone as a circle, and consider $s_j : \mathbb{S}^1 \to P(L_j^2) \subset P(L^5)$.
3. Since $(S_1, S_2) = 0$ for $S_i \in \varGamma s_i$, we have that $\varLambda := \mathrm{span}\{S_1, S_2\} : \mathbb{S}^1 \times \mathbb{S}^1 \to \mathcal{Z}$. We call such $\varLambda$ a *Dupin cyclide*.

For a Dupin cyclide $\varLambda = \mathrm{span}\{S_1, S_2\} : \mathbb{S}^1 \times \mathbb{S}^1 \to \mathcal{Z}$, assume that $X_1 \in \varGamma S_1$ is parametrized by u while $X_2 \in \varGamma S_2$ is parametrized by v. For some choice of a point sphere complex $\mathfrak{p}$ and space form vector $\mathfrak{q}$, let X and N denote the point sphere map and tangent plane congruence of $\varLambda$, respectively. At any fixed v_0, we then have $(X(u, v_0), X_2(v_0)) = 0$ so that X_2 projects to a 1-parameter family of spheres tangent to the surface, that is, X_2 envelops X. Similarly, we can see that X_1 also projects to a 1-parameter family of spheres enveloping X.

Therefore, Dupin cyclides are channel surfaces in two ways:

Definition 5.5.34 A surface is called a *channel surface* if it envelops a 1-parameter family of spheres.

Furthermore, since we have that

$$(X_1)_v = 0 \in \varGamma\varLambda \quad \text{and} \quad (X_2)_u = 0 \in \varGamma\varLambda,$$

Lemma 5.4.29 implies that $X_1 \in \varGamma S_1$ represents the curvature sphere congruence corresponding to the v-direction, while $X_2 \in \varGamma S_2$ represents the curvature sphere congruence corresponding to the u-direction.

Then without loss of generality, let

$$X_1 = K_2 = N + k_2 X \quad \text{and} \quad X_2 = K_1 = N + k_1 X$$

for principal curvatures k_1 and k_2. Using the Rodrigues' equation,

$$0 = (X_1)_v = N_v + k_2 X_v + (k_2)_v X = (k_2)_v X,$$

so that k_2 is constant in the corresponding curvature direction v. Similarly, one can show that k_1 is constant in the corresponding curvature direction u.

In fact, Dupin cyclides can be characterized via the following properties:

1. Principal curvatures along corresponding lines of curvature are constant.
2. Principal curvature spheres along lines of curvature are constant.
3. All lines of curvature are circular arcs.

Exercise 5.5.35 Let $\Lambda : \Sigma \to \mathcal{Z}$ project to a channel surface in some space form, with coordinates $(u, v) \in \Sigma$. Without loss of generality, assume that v parametrizes the enveloping 1-parameter family of spheres.

(1) Show that if the enveloping 1-parameter family of spheres is represented by $X_2 : I \to L^5$ for some interval on the reals I parametrized by v, then $X_2 \in \Gamma \Lambda$.
(2) Use Lemma 5.4.29 to show that X_2 represents the curvature sphere congruence corresponding to the u-direction.
(3) Deduce that if k_1 is the principal curvature corresponding to the u-direction, then k_1 is constant along the u-curvature lines.

The construction method of Dupin cyclides implies that a Dupin cyclide is essentially determined by the orthogonal splitting of $\mathbb{R}^{4,2} = V_1 \oplus_\perp V_2$, and any other splitting would differ by an orthogonal transformation of $\mathbb{R}^{4,2}$. Thus we can deduce the following result:

Theorem 5.5.36 ([14, §55]) *Up to the freedom of Lie sphere transformations, there is only one Dupin cyclide. In other words, all Dupin cyclides are Lie sphere equivalent.*

Example 5.5.37 Consider the Clifford torus

$$x(u, v) = \left(\cos v(\sqrt{2} + \cos u), \sin v(\sqrt{2} + \cos u), \sin u\right), \tag{5.5}$$

considered in $\mathbb{R}^3$. The surface envelops two 1-parameter families of spheres, one family whose centers are given by $(\sqrt{2}\cos v, \sqrt{2}\sin v, 0)^t$ with radius 1, and the other family with centers given by $(0, 0, -\sqrt{2}\tan u)^t$ and radius $1 + \frac{\sqrt{2}}{\cos u}$. Therefore, one has that

$$X_1(u) = \left(-1, 0, 0, \frac{-\sqrt{2}\sin u}{\cos u + \sqrt{2}}, \frac{2\cos u + \sqrt{2}}{\cos u + \sqrt{2}}, 1\right)^t$$

while

$$X_2(v) = \left(1, \sqrt{2}\cos v, \sqrt{2}\sin v, 0, 0, 1\right)^t.$$

Furthermore, we have that $(X_1, X_2) = 0$ so that the induced Legendre immersion Λ is given by

$$\Lambda = \mathrm{span}\{X_1, X_2\}.$$

Now

$$X_1(u) \in V_1 := \left\{ \begin{pmatrix} -A \\ 0 \\ 0 \\ B \\ C \\ A \end{pmatrix} \;\middle|\; A, B, C \in \mathbb{R} \right\},$$

$$X_2(v) \in V_2 := \left\{ \begin{pmatrix} A \\ B \\ C \\ 0 \\ 0 \\ A \end{pmatrix} \;\middle|\; A, B, C \in \mathbb{R} \right\},$$

while $V_1 \cong \mathbb{R}^{2,1}$, $V_2 \cong \mathbb{R}^{2,1}$ and $V_1 \perp V_2$. Then, for any $A \in O_{4,2}$, the projection of $A\Lambda : \Sigma \to \mathcal{Z}$ to $\mathbb{R}^3$ gives any arbitrary Dupin cyclide in $\mathbb{R}^3$ by Theorem 5.5.36.

5.6 Lie Cyclides

Assume that a surface x in some space form determined by the point sphere complex $\mathfrak{p}$ and space form vector $\mathfrak{q}$ has no umbilics, i.e. $k_1 \neq k_2$ everywhere. Thus the curvature spheres K_i can be used to determine the contact lift $\Lambda = \mathrm{span}\{K_1, K_2\}$. Set

$$V = \mathrm{span}\{K_1, \partial_2 K_1, \partial_2^2 K_1\}, \quad W = \mathrm{span}\{K_2, \partial_1 K_2, \partial_1^2 K_2\}.$$

Proposition 5.6.38 ([14, §84]) *At each point in the domain, V and W are subspaces orthogonal to each other with $(++-)$ signature.*

Proof First, from the contact condition of Λ and the fact that K_i is a curvature sphere congruence, we see that $dK_i \perp \Lambda$ while $\partial_i K_i \in \Gamma\Lambda$. Hence, let a_i and b_i be scalar functions so that

$$\partial_i K_i = a_i K_1 + b_i K_2.$$

Then we can deduce that

- $(\partial_2 K_1, K_2) = -(K_1, \partial_2 K_2) = 0$, so that
- $(\partial_2^2 K_1, K_2) = -(\partial_2 K_1, \partial_2 K_2) = 0$.

Also, for $j = 1, 2$, we have

$$(K_j, \partial_2 \partial_1 K_1) = (K_j, \partial_2 (a_1 K_1 + b_1 K_2))$$
$$= a_1 (K_j, \partial_2 K_1) + b_1 (K_j, \partial_2 K_2) = 0.$$

Similarly, we can further deduce the following facts:

- $(\partial_2 K_1, \partial_j K_j) = 0$.
- $(\partial_1 K_2, \partial_2 K_1) = \partial_1(K_2, \partial_2 K_1) - (K_2, \partial_1 \partial_2 K_2) = -(K_2, \partial_2 \partial_1 K_1) = 0$.
- $(\partial_2^2 K_1, \partial_1 K_2) = \partial_2(\partial_2 K_1, \partial_1 K_2) - (\partial_2 K_1, \partial_2 \partial_1 K_2)$, and then

$$(\partial_2^2 K_1, \partial_1 K_2) = -(\partial_2 K_1, \partial_1(a_2 K_1 + b_2 K_2))$$
$$= -a_2(\partial_2 K_1, \partial_1 K_1) - b_2(\partial_2 K_1, \partial_1 K_2) = 0.$$

- $(\partial_2^2 K_1, \partial_1^2 K_2) = \partial_2(\partial_2 K_1, \partial_1^2 K_2) - (\partial_2 K_1, \partial_2 \partial_1^2 K_2) = -(\partial_2 K_1, \partial_1^2 \partial_2 K_2)$, so that

$$(\partial_2^2 K_1, \partial_1^2 K_2) = -(\partial_2 K_1, \partial_1^2(a_2 K_1 + b_2 K_2))$$
$$= -\partial_1(\partial_2 K_1, \partial_1(a_2 K_1 + b_2 K_2)) + (\partial_2 \partial_1 K_1, \partial_1(a_2 K_1 + b_2 K_2))$$
$$= -\partial_1(a_2(\partial_2 K_1, \partial_1 K_1) + b_2(\partial_2 K_1, \partial_1 K_2))$$
$$+ (\partial_2(a_1 K_1 + b_1 K_2), \partial_1(a_2 K_1 + b_2 K_2))$$
$$= (\partial_2(a_1 K_1 + b_1 K_2), \partial_1(a_2 K_1 + b_2 K_2))$$
$$= (a_1 \partial_2 K_1 + b_1 \partial_2 K_2, a_2 \partial_1 K_1 + b_2 \partial_1 K_2) = 0.$$

- $\Lambda \subset \{\partial_2 K_1\}^{\perp}$ and Λ is 2-dimensional and totally null, so the metric of $\mathbb{R}^{4,2}$ restricted to $\{\partial_2 K_1\}^{\perp}$ cannot be of signature $(+ + + + -)$. So $\partial_2 K_1$ is either spacelike or lightlike.
- We now know

$$(\partial_2 K_1, \partial_2 K_1) \geq 0,$$

but it is actually strictly positive, seen as follows: If $(\partial_2 K_1, \partial_2 K_1) = 0$, then $\partial_2 K_1 \in \Gamma \Lambda$, since the maximal dimension of a null subspace of $\mathbb{R}^{4,2}$ is two, but this implies K_1 and K_2 are parallel, contradicting the fact that we have excluded umbilics.

Using the above properties, we can show the following:

- $(K_j, K_j) = 0, \quad j = 1, 2,$
- $(K_1, \partial_2 K_1) = (K_2, \partial_1 K_2) = 0,$
- $(\partial_2 K_1, \partial_2 K_1) = a > 0,$
- $(\partial_1 K_2, \partial_1 K_2) = b > 0,$
- $(K_1, \partial_2^2 K_1) = -a,$
- $(K_2, \partial_1^2 K_2) = -b,$
- $(K_1, K_2) = (K_1, \partial_1 K_2) = (K_1, \partial_1^2 K_2) = 0,$
- $(\partial_2 K_1, K_2) = (\partial_2 K_1, \partial_1 K_2) = (\partial_2 K_1, \partial_1^2 K_2) = 0,$
- $(\partial_2^2 K_1, K_2) = (\partial_2^2 K_1, \partial_1 K_2) = (\partial_2^2 K_1, \partial_1^2 K_2) = 0.$

Here, a and b are functions (not constants).

Now, to complete the proof, note that the matrix

$$\begin{pmatrix} (K_1, K_1) & (\partial_2 K_1, K_1) & (\partial_2^2 K_1, K_1) \\ (K_1, \partial_2 K_1) & (\partial_2 K_1, \partial_2 K_1) & (\partial_2^2 K_1, \partial_2 K_1) \\ (K_1, \partial_2^2 K_1) & (\partial_2 K_1, \partial_2^2 K_1) & (\partial_2^2 K_1, \partial_2^2 K_1) \end{pmatrix} = \begin{pmatrix} 0 & 0 & -a \\ 0 & a & * \\ -a & * & * \end{pmatrix}$$

has strictly negative determinant, so V has a nondegenerate metric. Thus the metric on V has either $(+ + -)$ or $(- - -)$ signature. But $(- - -)$ signature is not possible in $\mathbb{R}^{4,2}$, so V has $(+ + -)$ signature, and similarly so does W. $\qquad\square$

Remark 5.6.39 Note that everything in Proposition 5.6.38 and its proof still holds if we replace K_1 and K_2 by any scalar multiples of them.

Remark 5.6.40 Since V and W are perpendicular, the Lie cyclide splitting of $\mathbb{R}^{4,2}$ is sometimes written as $V \oplus V^{\perp}$.

At each point, the splitting $V \oplus W$ of $\mathbb{R}^{4,2}$ by the perpendicular 3-planes V and W is called a *Lie cyclide*, giving a congruence of Lie cyclides over the surface. Furthermore, each Lie cyclide produces a Dupin cyclide (as explained at the beginning of Sect. 5.5), which makes second order contact with the surface in both principal directions. (This is somewhat akin to attaching quadratic surfaces with second order tangential contact at points of a surface in $\mathbb{R}^3$.)

Proposition 5.6.41 *The congruence of Lie cyclides along a surface is constant if and only if the surface is a Dupin cyclide.*

Proof Lemma 5.4.29 implies there exist real scalar functions a and b so that

$$\partial_1 K_1 = a K_1 + b K_2,$$

and the Lie cyclide congruence is constant along u-curvature lines, i.e.

$$\partial_1 K_1 \in \mathrm{span}\{K_1, \partial_2 K_1, \partial_2^2 K_1\},$$

if and only if $b = 0$. Equivalently,

$$K_1 \parallel \partial_1 K_1 = \partial_1 N + k_1 \partial_1 X + \partial_1 k_1 X = \partial_1 k_1 X,$$

which boils down to the fact that $\partial_1 k_1 = 0$. A symmetric argument shows that Lie cyclide congruence is constant along v-curvature lines if and only if $\partial_2 k_2 = 0$, and we have a Dupin cyclide. $\qquad\square$

In fact, one can consider a more general case and treat channel surfaces similarly:

Lemma 5.6.42 *A surface is channel if and only if the Lie cyclide congruence is constant along the one principal direction of the surface, and thus the congruence consists of only a 1-parameter family of Lie cyclides.*

Exercise 5.6.43 A surface of revolution is a channel surface, and hence its Lie cyclide congruence is also constant along the direction of the rotation. In this exercise, we verify this directly through explicit parametrization. Let $x : \Sigma \to \mathbb{R}^3$ be a surface of revolution so that

$$x(u, v) = (f(u)\cos v,\ f(u)\sin v,\ g(u)),$$

where

$$f_u^2 + g_u^2 = f^2.$$

(1) Verify that the unit normal is given by

$$n(u, v) = f^{-1}(-g_u\cos v,\ -g_u\sin v,\ f_u),$$

   and see that the principal curvatures are

$$k_1 = f^{-3}(f_u g_{uu} - f_{uu}g_u),\quad k_2 = f^{-2}g_u.$$

(2) Denoting the lift of x and n as X and N, respectively, show that $K_2 = k_2 X + N$ is independent of v.

5.7   Normal Bundle for Surfaces in $P(L^4)$

So far, we have seen that any sphere congruence defined on Σ can be described in Lie sphere geometry via

$$s : \Sigma \to P(L^5) \subset P(\mathbb{R}^{4,2}).$$

We view s both as a map into the projectivized light cone, and a 1-dimensional null line subbundle of the trivial bundle $\Sigma \times \mathbb{R}^{4,2}$. Now we wish to see when such a map s is *isothermic*; for this, we must first describe the normal bundle for surfaces in $P(L^5)$. To serve as an introduction, we will first consider the normal bundle for surfaces in $P(L^4)$, making connection with how isothermicity was defined in the context of Möbius geometry in Sects. 2.6 and 2.8.

Taking an arbitrary $\mathbf{v} \in \mathbb{R}^{4,1} \setminus \{0\}$, the projective space $P(\mathbb{R}^{4,1})$ can be separated into a disjoint union as follows:

$$P(\mathbb{R}^{4,1}) = \mathcal{U} \cup \mathcal{V},$$

$$\mathcal{V} := \{\mathrm{span}\{X\} \in P(\mathbb{R}^{4,1}) \mid \langle X, \mathbf{v}\rangle = 0\},\quad \mathcal{U} := \{\mathrm{span}\{X\} \in P(\mathbb{R}^{4,1}) \mid \langle X, \mathbf{v}\rangle \neq 0\}.$$

Then, we can identify $\mathcal{U}$ with the hyperplane

$$\mathcal{P} := \{X \in \mathbb{R}^{4,1} \mid \langle X, \mathbf{v} \rangle = -1\},$$

and $\mathcal{V}$ is equal to the set of lines at the origin parallel to the hyperplane $\mathcal{P}$. We denote this identification by $\varphi : \mathcal{U} \to \mathcal{P}$, and $(\mathcal{U}, \varphi)$ form an affine coordinate chart for $P(\mathbb{R}^{4,1})$ at any point in $\mathcal{U}$.

Let $L \in P(L^4)$, and fix any vector $\mathbf{v} \in \mathbb{R}^{4,1}$ so that $L \in \mathcal{U}$. Then we can find $X \in L^4$ with $L = \operatorname{span}\{X\}$ so that $\langle X, \mathbf{v} \rangle = -1$, i.e. $X = \varphi(L) \in \mathcal{P} \cap L^4$. Since $\mathcal{U} \cap P(L^4) \cong \mathcal{P} \cap L^4$ via φ, we have

$$\mathrm{d}\varphi(T_L P(L^4)) = T_X(\mathcal{P} \cap L^4) = \{A \in T_X \mathbb{R}^{4,1} \mid \langle A, \mathbf{v} \rangle = \langle A, X \rangle = 0\}$$

$$= \operatorname{span}\{X, \mathbf{v}\}^\perp \cong \operatorname{span}\{X\}^\perp / \operatorname{span}\{X\} = L^\perp / L,$$

where the isomorphism above is obtained by applying the homomorphism theorem to the linear map

$$\pi : \operatorname{span}\{X\}^\perp \to \operatorname{span}\{X, \mathbf{v}\}^\perp, \quad B \mapsto B + \langle B, \mathbf{v} \rangle X.$$

Let us now consider $L : \Sigma \to P(L^4)$ as a surface. Considering L as a null line subbundle of the trivial bundle $\Sigma \times \mathbb{R}^{4,1}$, let $X \in \Gamma L$ and choose $\mathbf{v} \in \mathbb{R}^{4,1}$ such that $\langle X, \mathbf{v} \rangle = -1$. Thus, we have $X : \Sigma \to \mathcal{P} \cap L^4$, equivalently, we have $X = \varphi(L)$. Since $(\mathcal{U}, \varphi)$ is a chart containing L of $P(\mathbb{R}^{4,1})$, we can identify the tangent space of the surface L at each point $p = (u, v) \in \Sigma$ with

$$T_{L(p)} L \cong \operatorname{span}\{X_u, X_v\} \cong \operatorname{span}\{X_u, X_v, X\} / \operatorname{span}\{X\},$$

where the linear spans are evaluated at $p \in \Sigma$ throughout the section. Therefore, viewing the normal space of L at each point in the domain as the subspace of

$$T_{L(p)} P(L^4) \cong \operatorname{span}\{X\}^\perp / \operatorname{span}\{X\}$$

that is orthogonal to $T_{L(p)} L \cong \operatorname{span}\{X_u, X_v, X\} / \operatorname{span}\{X\}$, we see that

$$\mathcal{N}_{L(p)} L \cong \operatorname{span}\{X_u, X_v, X\}^\perp / \operatorname{span}\{X\}$$

is a 1-dimensional subspace of the tangent space of the projective light cone.

Now, for any non-zero $\tilde{X} \in \Gamma L$, there is some λ such that $\tilde{X} = \lambda X$, implying that $\tilde{X}_u = \lambda_u X + \lambda X_u$ and $\tilde{X}_v = \lambda_v X + \lambda X_v$. Hence,

$$\mathcal{N}_{L(p)} L \cong \operatorname{span}\{\tilde{X}_u, \tilde{X}_v, \tilde{X}\}^\perp / \operatorname{span}\{\tilde{X}\} = \{Y + \operatorname{span}\{\tilde{X}\} \mid Y \perp \tilde{X}_u, \tilde{X}_v, \tilde{X}\}.$$

Therefore, we see that the conditions for L to be isothermic with respect to the coordinates (u, v) are

$$\langle X_u, X_u \rangle = \langle X_v, X_v \rangle, \quad \langle X_u, X_v \rangle = 0, \quad \langle X_{uv}, Y \rangle = 0 \qquad (5.6)$$

for any $Y + \mathrm{span}\{X\} \in \mathcal{N}_{L(p)}L$, or equivalently,

$$\langle \tilde{X}_u, \tilde{X}_u \rangle = \langle \tilde{X}_v, \tilde{X}_v \rangle, \quad \langle \tilde{X}_u, \tilde{X}_v \rangle = 0, \quad \langle \tilde{X}_{uv}, Y \rangle = 0$$

since $\tilde{X}_{uv} = \lambda_{uv} X + \lambda_u X_v + \lambda_v X_u + \lambda X_{uv}$.

In fact, we have seen such treatment of isothermicity in Sect. 2.6: recall that for a space form vector $\mathfrak{q}_\kappa$, we have $X_\kappa \in \Gamma L$ with $X_\kappa : \Sigma \to M_\kappa$ where

$$M_\kappa = \{X \in L^4 \mid \langle X, \mathfrak{q}_\kappa \rangle = -1\}.$$

Letting $\mathbf{v} = \mathfrak{q}_\kappa$, we see that now $\mathcal{P} \cap L^4 = M_\kappa$. Therefore, we have that $T_X(\mathcal{P} \cap L^4) = T_X M_\kappa = \{\mathcal{T}_a \mid a \in \mathbb{R}^3\}$. Hence, it is not difficult to see that $\mathcal{T}_n + \mathrm{span}\{X\} \in \mathcal{N}_{L(p)}L$, and the isothermicity of X is given by (5.6) (see Exercise 2.8.72). To see that the isothermicity does not depend on the choice of the section $\tilde{X} \in \Gamma L$, we needed to involve the cross ratios characterization as in Lemma 2.8.71, which is a Möbius geometric notion. However, now using the projective geometry tools, we see that one can obtain a notion of isothermicity for maps into the projective light cone $P(L^4)$ directly. Such a projective geometric approach allows us to sidestep the lack of cross ratios in Lie sphere geometry (in general), and discuss the isothermicity for maps into the projective light cone $P(L^5)$ of $\mathbb{R}^{4,2}$.

5.8   Normal Bundle for Surfaces in $P(L^5)$

As in the previous section, take an arbitrary $\mathbf{v} \in \mathbb{R}^{4,2} \backslash \{0\}$, and separate the projective space $P(\mathbb{R}^{4,2})$ into a disjoint union as follows:

$$P(\mathbb{R}^{4,2}) = \mathcal{U} \cup \mathcal{V},$$

$$\mathcal{V} := \{\mathrm{span}\{X\} \in P(\mathbb{R}^{4,2}) \mid (X, \mathbf{v}) = 0\}, \quad \mathcal{U} := \{\mathrm{span}\{X\} \in P(\mathbb{R}^{4,2}) \mid (X, \mathbf{v}) \neq 0\}.$$

We let φ be an affine coordinate chart of $\mathcal{U}$ via the hyperplane

$$\mathcal{P} := \{X \in \mathbb{R}^{4,2} \mid (X, \mathbf{v}) = -1\}.$$

Exercise 5.8.44   For a point $s \in P(L^5)$, fix $\mathbf{v}$ so that $s \in \mathcal{U}$, and set $S := \varphi(s) \in \mathcal{P} \cap L^5$. Show that

$$d\varphi(T_s P(L^5)) = T_S(\mathcal{P} \cap L^5) = \{A \in T_S \mathbb{R}^{4,2} \mid (A, \mathbf{v}) = (A, S) = 0\}$$

$$= \mathrm{span}\{S, \mathbf{v}\}^\perp \cong \mathrm{span}\{S\}^\perp / \mathrm{span}\{S\} = s^\perp / s.$$

Now let $s : \Sigma \to P(L^5)$, and choose $S \in \Gamma s$ such that

$$S : \Sigma \to \mathcal{P} \cap L^5.$$

Then for any $p \in \Sigma$, the tangent space of the surface s in $P(L^5)$ can be identified via φ as

$$T_{s(p)}s \cong \operatorname{span}\{S_u, S_v\} \cong \operatorname{span}\{S_u, S_v, S\}/\operatorname{span}\{S\}.$$

Hence, the normal space becomes

$$\mathcal{N}_{s(p)}s \cong \operatorname{span}\{S, S_u, S_v\}^{\perp}/\operatorname{span}\{S\},$$

a 2-dimensional subspace of the tangent space of the projectivized light cone. We will denote the normal bundle of $s : \Sigma \to P(L^5)$ as $\mathcal{N}s$.

Therefore, the conditions for s to be isothermic with respect to the coordinates (u, v) are

$$(S_u, S_u) = (S_v, S_v), \quad (S_u, S_v) = 0, \quad (S_{uv}, Y) = 0 \tag{5.7}$$

for any smooth vector field $Y + s \in \Gamma \mathcal{N}s$.

Remark 5.8.45   Due to the definition of normal bundles, one sees that the condition $(S_{uv}, Y) = 0$ for $S \in L^5$ is equivalent to

$$S_{uv} \in \operatorname{span}\{S, S_u, S_v\}.$$

Exercise 5.8.46   Using (5.7), show that if s is isothermic then any non-zero $S \in \Gamma s$ satisfies

$$(S_u, S_u) = (S_v, S_v), \quad (S_u, S_v) = 0, \quad S_{uv} \in \operatorname{span}\{S, S_u, S_v\}. \tag{5.8}$$

Conversely, show that s is isothermic if there is some non-zero $S \in \Gamma s$ satisfying (5.8).

Now, taking the timelike point sphere complex $\mathfrak{p}$ to project to Möbius geometry $\operatorname{span}\{\mathfrak{p}\}^{\perp} \cong \mathbb{R}^{4,1}$, assume that a sphere congruence $s : \Sigma \to P(L^5)$ satisfies $(S, \mathfrak{p}) \neq 0$ for any $S \in \Gamma s$. Then s gives a sphere congruence with spheres of non-zero radius in resulting space forms. For some non-zero $S \in \Gamma s$, assume

$$(S_u, S_u) > 0, \quad (S_v, S_v) > 0 \quad \text{and} \quad (S_u, S_v) = 0,$$

so that

$$\operatorname{span}\{S, S_u, S_v\}^{\perp} \cap \operatorname{span}\{\mathfrak{p}\}^{\perp}$$

has signature $(+-)$. Then there exist precisely two independent vectors X, $\hat{X}$ (up to scalar factors) at every point in Σ so that

$$X, \hat{X} \in \mathrm{span}\{S, S_u, S_v\}^\perp \cap \mathrm{span}\{\mathfrak{p}\}^\perp \cap L^5,$$

and we have that

$$\mathrm{Proj}_{\mathfrak{p}^\perp}(\mathcal{N}_{s(p)}s) \cong \mathrm{span}\{X, \hat{X}\}. \tag{5.9}$$

Therefore, defining

$$\Lambda := \mathrm{span}\{S, X\}, \quad \hat{\Lambda} := \mathrm{span}\{S, \hat{X}\}, \tag{5.10}$$

we see that these two are maps into the set of contact elements. Viewing $X, \hat{X}$: $\Sigma \to P(L^4)$, they are the two surfaces that envelop the sphere congruence given by s.

In fact, (5.9) implies that the conditions for s to be isothermic with respect to the coordinates (u, v) can be written in terms of X and $\hat{X}$ as:

$$(S_u, S_u) = (S_v, S_v), \quad (S_u, S_v) = 0, \quad \text{and} \quad (S_{uv}, S) = (S_{uv}, X) = (S_{uv}, \hat{X}) = 0.$$

Remark 5.8.47 Should $(S, \mathfrak{p}) = 0$ on Σ for any $S \in \Gamma s$, the directions of X and $\hat{X}$ would coincide, and in fact would become the direction of s itself. In fact, one could reinterpret s as a map into $P(L^4)$, and the isothermicity of s would imply that s is an isothermic surface in the Möbius geometry defined by $\mathfrak{p}$.

5.9   Definition of Ω-surfaces

Suppose we have a Legendre immersion $\Lambda : \Sigma \to \mathcal{Z}$, and denote by X and N the point sphere map and tangent plane congruence of Λ with respect to some choice of point sphere complex $\mathfrak{p}$ and space form vector $\mathfrak{q}$. Let $s : \Sigma \to P(L^5)$ be a sphere congruence enveloped by Λ, so that s is a subbundle of Λ by Proposition 5.4.27. Therefore, by Corollary 5.4.28, we may assume $s = \mathrm{span}\{S\}$ where

$$S = \alpha X + N,$$

for some real-valued function $\alpha : \Sigma \to \mathbb{R} \cup \{\infty\}$. Note that the case $\alpha = \infty$ corresponds to the case that $s = \mathrm{span}\{X\}$.

Exercise 5.9.48 Show that $X, N \in \mathrm{span}\{S, S_u, S_v\}^\perp$.

Now we arrive at the main definition of this chapter.

Definition 5.9.49 ([87–89]) A Legendre immersion $\Lambda : \Sigma \to \mathcal{Z}$ is called an Ω-*surface* if there exists an isothermic sphere congruence $s : \Sigma \to P(L^5)$ enveloped by Λ (congruences containing principal curvature spheres are excluded). We also call the surface $x : \Sigma \to \mathfrak{R}_\kappa$ in a space form an Ω-surface if its contact lift is Ω.

Lemma 5.9.50 *Let $\Lambda : \Sigma \to \mathcal{Z}$ be an Ω-surface with enveloped isothermic sphere congruence s. If (u, v) are isothermic coordinates for s, then (u, v) are curvature line coordinates for some space form projection X.*

Proof Let X and N be the point sphere map and tangent plane congruence of Λ, respectively, for some choice of point sphere complex and space form vector. Writing $s = \mathrm{span}\{S\}$ for $s = \alpha X + N$, if $\alpha = \infty$ so that $s = \mathrm{span}\{X\}$, then we have that X is isothermic with isothermic coodinates (u, v).

Now assume that α is non-zero, and let the fundamental forms of X with normal N in the appropriate space form be given by

$$g = \begin{pmatrix} g_{11} & g_{12} \\ g_{12} & g_{22} \end{pmatrix} = \begin{pmatrix} (X_u, X_u) & (X_u, X_v) \\ (X_u, X_v) & (X_v, X_v) \end{pmatrix},$$

$$b = \begin{pmatrix} b_{11} & b_{12} \\ b_{12} & b_{22} \end{pmatrix} = -\begin{pmatrix} (X_u, N_u) & (X_u, N_v) \\ (X_v, N_u) & (X_v, N_v) \end{pmatrix},$$

$$\mathrm{I\!I\!I} = \begin{pmatrix} \mathrm{I\!I\!I}_{11} & \mathrm{I\!I\!I}_{12} \\ \mathrm{I\!I\!I}_{12} & \mathrm{I\!I\!I}_{22} \end{pmatrix} = \begin{pmatrix} (N_u, N_u) & (N_u, N_v) \\ (N_v, N_u) & (N_v, N_v) \end{pmatrix}.$$

Since (u, v) are isothermic coordinates for s, we have that

$$0 = (S_u, S_v) = \mathrm{I\!I\!I}_{12} - 2\alpha b_{12} + \alpha^2 g_{12}.$$

On the other hand, by the Cayley-Hamilton identity (2.36), we have

$$\mathrm{I\!I\!I}_{12} - 2H b_{12} + K g_{12} = 0$$

for the Gaussian curvature K and mean curvature H of the point sphere map X. Therefore, we see that

$$(-2H + 2\alpha)b_{12} + (K - \alpha^2)g_{12} = 0. \tag{5.11}$$

Now noting that $X, N \in \mathrm{span}\{S, S_u, S_v\}^\perp$, the isothermicity of s implies that

$$0 = (S_{uv}, X) = -(S_u, X_v) = -\alpha g_{12} + b_{12}. \tag{5.12}$$

Thus, (5.11) now reads

$$(\alpha^2 - 2H\alpha + K)g_{12} = 0.$$

Since we are assuming that s is not a curvature sphere congruence, that is, $\alpha \neq k_1, k_2$, we have that

$$\alpha^2 - 2H\alpha + K \neq 0,$$

and thus $g_{12} = 0$. Then, (5.12) tells us that $b_{12} = 0$, so (u, v) are curvature line coordinates for X.

Finally, if $\alpha = 0$, then the isothermicity of s implies that $b_{12} = \mathrm{III}_{12} = 0$, and the Cayley-Hamilton identity (2.36) gives us $g_{12} = 0$.   □

Remark 5.9.51 Suppose that a Legendre immersion $\Lambda = \mathrm{span}\{X, N\}$ is Ω with an enveloping isothermic sphere congruence S. Then from (5.10), the sphere congruence resulting from S is enveloped by another surface $\hat{X}$. In fact, Lemma 5.9.50 implies that the curvature lines on X and $\hat{X}$ also correspond, making the two surfaces a *Ribaucour pair* [187].

Exercise 5.9.52 Show that a Lie sphere transformation of an Ω-surface is again an Ω-surface. (Hint: Given an Ω-surface Λ with enveloping isothermic sphere congruence s, show that $\tilde{s} := \mathrm{span}\{AS\}$ for any $S \in \Gamma s$ is an enveloping isothermic sphere congruence of $\tilde{\Lambda} := A\Lambda$.)

5.10   The First Fundamental Form of Sphere Congruences

With an eye on obtaining Demoulin's equation [88] which characterizes Ω-surfaces in space forms, let $s : \Sigma \to P(L^5)$ be a sphere congruence enveloped by a Legendre immersion $\Lambda : \Sigma \to \mathcal{Z}$ so that s is a 1-dimensional subbundle of Λ. Let X and N be the point sphere map and tangent plane congruence of Λ with respect to the point sphere complex $\mathfrak{p}$ and space form vector $\mathfrak{q}_0$ so that $X : \Sigma \to M_0$. Assuming that X is umbilic-free on Σ, we will use Lemma 5.9.50 and suppose that $(u, v) \in \Sigma$ are curvature line coordinates of X.

As observed in Sect. 5.8, isothermicity of a sphere congruence requires us to investigate both the conformality of first fundamental form and the diagonality of the second fundamental forms; in this section, we identify the condition that is equivalent to the conformality of the first fundamental form. Since we have s is a subbundle of Λ, let $S \in \Gamma s$ so that $S = \alpha X + N$, and we can compute the first fundamental form of S:

$$(\mathrm{d}S, \mathrm{d}S) = (\alpha - k_1)^2 g_{11}\, \mathrm{d}u^2 + (\alpha - k_2)^2 g_{22}\, \mathrm{d}v^2. \tag{5.13}$$

Note that the first fundamental form cannot be conformal if α equals one of k_1 or k_2 since we are avoiding umbilic points. Thus we have that

$$\alpha \neq k_1 \quad \text{and} \quad \alpha \neq k_2.$$

Then, $(S_u, S_u) = (S_v, S_v)$ if and only if

$$\alpha = \frac{k_1\sqrt{g_{11}} \mp k_2\sqrt{g_{22}}}{\sqrt{g_{11}} \mp \sqrt{g_{22}}} \tag{5.14}$$

For such α, we can calculate that

$$\alpha - k_1 = \pm(k_1 - k_2)\frac{\sqrt{g_{22}}}{\sqrt{g_{11}} \mp \sqrt{g_{22}}}, \quad \alpha - k_2 = (k_1 - k_2)\frac{\sqrt{g_{11}}}{\sqrt{g_{11}} \mp \sqrt{g_{22}}},$$

$$\frac{\alpha - k_1}{\alpha - k_2} = \pm\frac{\sqrt{g_{22}}}{\sqrt{g_{11}}}, \quad (\alpha - k_1)(\alpha - k_2) = \pm(k_1 - k_2)^2\frac{\sqrt{g_{11}}\sqrt{g_{22}}}{(\sqrt{g_{11}} \mp \sqrt{g_{22}})^2}.$$

Using the Codazzi equations (2.35), we can compute that

$$\frac{\alpha_u}{k_1 - \alpha} = (\log(\sqrt{g_{11}} \mp \sqrt{g_{22}}))_u \mp \frac{\sqrt{g_{11}}}{\sqrt{g_{22}}}\frac{k_{1,u}}{k_1 - k_2},$$

$$\frac{\alpha_v}{k_2 - \alpha} = (\log(\sqrt{g_{11}} \mp \sqrt{g_{22}}))_v \pm \frac{\sqrt{g_{22}}}{\sqrt{g_{11}}}\frac{k_{2,v}}{k_1 - k_2}.$$

Therefore, we conclude that

$$\left(\frac{\alpha_u}{k_1 - \alpha}\right)_v - \left(\frac{\alpha_v}{k_2 - \alpha}\right)_u = \mp\left(\left(\frac{\sqrt{g_{11}}}{\sqrt{g_{22}}}\frac{k_{1,u}}{k_1 - k_2}\right)_v + \left(\frac{\sqrt{g_{22}}}{\sqrt{g_{11}}}\frac{k_{2,v}}{k_1 - k_2}\right)_u\right). \tag{5.15}$$

Remark 5.10.53 If we have $g_{11} = g_{22}$, then we have that either $\alpha = \frac{1}{2}(k_1 + k_2)$, or $\alpha = \infty$. However, since s is projectively defined, when $\alpha = \infty$, we have $s = \mathrm{span}\{X\}$.

5.11   The Second Fundamental Form of Sphere Congruences

Now we look at the condition on α that is equivalent to the diagonality of the second fundamental form of a sphere congruence. For this, let $\mathcal{N}s$ denote the normal bundle of $s = \mathrm{span}\{S\} : \Sigma \to P(L^5)$, and let $Y + s \in \Gamma\mathcal{N}s$ so that $(Y, S) = (Y, S_u) = (Y, S_v) = 0$.

Case of α neither 0 nor ∞ Since we have that $S = \alpha X + N$, the conditions $(Y, S) = (Y, S_u) = (Y, S_v) = 0$ imply that

$$(Y, X) = -\alpha^{-1}(Y, N),$$

$$(Y, X_u) = \frac{\alpha_u}{\alpha(\alpha - k_1)}(Y, N),$$

$$(Y, X_v) = \frac{\alpha_v}{\alpha(\alpha - k_2)}(Y, N),$$

where we have used the Rodrigues' equation $N_u = -k_1 X_u$ and $N_v = -k_2 X_v$. Using Exercise 5.4.31, we calculate

$$0 = (Y, S_{uv}) = (Y, \alpha_{uv} X + \alpha_u X_v + \alpha_v X_u + \alpha X_{uv} + N_{uv}) = \beta(Y, N), \qquad (5.16)$$

where

$$\beta = -\frac{\alpha_{uv}}{\alpha} + \frac{\alpha_u}{\alpha(\alpha - k_1)}\left(\alpha_v - \frac{(\alpha - k_2)k_{1,v}}{k_1 - k_2}\right) + \frac{\alpha_v}{\alpha(\alpha - k_2)}\left(\alpha_u - \frac{(\alpha - k_1)k_{2,u}}{k_2 - k_1}\right).$$

In particular, (5.16) must hold for any $Y + s \in \Gamma \mathcal{N} s$; therefore, choosing Y so that $(Y, N) \neq 0$, we see that $\beta = 0$, or equivalently,

$$\frac{\alpha_u}{\alpha - k_1}\left(\alpha_v - \frac{(\alpha - k_2)k_{1,v}}{k_1 - k_2}\right) + \frac{\alpha_v}{\alpha - k_2}\left(\alpha_u - \frac{(\alpha - k_1)k_{2,u}}{k_2 - k_1}\right) = \alpha_{uv}. \qquad (5.17)$$

We now rewrite (5.17) in a cleaner form:

Lemma 5.11.54 *We have that (5.17) is equivalent to*

$$\left(\frac{\alpha_u}{k_1 - \alpha}\right)_v = \left(\frac{\alpha_v}{k_2 - \alpha}\right)_u. \qquad (5.18)$$

Proof Noting that

$$\frac{1}{k_1 - k_2}\frac{\alpha - k_2}{\alpha - k_1} + \frac{1}{k_2 - k_1}\frac{\alpha - k_1}{\alpha - k_2} = \frac{1}{\alpha - k_1} + \frac{1}{\alpha - k_2},$$

(5.17) becomes

$$\begin{aligned}
\alpha_{uv} &= \left(\alpha_v \frac{\alpha - k_2}{k_1 - k_2} - (\alpha - k_2)\frac{k_{1,v}}{k_1 - k_2}\right)\frac{\alpha_u}{\alpha - k_1} \\
&\quad + \left(\alpha_u \frac{\alpha - k_1}{k_2 - k_1} - (\alpha - k_1)\frac{k_{2,u}}{k_2 - k_1}\right)\frac{\alpha_v}{\alpha - k_2} \\
&= \frac{(\alpha_v - k_{1,v})\alpha_u}{k_1 - k_2}\frac{k_2 - \alpha}{k_1 - \alpha} - \frac{(\alpha_u - k_{2,u})\alpha_v}{k_1 - k_2}\frac{k_1 - \alpha}{k_2 - \alpha},
\end{aligned}$$

so

$$\frac{\alpha_{uv}}{k_1 - \alpha} + (\alpha - k_1)_v \frac{\alpha_u}{(k_1 - \alpha)^2} = \frac{\alpha_{uv}}{k_2 - \alpha} + (\alpha - k_2)_u \frac{\alpha_v}{(k_2 - \alpha)^2},$$

giving us the desired conclusion.   □

Combining the conditions on α equivalent to the conformality of the first fundamental form (5.14), (5.15), and the diagonality of the second fundamental form (5.18), we have now proven the following:

Theorem 5.11.55 ([88]) *Let $\Lambda : \Sigma \to \mathcal{Z}$ be a Legendre immersion parametrized by curvature line coordinates $(u, v) \in \Sigma$ with point sphere map X and tangent plane congruence N so that $\Lambda = \mathrm{span}\{X, N\}$. The sphere congruences $s^{\pm} = \mathrm{span}\{\alpha^{\pm}X + N\}$, with $\alpha^{\pm} : \Sigma \to \mathbb{R} \setminus \{0\}$, enveloped by Λ are isothermic with respect to the curvature line coordinates (u, v) of X if and only if*

$$\alpha^{\pm} = \frac{k_1\sqrt{g_{11}} \mp k_2\sqrt{g_{22}}}{\sqrt{g_{11}} \mp \sqrt{g_{22}}},$$

and the Demoulin's equation holds, that is,

$$\left(\frac{\sqrt{g_{11}}}{\sqrt{g_{22}}}\frac{k_{1,u}}{k_1 - k_2}\right)_v + \left(\frac{\sqrt{g_{22}}}{\sqrt{g_{11}}}\frac{k_{2,v}}{k_1 - k_2}\right)_u = 0. \tag{5.19}$$

In fact, we have also shown the following:

Corollary 5.11.56 *Whenever we have one isothermic sphere congruence $s = \mathrm{span}\{\alpha X + N\}$ with $\alpha \in \mathbb{R} \setminus \{0\}$ for a Legendre immersion Λ, there is a second isothermic sphere congruence enveloping Λ. We call these two isothermic sphere congruences an Ω-pair.*

Case of $\alpha = 0$ Because we have divided by α at some places in the above computations, we now deal with the case $\alpha = 0$. Suppose we have that the sphere congruence under consideration satisfies

$$s = \mathrm{span}\{N\}.$$

For any $Y + s \in \Gamma \mathcal{N}s$, we have in particular that

$$(Y, N) = (Y, N_u) = (Y, N_v) = 0,$$

and thus Exercise 5.4.31 together with Rodrigues' equation implies

$$(Y, N_{uv}) = 0.$$

Thus, $s = \mathrm{span}\{N\}$ is isothermic if and only if

$$(N_u, N_u) = (N_v, N_v), \quad (N_u, N_v) = 0,$$

or, equivalently, the third fundamental form of the space form projection is conformal, i.e.

$$k_1^2 g_{11} = k_2^2 g_{22}. \tag{5.20}$$

If k_1 were zero, then (5.20) implies k_2 is also zero, so we would have umbilic points, which have been excluded, so that

$$k_1 \neq k_2, \quad K := k_1 k_2 \neq 0.$$

Remark 5.11.57 Surfaces satisfying (5.20) are called *L-isothermic surfaces* (see, for example, [164, 166, 167]).

In view of Corollary 5.11.56, we now seek the second isothermic sphere congruence enveloping an *L*-isothermic Λ:

Lemma 5.11.58 *Suppose that $\Lambda : \Sigma \to \mathcal{Z}$ is an L-isothermic surface with space form projection $X : \Sigma \to M_0$ and tangent plane congruence N. If X is not a minimal surface, i.e. $H_0 \neq 0$, then X satisfies Demoulin's equation (5.19).*

Proof Since we have that Λ is L-isothermic, we know that (5.20) holds. Thus, we have that

$$k_1 \sqrt{g_{11}} = \pm k_2 \sqrt{g_{22}}.$$

As we are assuming $H_0 \neq 0$, both cases are proven similarly; hence, we will assume that

$$k_1 \sqrt{g_{11}} = k_2 \sqrt{g_{22}}. \tag{5.21}$$

Differentiating (5.21) with respect to u, we obtain

$$k_{1,u} \sqrt{g_{11}} = k_{2,u} \sqrt{g_{22}} + k_2 (\sqrt{g_{22}})_u - k_1 (\sqrt{g_{11}})_u,$$

and Codazzi equations (2.35) imply

$$k_{1,u} \frac{\sqrt{g_{11}}}{\sqrt{g_{22}}} = \frac{1}{2} \frac{g_{22,u}}{g_{22}} (k_1 - k_2) + \frac{1}{2} k_2 \frac{g_{22,u}}{g_{22}} - k_1 \frac{(\sqrt{g_{11}})_u}{\sqrt{g_{22}}}$$

$$= \frac{k_1}{\sqrt{g_{22}}} (\sqrt{g_{22}} - \sqrt{g_{11}})_u.$$

Similarly, taking the v-derivative of (5.21) and using Codazzi equations (2.35) tells us that

$$k_{2,v}\frac{\sqrt{g_{22}}}{\sqrt{g_{11}}} = \frac{k_2}{\sqrt{g_{11}}}(\sqrt{g_{11}} - \sqrt{g_{22}})_v.$$

Now, from (5.21), we can show

$$\frac{k_2}{k_1 - k_2} = \frac{\sqrt{g_{11}}}{\sqrt{g_{22}} - \sqrt{g_{11}}}, \qquad \frac{k_1}{k_1 - k_2} = \frac{\sqrt{g_{22}}}{\sqrt{g_{22}} - \sqrt{g_{11}}},$$

so that

$$\frac{k_{1,u}}{k_1 - k_2}\frac{\sqrt{g_{11}}}{\sqrt{g_{22}}} = \frac{(\sqrt{g_{22}} - \sqrt{g_{11}})_u}{\sqrt{g_{22}} - \sqrt{g_{11}}} = (\log(\sqrt{g_{22}} - \sqrt{g_{11}}))_u,$$

$$\frac{k_{2,v}}{k_1 - k_2}\frac{\sqrt{g_{22}}}{\sqrt{g_{11}}} = \frac{(\sqrt{g_{11}} - \sqrt{g_{22}})_v}{\sqrt{g_{22}} - \sqrt{g_{11}}} = -(\log(\sqrt{g_{22}} - \sqrt{g_{11}}))_v.$$

Therefore,

$$\left(\frac{k_{1,u}}{k_1 - k_2}\frac{\sqrt{g_{11}}}{\sqrt{g_{22}}}\right)_v + \left(\frac{k_{2,v}}{k_1 - k_2}\frac{\sqrt{g_{22}}}{\sqrt{g_{11}}}\right)_u = 0,$$

that is, Demoulin's equation (5.19) holds. $\square$

Corollary 5.11.59 *Let $\Lambda : \Sigma \to \mathcal{Z}$ be an L-isothermic surface with space form projection $X : \Sigma \to M_0$ and tangent plane congruence N. If X is not a minimal surface, i.e. $H_0 \neq 0$, then the two isothermic sphere congruences are given by*

$$s^+ = \mathrm{span}\{N\}, \quad \text{and} \quad s^- = \mathrm{span}\{K_0 H_0^{-1} X + N\}.$$

On the other hand, if $H_0 = 0$, then since $k_1 = -k_2$, we have $g_{11} = g_{22}$ implying that the second isothermic sphere congruence must fall into the case of $\alpha = \infty$, bringing us to the next final case.

Case of $\alpha = \infty$ In this case, $s = \mathrm{span}\{X\}$, with $(X, \mathfrak{p}) = 0$ so that $\mathfrak{p}$ is a constant section in the normal bundle of s. Therefore, the isothermicity of s in $P(L^5)$ is equivalent to the isothermicity of X viewed as a map into Möbius geometry $\mathrm{span}\{\mathfrak{p}\}^\perp \cong \mathbb{R}^{4,1}$. Thus, we also conclude, conversely, that the following lemma holds:

Lemma 5.11.60 ([87]) *All isothermic surfaces are Ω-surfaces.*

Now assume that X is isothermic so that $g_{11} = g_{22}$. Corollary 5.11.56 implies that we have another isothermic sphere congruence S; the relation (5.14) allows us to see that $\alpha = \frac{1}{2}(k_1 + k_2)$. When $\alpha = H_0 \neq 0$, we do indeed have this second isothermic sphere congruence, by the following Lemma 5.11.61.

Lemma 5.11.61 *Suppose that X is isothermic. Then the mean curvature sphere congruence (or the central sphere congruence) is also isothermic.*

Proof Using the Codazzi equations for curvature line coordinates as in (2.35), we can verify that

$$k_{1,uv} = \frac{g_{11,uv}}{2g_{11}}(k_2 - k_1) - \frac{g_{11,u}g_{11,v}}{2g_{11}^2}(k_2 - k_1) + \frac{g_{11,v}}{2g_{11}}(k_{2,u} - k_{1,u})$$

$$= \frac{g_{11,uv}}{2g_{11}}(k_2 - k_1) - \frac{g_{11,u}g_{11,v}}{2g_{11}^2}(k_2 - k_1) + \frac{k_{1,v}}{k_2 - k_1}(k_{2,u} - k_{1,u}),$$

$$k_{2,uv} = \frac{g_{22,uv}}{2g_{22}}(k_1 - k_2) - \frac{g_{22,u}g_{22,v}}{2g_{22}^2}(k_1 - k_2) + \frac{k_{2,u}}{k_1 - k_2}(k_{1,v} - k_{2,v}).$$

Now,

$$-\tfrac{1}{2}(k_{1,uv} + k_{2,uv})(k_2 - k_1)^2 = -\tfrac{1}{2}(k_2 - k_1)(-k_{2,u}(k_{1,v} - k_{2,v}) + k_{1,v}(k_{2,u} - k_{1,u})),$$

so we have

$$k_{1,v}H_{0,u} - k_{2,u}H_{0,v} = H_{0,uv}(k_1 - k_2) \ .$$

and confirms Demoulin's equation (5.19). $\square$

Finally, for $H_0 = 0$ case, we have the following:

Corollary 5.11.62 *Let $\Lambda : \Sigma \to \mathcal{Z}$ be a Legendre immersion that projects to a minimal surface in Euclidean 3-space. Then it is an Ω-surface, with the pair of isothermic sphere congruences given by the point sphere map and the tangent plane congruence.*

We have so far computed Demoulin's equation depending on the space form projection to Euclidean space M_0. However, the form of Demoulin's equation (5.19) does not depend on the choice of the space form M_κ:

Lemma 5.11.63 *Demoulin's equation is independent of the choice of space form.*

Proof Let $\Lambda : \Sigma \to \mathcal{Z}$ be a Legendre immersion and let $X_0 \in \Gamma\Lambda$ be the point sphere map with respect to the point sphere complex $\mathfrak{p}$ and space form vector $\mathfrak{q}_0$ so that $X : \Sigma \to M_0$. Suppose now that $\hat{X} \in \Gamma\Lambda$ so that $\hat{X} : \Sigma \to M_\kappa$ is defined by a different space form vector $\mathfrak{q}_\kappa$. Denoting the first fundamental form and principal curvatures of X by g and k_j, respectively, further assume that $x : \Sigma \to \mathfrak{R}_0 \cong \mathbb{R}^3$ is the space form projection with unit normal n_0. If we let $\hat{g}$ and $\hat{k}_j$ be the first fundamental form and principal curvatures of $\hat{X}$, respectively, then we have from (2.31) and (2.33) that

$$\hat{k}_j = t k_j + 2\kappa(x \cdot n_0), \quad \hat{g}_{11} = t^{-2} g_{11}, \quad \hat{g}_{22} = t^{-2} g_{22}, \qquad (5.22)$$

with

$$t = 1 + \kappa |x|^2.$$

Therefore,

$$\left(\frac{\sqrt{\hat{g}_{11}}}{\sqrt{\hat{g}_{22}}} \frac{\hat{k}_{1,u}}{\hat{k}_1 - \hat{k}_2} \right)_v + \left(\frac{\sqrt{\hat{g}_{22}}}{\sqrt{\hat{g}_{11}}} \frac{\hat{k}_{2,v}}{\hat{k}_1 - \hat{k}_2} \right)_u$$

$$= \left(\frac{\sqrt{g_{11}}}{\sqrt{g_{22}}} \frac{2\kappa(x \cdot x_u)k_1 + 2\kappa(x \cdot n_{0,u}) + t k_{1,u}}{t(k_1 - k_2)} \right)_v$$

$$+ \left(\frac{\sqrt{g_{22}}}{\sqrt{g_{11}}} \frac{2\kappa(x \cdot x_v)k_2 + 2\kappa(x \cdot n_{0,v}) + t k_{2,v}}{t(k_1 - k_2)} \right)_u$$

$$= \left(\frac{\sqrt{g_{11}}}{\sqrt{g_{22}}} \frac{0 + t k_{1,u}}{t(k_1 - k_2)} \right)_v + \left(\frac{\sqrt{g_{22}}}{\sqrt{g_{11}}} \frac{0 + t k_{2,v}}{t(k_1 - k_2)} \right)_u .$$

Hence, if Demoulin's equation holds for one space form projection then it holds for any space form projection. $\qquad \square$

Finally, as in Sect. 2.8, one can always stretch the coordinates and obtain new curvature line coordinates $(\tilde{u}, \tilde{v})$ from given curvature line coordinates (u, v) via taking $\Lambda(u, v) = \Lambda(u(\tilde{u}), v(\tilde{v}))$ for any strictly monotonic functions u depending only on $\tilde{u}$ and v depending only on $\tilde{v}$. Denoting the first fundamental form with respect to $(\tilde{u}, \tilde{v})$ by $\tilde{g}$, we then have

$$g_{11} = U^2 \tilde{g}_{11}, \quad g_{22} = V^2 \tilde{g}_{22},$$

where $U := \frac{d\tilde{u}}{du}$ and $V := \frac{d\tilde{v}}{dv}$ are functions of u and v alone, respectively. Then the Demoulin's equation is equivalent to

$$\left(\frac{U}{V} \frac{\sqrt{\tilde{g}_{11}}}{\sqrt{\tilde{g}_{22}}} \frac{k_{1,\tilde{u}}}{k_1 - k_2} \right)_{\tilde{v}} + \left(\frac{V}{U} \frac{\sqrt{\tilde{g}_{22}}}{\sqrt{\tilde{g}_{11}}} \frac{k_{2,\tilde{v}}}{k_1 - k_2} \right)_{\tilde{u}} = 0.$$

Summarizing:

Theorem 5.11.64 *Let $\Lambda : \Sigma \to \mathcal{Z}$ be a Legendre immersion with curvature line coordinates $(u, v) \in \Sigma$. Then Λ is an Ω-surface if and only if its space form projection satisfies Demoulin's equation, that is,*

$$\left(\frac{U}{V}\frac{\sqrt{g_{11}}}{\sqrt{g_{22}}}\frac{k_{1,u}}{k_1 - k_2}\right)_v + \left(\frac{V}{U}\frac{\sqrt{g_{22}}}{\sqrt{g_{11}}}\frac{k_{2,v}}{k_1 - k_2}\right)_u = 0,$$

where U and V are functions of u and v alone, respectively.

5.12   Harmonic Separation of Curvature Sphere Congruences

To discuss harmonic separation of sphere congruences, we use cross ratios in $\mathbb{R}^{4,2}$ as follows: When four points $\mathbf{v}_1, \mathbf{v}_2, \mathbf{v}_3, \mathbf{v}_4$ lie in a line in $\mathbb{R}^{4,2}$, there exist vectors $\mathbf{v}$ and $\mathbf{w}$ (independent of the counter j) so that

$$\mathbf{v}_j = \mathbf{v} + \alpha_j \mathbf{w}$$

for some $\alpha_j \in \mathbb{R}$. We can then take the cross ratio of $\mathbf{v}_1, \mathbf{v}_2, \mathbf{v}_3, \mathbf{v}_4$ to be

$$\mathrm{cr}(\mathbf{v}_1, \mathbf{v}_2, \mathbf{v}_3, \mathbf{v}_4) = \frac{\alpha_2 - \alpha_1}{\alpha_3 - \alpha_2}\frac{\alpha_4 - \alpha_3}{\alpha_1 - \alpha_4}.$$

The four vectors $\mathbf{v}_1, \mathbf{v}_2, \mathbf{v}_3, \mathbf{v}_4$ are said to be *harmonically separated* if their cross ratio is -1, i.e.

$$\mathrm{cr}(\mathbf{v}_1, \mathbf{v}_2, \mathbf{v}_3, \mathbf{v}_4) = -1.$$

Lemma 5.12.65 *The two isothermic sphere congruences of an Ω-surface separate the curvature sphere congruences harmonically.*

Proof Since sphere congruences are projectively invariant, we can write any sphere congruence enveloping an Ω-surface $\Lambda = \mathrm{span}\{X, N\}$ as

$$\alpha X + N.$$

Therefore, we can compute the cross ratios between sphere congruences enveloping Λ. Now, the four values of α under consideration, corresponding to the pair of isothermic sphere congruences and the curvature sphere congruences, are

$$\alpha_1 = k_1, \quad \alpha_2 = \frac{k_1\sqrt{g_{11}} + k_2\sqrt{g_{22}}}{\sqrt{g_{11}} + \sqrt{g_{22}}}, \quad \alpha_3 = k_2, \quad \alpha_4 = \frac{k_1\sqrt{g_{11}} - k_2\sqrt{g_{22}}}{\sqrt{g_{11}} - \sqrt{g_{22}}}.$$

Then the cross ratio is

$$(\alpha_2 - \alpha_1)(\alpha_3 - \alpha_2)^{-1}(\alpha_4 - \alpha_3)(\alpha_1 - \alpha_4)^{-1}$$

$$= \left\{ \left(-k_1 + \frac{k_1\sqrt{g_{11}} + k_2\sqrt{g_{22}}}{\sqrt{g_{11}} + \sqrt{g_{22}}} \right) \left(k_2 - \frac{k_1\sqrt{g_{11}} + k_2\sqrt{g_{22}}}{\sqrt{g_{11}} + \sqrt{g_{22}}} \right)^{-1} \right.$$

$$\left. \left(\frac{k_1\sqrt{g_{11}} - k_2\sqrt{g_{22}}}{\sqrt{g_{11}} - \sqrt{g_{22}}} - k_2 \right) \left(k_1 - \frac{k_1\sqrt{g_{11}} - k_2\sqrt{g_{22}}}{\sqrt{g_{11}} - \sqrt{g_{22}}} \right)^{-1} \right\}$$

$$= -1.$$

$\square$

The harmonic separation of the two curvature sphere congruences by the pair of isothermic sphere congruences gives us the next results:

Lemma 5.12.66 *For the curvature sphere congruences $s_1 = \mathrm{span}\{K_1\}$ and $s_2 = \mathrm{span}\{K_2\}$, $\mathrm{cr}(K_1, \alpha_1 K_1 + \alpha_2 K_2, K_2, \beta_1 K_1 - \beta_2 K_2) = -1$ implies that the vectors (α_1, α_2) and (β_1, β_2) are parallel.*

Proof Using the properties of cross ratios,

$$-1 = \mathrm{cr}(K_1, \alpha_1 K_1 + \alpha_2 K_2, K_2, \beta_1 K_1 - \beta_2 K_2)$$

$$= \mathrm{cr}(K_1, K_1 + (\alpha_2/\alpha_1)K_2, K_2, K_1 + (-\beta_2/\beta_1)K_2)$$

$$= \mathrm{cr}(0, \alpha_2/\alpha_1, \infty, -\beta_2/\beta_1) = -\alpha_2\beta_1/(\alpha_1\beta_2) .$$

$\square$

Corollary 5.12.67 *For an Ω-surface Λ, let K_1, K_2 denote arbitrary lifts of the curvature sphere congruences, and $s^{\pm}$ the pair of isothermic sphere congruences enveloping Λ. Then we can choose $S^{\pm} \in \Gamma s^{\pm}$ so that*

$$S^+ = \alpha_1 K_1 + \alpha_2 K_2 \quad and \quad S^- = \alpha_1 K_1 - \alpha_2 K_2$$

for some functions α_1 and α_2.

Chapter 6
Integrability of Ω-surfaces via Isothermicity

6.1 Isothermic Sphere Congruences as Isothermic Surfaces

The integrable structure of Ω-surfaces is induced from that of the enveloping isothermic sphere congruences. To better understand the integrability of isothermic sphere congruences, we make comparison between isothermic surfaces and isothermic sphere congruences. Recall that a point $x \in \mathbb{R}^3$ can be lifted into $X \in M_0 \subset L^4 \subset \mathbb{R}^{4,1}$ via

$$X = \left(\tfrac{1}{2}(1 + |x|^2), x^t, \tfrac{1}{2}(1 - |x|^2) \right)^t,$$

where M_0 is determined by the space form vector $\mathfrak{q}_0 = (1, 0, 0, 0, -1)^t$. Choosing $\mathfrak{o} = \tfrac{1}{2}(1, 0, 0, 0, 1)^t$ so that $\langle \mathfrak{o}, \mathfrak{q}_\mathfrak{o} \rangle = -1$, and viewing x as a vector in $\mathbb{R}^{4,1}$ via $(0, x^t, 0)^t$, we see that

$$X = x + \mathfrak{o} + \frac{1}{2}(x \cdot x)\, \mathfrak{q}_0.$$

We have further seen that the isothermicity of $L = \mathrm{span}\{X\}$ in $P(L^4)$ is equivalent to the isothermicity of x in $\mathbb{R}^3$.

Now, let $x \in \mathbb{R}^{3,1}$, where $\mathbb{R}^{3,1}$ is the Minkowski 4-space with signature $(- + ++)$. Writing $x = (r, c_1, c_2, c_3)^t$ for $c = (c_1, c_2, c_3) \in \mathbb{R}^3$, we can lift x into $L^5 \subset \mathbb{R}^{4,2}$ analogously as follows: For $\mathfrak{q}_0 = (1, 0, 0, 0, -1, 0)^t$, let $\mathfrak{o} = \tfrac{1}{2}(1, 0, 0, 0, 1, 0)^t$ so that $(\mathfrak{o}, \mathfrak{q}_0) = -1$. Viewing $x = (r, c_1, c_2, c_3)^t$ as a vector in $\mathbb{R}^{4,2}$ via $(0, c^t, 0, r)^t \in \mathbb{R}^{4,2}$, we see that

J. Cho et al., *Discrete Isothermic Surfaces in Lie Sphere Geometry*, Lecture Notes in Mathematics 2375, https://doi.org/10.1007/978-3-031-95592-1_6

$$S = x + \mathfrak{o} + \tfrac{1}{2}\langle x, x\rangle_{3,1}\, \mathfrak{q}_0$$

$$= \begin{pmatrix} 0 \\ c \\ 0 \\ r \end{pmatrix} + \begin{pmatrix} \tfrac{1}{2} \\ \mathbf{0} \\ \tfrac{1}{2} \\ 0 \end{pmatrix} + \frac{1}{2}\begin{pmatrix} \langle x, x\rangle_{3,1} \\ \mathbf{0} \\ -\langle x, x\rangle_{3,1} \\ 0 \end{pmatrix} = \begin{pmatrix} \tfrac{1}{2}(1 + |c|^2 - r^2) \\ c \\ \tfrac{1}{2}(1 - |c|^2 + r^2) \\ r \end{pmatrix},$$

where $\mathbf{0}$ denotes the zero vector. Comparing to the lift of spheres in $\mathbb{R}^3$ to Lie sphere geometry in (5.2), we understand that S projects to a sphere with center $c \in \mathbb{R}^3$ and radius r in the Euclidean space determined by taking $\mathfrak{q}_0$ as the space form vector and $\mathfrak{p} = (0, 0, 0, 0, 0, 1)^t$ as the point sphere complex. Therefore, viewing $\mathfrak{p}$ as a vector in $\mathbb{R}^{3,1}$, we see that $x = c + r\mathfrak{p}$ represents a sphere with center $c \in \mathbb{R}^3$ and radius r in the Euclidean space. This is called the *isotropy projection*, and is the basis for *Laguerre geometry* (see, for example, [65, §3.4]).

Any map $x : \Sigma \to \mathbb{R}^{3,1}$ represents a sphere congruence, corroborated by the fact that its lift $S : \Sigma \to L^5$ also represents the same sphere congruence. Hence, as in the isothermic surfaces case, one can deduce that the isothermicity of $s = \mathrm{span}\{S\}$ in $P(L^5)$ is equivalent to the isothermicity of x in $\mathbb{R}^{3,1}$. Accordingly, many of the characterizations of isothermic surfaces in $P(L^4)$ also apply to the isothermic sphere congruences in $P(L^5)$, some of which we recover in the following discussions.

Moutard Lifts Let $\Lambda : \Sigma \to \mathcal{Z}$ be a Legendre immersion. We say that the 1-dimensional subbundle $s : \Sigma \to P(L^5)$ of Λ (so that s represents a sphere congruence enveloping Λ) is *nondegenerate* if s never coincides with a principal curvature sphere. Due to Lemma 5.4.29, for any section $S \in \Gamma s$, neither S_u nor S_v lie in the null plane given by Λ, so in particular,

$$(S_u, S_u) > 0 \quad \text{and} \quad (S_v, S_v) > 0.$$

The following lemma can be proved analogously to Theorem 2.9.74.

Lemma 6.1.1 *Let s be a nondegenerate 1-dimensional subbundle of a Legendre immersion $\Lambda : \Sigma \to \mathcal{Z}$. Then s is isothermic if and only if s admits a Moutard lift $\varsigma \in \Gamma s$, i.e. $\varsigma_{uv} \parallel \varsigma$.*

Exercise 6.1.2 Prove Lemma 6.1.1 using Exercise 5.8.46.

We have seen in Corollary 5.11.56 that, generically, existence of one isothermic sphere congruence implies existence of a second one. We now introduce an alternate proof of this fact using the Moutard lifts characterization of isothermicity:

Lemma 6.1.3 *Let $s_1 = \mathrm{span}\{K_1\}$ and $s_2 = \mathrm{span}\{K_2\}$ denote the (principal) curvature sphere congruences of a Legendre immersion Λ. If $\alpha K_1 + \beta K_2$ is Moutard for some functions α and β, then so is $\alpha K_1 - \beta K_2$.*

Proof By Lemma 5.4.29, there exist scalars A, B, C and D such that

$$K_{1,u} = AK_1 + BK_2, \quad K_{2,v} = CK_1 + DK_2.$$

By assumption, we have that $\alpha K_1 + \beta K_2$ and $(\alpha K_1 + \beta K_2)_{uv}$ are parallel, and we also have

$$(\alpha K_1 + \beta K_2)_{uv} = (\alpha_{uv} + \alpha_v A + \beta_u C + \alpha A_v + \beta C_u + \alpha BC + \beta AC)K_1$$
$$+ (\beta_{uv} + \alpha_v B + \beta_u D + \alpha B_v + \beta D_u + \alpha BD + \beta BC)K_2$$
$$+ (\alpha A + \alpha_u)K_{1,v} + (\beta D + \beta_v)K_{2,u}.$$

Then Proposition 5.6.38 and the fact that this is a Moutard lift imply $\alpha A + \alpha_u = \beta D + \beta_v = 0$, so

$$(\alpha K_1 + \beta K_2) \parallel (\alpha K_1 + \beta K_2)_{uv}$$
$$= ((\beta C)_u + C(\alpha B + \beta A))K_1 + ((\alpha B)_v + B(\alpha D + \beta C))K_2,$$

which gives

$$\beta(\beta C)_u + \beta^2 AC = \alpha(\alpha B)_v + \alpha^2 BD. \tag{6.1}$$

Now, changing $\alpha K_1 + \beta K_2$ to $\alpha K_1 - \beta K_2$, we obtain

$$(\alpha K_1 - \beta K_2)_{uv} = \big(-(\beta C)_u + C(\alpha B - \beta A)\big)K_1 + \big((\alpha B)_v + B(\alpha D - \beta C)\big)K_2,$$

and we want this to be parallel to $\alpha K_1 - \beta K_2$. For this, we need

$$(-\beta)(-\beta C)_u + (-\beta)^2 AC = \alpha(\alpha B)_v + \alpha^2 BD,$$

which holds, as it is equivalent to (6.1). $\qquad\square$

Remark 6.1.4 If we rescale K_1 and K_2 so that $\alpha = \beta = 1$ in Lemma 6.1.3, then the argument in the proof of Lemma 6.1.3 shows $A = D = 0$, so

$$K_{1,u} \parallel K_2 \quad \text{and} \quad K_{2,v} \parallel K_1.$$

Flat Connections Again, similar to the isothermic surfaces case seen in Theorem 3.3.15, isothermic sphere congruences also admit a 1-parameter family of flat connections defined on the trivial vector bundle $\underline{\mathbb{R}}^{4,2} := \Sigma \times \mathbb{R}^{4,2}$. Using the analogous definition of the wedge product from Sect. 3.1, we use the identification

$$\wedge^2 \mathbb{R}^{4,2} \ni A \wedge B \sim A \wedge B \in \mathfrak{o}_{4,2}$$

for $A, B \in \mathbb{R}^{4,2}$ where $A \wedge B$ acts on any $X \in \mathbb{R}^{4,2}$ via

$$(A \wedge B)X = (A, X)B - (B, X)A.$$

Theorem 6.1.5 *For a sphere congruence* $s : \Sigma \to P(L^5)$, *s is isothermic with isothermic coordinates* $(u, v) \in \Sigma$ *if and only if* $\Gamma^\lambda = \mathrm{d} + \lambda\tau$ *is flat for any choice of* λ, *where the retraction form* τ *of* s *is*

$$\tau := -\frac{2}{(S_u, S_u)} S \wedge S_u \mathrm{d}u + \frac{2}{(S_v, S_v)} S \wedge S_v \mathrm{d}v \tag{6.2}$$

for any section $S \in \Gamma s$.

Remark 6.1.6 Note that τ is invariant of the choice of section $S \in \Gamma s$ (see Exercise 3.3.14).

Proof We first note that for τ defined as in (6.2), we always have $[\tau \wedge \tau] = 0$; therefore, the flatness of $\mathrm{d} + \lambda\tau$ is equivalent to the closure of τ, i.e. $\mathrm{d}\tau = 0$.

Suppose that s is isothermic with isothermic coordinates $(u, v) \in \Sigma$. Letting $\varsigma \in \Gamma s$ be the Moutard lift so that $\varsigma_{uv} = \alpha\varsigma$ for some function α, we have

$$2(\varsigma_u, \varsigma_u)_v = (\varsigma_{uv}, \varsigma_u) = \alpha(\varsigma, \varsigma_u) = 0,$$
$$2(\varsigma_v, \varsigma_v)_u = (\varsigma_{uv}, \varsigma_v) = \alpha(\varsigma, \varsigma_v) = 0.$$

Therefore, we may scale u and v independently and assume without loss of generality that

$$(\varsigma_u, \varsigma_u) = (\varsigma_v, \varsigma_v) = 2,$$

allowing us to write the retraction form in terms of ς as

$$\tau = \varsigma \wedge (-\varsigma_u \mathrm{d}u + \varsigma_v \mathrm{d}v).$$

Then we have

$$\mathrm{d}\tau = (\varsigma_v \wedge \varsigma_u + \varsigma \wedge \varsigma_{uv} + \varsigma_u \wedge \varsigma_v + \varsigma \wedge \varsigma_{uv})(\mathrm{d}u \wedge \mathrm{d}v) = 0,$$

since ς is a Moutard lift.

On the other hand, let $\mathrm{d} + \lambda\tau$ be a 1-parameter family of flat connections. Then since $\mathrm{d}\tau = 0$, we have

$$0 = -\frac{2}{(S_u, S_u)}(S_v \wedge S_u + S \wedge S_{uv}) + \frac{4(S_u, S_{uv})}{(S_u, S_u)^2} S \wedge S_u \tag{6.3}$$

$$-\frac{2}{(S_v, S_v)}(S_u \wedge S_v + S \wedge S_{uv}) + \frac{4(S_v, S_{uv})}{(S_v, S_v)^2} S \wedge S_v,$$

so that

$$S_{uv} \in \mathrm{span}\{S, S_u, S_v\}.$$

In particular, we then have $(S_{uv}, S) = 0$, or equivalently,

$$(S_u, S_v) = 0,$$

and evaluating (6.3) on both sides with S_u implies

$$2S_v + \tfrac{2(S_u,S_{uv})}{(S_u,S_u)} S - \tfrac{4(S_u,S_{uv})}{(S_u,S_u)} S - \tfrac{2(S_u,S_u)}{(S_v,S_v)} S_v + \tfrac{2(S_u,S_{uv})}{(S_v,S_v)} S = 0.$$

Finally, taking the inner product with S_v on both sides tells us

$$2(S_v, S_v) - 2(S_u, S_u) = 0,$$

allowing us to use Exercise 5.8.46 to conclude that s is isothermic. $\square$

The existence of a 1-parameter family of flat connections implies the existence of the trivializing gauge transformation via Proposition 3.2.11:

Definition 6.1.7 Let $s : \Sigma \to P(L^5)$ be an isothermic sphere congruence with associated flat connection Γ^λ. A *Calapso transformation* $T^\lambda : \Sigma \to O_{4,2}$ is the orthogonal trivializing gauge transformation of Γ^λ, i.e.

$$T^\lambda \bullet \Gamma^\lambda = \mathrm{d}.$$

Then $s^\mu := \mathrm{span}\{T^\mu S\} : \Sigma \to P(L^5)$ for $S \in \Gamma s$ is called a *Calapso transform* (with respect to parameter μ) of s.

We leave the checking of the following analogous results regarding Calapso transformations of an isothermic sphere congruences as exercises (see Sect. 3.4):

Exercise 6.1.8 Show that $(T^\lambda)^{-1}\mathrm{d}T^\lambda = \lambda\tau$.

Exercise 6.1.9 Let $s : \Sigma \to P(L^5)$ be an isothermic sphere congruence so that it admits a Moutard lift $\varsigma \in \Gamma s$. Using Exercise 6.1.8, show that $T^\mu\varsigma$ is also Moutard, and conclude that $s^\mu : \Sigma \to P(L^5)$ is also isothermic.

Exercise 6.1.10 Let τ^μ be the retraction form of s^μ. Show that we have $\tau^\mu = T^\mu\tau(T^\mu)^{-1}$.

6.2 Flat Connections of Ω-surfaces

Assuming that $\Lambda : \Sigma \to \mathcal{Z}$ is an Ω-surface, let $s^\pm : \Sigma \to P(L^5)$ be the pair of isothermic sphere congruences enveloping Λ. Denote the corresponding retraction forms by $\tau^\pm$, and the flat connections by $\Gamma^{\pm,\lambda} = \mathrm{d}+\lambda\tau^\pm$. First, we investigate how the two retraction forms are related.

Lemma 6.2.11 *For the right choices of Moutard lifts $\varsigma^{\pm} \in \Gamma s^{\pm}$,*

$$\tau^+ - \tau^- = d(\varsigma^+ \wedge \varsigma^-).$$

Proof By Corollary 5.12.67, Lemma 6.1.3 and Remark 6.1.4, we can normalize the lifts of the principal curvature spheres K_1, $K_2 \in \Gamma\Lambda$ so that

$$\varsigma^+ = \frac{1}{\sqrt{2}}(K_1 + K_2), \quad \varsigma^- = \frac{1}{\sqrt{2}}(K_1 - K_2) \tag{6.4}$$

are Moutard lifts of the isothermic sphere congruences, and then

$$(K_1)_u = \beta K_2 \quad \text{and} \quad (K_2)_v = \gamma K_1$$

for some functions β and γ. Since $\varsigma^{\pm}$ are Moutard lifts, $(\varsigma_u^{\pm}, \varsigma_u^{\pm})$ can be taken to be constant (see the proof of Theorem 6.1.5), so without loss of generality assume that $(\varsigma_u^{\pm}, \varsigma_u^{\pm}) = 2$. Then expressing the retraction forms $\tau^{\pm}$ using the respective Moutard lifts $\varsigma^{\pm}$, we have

$$\begin{aligned}
\tau^+ &- \tau^- - d(\varsigma^+ \wedge \varsigma^-) \\
&= \varsigma^+ \wedge (-\varsigma_u^+ du + \varsigma_v^+ dv) - \varsigma^- \wedge (-\varsigma_u^- du + \varsigma_v^- dv) \\
&\quad - \varsigma^+ \wedge (\varsigma_u^- du + \varsigma_v^- dv) + \varsigma^- \wedge (\varsigma_u^+ du + \varsigma_v^+ dv) \\
&= -\varsigma^+ \wedge ((\varsigma^+ + \varsigma^-)_u du + (\varsigma^- - \varsigma^+)_v dv) \\
&\quad + \varsigma^- \wedge ((\varsigma^+ + \varsigma^-)_u du + (\varsigma^+ - \varsigma^-)_v dv) \\
&= -\sqrt{2}(\varsigma^+ \wedge (K_{1,u} du - K_{2,v} dv) - \varsigma^- \wedge (K_{1,u} du + K_{2,v} dv)) \\
&= -\sqrt{2}(\varsigma^+ \wedge (\beta K_2 du - \gamma K_1 dv) - \varsigma^- \wedge (\beta K_2 du + \gamma K_1 dv)) \\
&= -\sqrt{2}\left(\beta(\varsigma^+ - \varsigma^-) \wedge K_2 du - \gamma(\varsigma^+ + \varsigma^-) \wedge K_1 dv\right) \\
&= -\sqrt{2}\left(\beta(\sqrt{2}K_2) \wedge K_2 du - \gamma(\sqrt{2}K_1) \wedge K_1 dv\right) = 0.
\end{aligned}$$

$\square$

With $\varsigma^{\pm}$ as in (6.4), we have that

$$\varsigma^+ \wedge \varsigma^- = -K_1 \wedge K_2, \tag{6.5}$$

and then

$$\tau^+ - \tau^- = -d(K_1 \wedge K_2). \tag{6.6}$$

Now, we examine the relationship between the pair of flat connections of isothermic sphere congruences enveloping an Ω-surface. To do this, fix the particular lifts of the curvature spheres $K_1, K_2 \in \Gamma \Lambda$ and Moutard lifts $\varsigma^{\pm} \in \Gamma s^{\pm}$ so that (6.4) holds. We will first prove the following useful lemma:

Lemma 6.2.12 *Let $\Lambda : \Sigma \to \mathcal{Z}$ be a Legendre immersion. For any $S_1, S_2 \in \Gamma \Lambda$, we have $(S_1 \wedge S_2)^2 = 0$ and $(\mathrm{d}(S_1 \wedge S_2))(S_1 \wedge S_2) = 0$.*

Proof This result follows from noting that at every point in the domain,

$$\mathrm{Image}(S_1 \wedge S_2) \subset \Lambda = \mathrm{span}\{S_1, S_2\},$$

and $S_1 \wedge S_2$ restricted to $\mathrm{span}\{S_1, S_2\}$ is zero. Furthermore, for $Z \in \mathbb{R}^{4,2}$, we have that $\hat{Z} := (S_1 \wedge S_2)Z \in \mathrm{span}\{S_1, S_2\}$, so that

$$(\mathrm{d}(S_1 \wedge S_2))(S_1 \wedge S_2)Z = (\mathrm{d}S_1 \wedge S_2 + S_1 \wedge \mathrm{d}S_2)\hat{Z} = 0,$$

by the contact condition of Λ. $\square$

We now define g^{λ} as

$$g^{\lambda} := \exp(\lambda K_1 \wedge K_2) = 1 + \lambda K_1 \wedge K_2 + \sum_{n=2}^{\infty} \frac{1}{n!} \lambda^n (K_1 \wedge K_2)^n = 1 + \lambda K_1 \wedge K_2,$$

where we used Lemma 6.2.12.

Exercise 6.2.13 Show that $(g^{\lambda})^{-1} = g^{-\lambda}$.

Then we can show the following:

Lemma 6.2.14 *We have that $g^{\lambda} \bullet \mathrm{d} = \mathrm{d} - \lambda \mathrm{d}(K_1 \wedge K_2)$.*

Proof We have

$$g^{\lambda} \circ \mathrm{d} \circ (g^{\lambda})^{-1} = \mathrm{d} + g^{\lambda} \mathrm{d}((g^{\lambda})^{-1}) = \mathrm{d} - (\mathrm{d}g^{\lambda})(g^{\lambda})^{-1}$$

$$= \mathrm{d} - \lambda(\mathrm{d}(K_1 \wedge K_2))(1 - \lambda K_1 \wedge K_2),$$

and the result follows from Lemma 6.2.12. $\square$

Lemma 6.2.14 can be used to show that g^{λ} is the gauge transformation between the 1-parameter family of flat connections $\Gamma^{\pm,\lambda}$ coming from the pair of isothermic sphere congruences enveloping Λ:

Theorem 6.2.15 *Suppose $\Lambda : \Sigma \to \mathbb{Z}$ is an Ω-surface enveloped by isothermic sphere congruences $s^{\pm} : \Sigma \to P(L^5)$ with associated flat connections $\Gamma^{\pm,\lambda}$. Then we have*

$$g^{\lambda} \bullet \Gamma^{-,\lambda} = \Gamma^{+,\lambda}.$$

Proof We have

$$g^\lambda \circ \Gamma^{-,\lambda} \circ (g^\lambda)^{-1} = g^\lambda \circ (\mathrm{d} + \lambda\tau^-) \circ (g^\lambda)^{-1}$$

$$= \mathrm{d} - \lambda\mathrm{d}(K_1 \wedge K_2) + \lambda g^\lambda \tau^- (g^\lambda)^{-1}$$

$$= \mathrm{d} + \lambda(\tau^+ - \tau^-) + \lambda g^\lambda \tau^- (g^\lambda)^{-1},$$

by Lemma 6.2.14 and (6.6). Because the image of τ^- is perpendicular to Λ, the contact condition gives

$$g^\lambda \tau^- (g^\lambda)^{-1} = (1 + \lambda K_1 \wedge K_2)\tau^-(1 - \lambda K_1 \wedge K_2) = \tau^-,$$

so

$$g^\lambda \bullet \Gamma^{-,\lambda} = \mathrm{d} + \lambda(\tau^+ - \tau^-) + \lambda\tau^- = \mathrm{d} + \lambda\tau^+,$$

and the result follows. $\qquad\qquad\qquad\qquad\qquad\qquad\qquad\qquad\qquad\qquad\qquad\square$

Exercise 6.2.16 Show that $[\tau^+ \wedge \tau^-] = 0$. (Hint: express $\tau^\pm$ using the Moutard lifts $\varsigma^\pm \in \Gamma s^\pm$, and use the contact condition of Λ.)

6.3   Calapso Transformations of Ω-surfaces

As seen in Definition 6.1.7, isothermic sphere congruences $s^\pm$ and the respective flat connections $\Gamma^{\pm,\lambda}$ admit Calapso transformations via the trivializing gauge transformations $T^{\pm,\lambda}$, i.e.

$$T^{\pm,\lambda} \bullet \Gamma^{\pm,\lambda} = \mathrm{d}.$$

Since the two flat connections coming from the pair of isothermic sphere congruences are related by a gauge transformation as in Theorem 6.2.15, we have the following lemma:

Lemma 6.3.17 *Under suitable initial conditions for* $T^{\pm,\lambda}$,

$$T^{-,\lambda} = T^{+,\lambda} g^\lambda. \qquad\qquad\qquad\qquad (6.7)$$

Proof Using the definition of trivializing gauge and Theorem 6.2.15, we have that

$$T^{-,\lambda} \bullet \Gamma^{-,\lambda} = \mathrm{d} = T^{+,\lambda} \bullet \Gamma^{+,\lambda} = T^{+,\lambda} \bullet (g^\lambda \bullet \Gamma^{-,\lambda}) = (T^{+,\lambda} g^\lambda) \bullet \Gamma^{-,\lambda}.$$

Thus, choosing the right initial conditions, the uniqueness of solutions to differential equations gives us the desired conclusion. $\qquad\qquad\qquad\qquad\qquad\qquad\square$

Noting that for any $S \in \Gamma\Lambda$, we have $g^\lambda S = S$, and (6.7) tells us the following:

Proposition 6.3.18 *Let $\Lambda : \Sigma \to \mathcal{Z}$ be an Ω-surface with enveloping isothermic sphere congruences $s^\pm$, with respective Calapso transformaions $T^{\pm,\lambda}$. Then we have*

$$T^{+,\lambda}\Lambda = T^{-,\lambda}\Lambda$$

at every point in the domain.

Hence, $\Lambda^\mu : \Sigma \to \mathcal{Z}$ defined via $\Lambda^\mu := T^{+,\mu}\Lambda = T^{-,\mu}\Lambda$ is well-defined. Furthermore, we know that $s^{\pm,\mu} := T^{\pm,\mu}s^\pm : \Sigma \to P(L^5)$ are isothermic sphere congruences enveloping Λ^μ, telling us that Λ^μ is again an Ω-surface:

Definition 6.3.19 We call $\Lambda^\mu := T^{+,\mu}\Lambda = T^{-,\mu}\Lambda$ a *Calapso transform* of Λ with parameter μ.

Thus, Calapso transformations of a given Ω-surface constitute a 1-parameter family of deformations of the given surface. In fact, analogous to the isothermic surfaces in Möbius geometry, it is known that these are second order deformations of the original surface in Lie sphere geometry (see [165]), making them *Lie applicable*.

6.4 Christoffel Dual Lifts of Ω-surfaces

The next lemma introduces another characterization of Ω-surfaces.

Lemma 6.4.20 (cf. [177, p. 60], [176, Proposition 3.17]) *Consider a Legendre immersion $\Lambda : \Sigma \to \mathcal{Z}$, and suppose there exist sections $\sigma^\pm \in \Gamma\Lambda$ such that*

$$\mathrm{d}\sigma^+ \wedge \mathrm{d}\sigma^- = 0.$$

Then

1. $s^\pm := \mathrm{span}\{\sigma^\pm\}$ are isothermic with the same isothermic coordinates $(u, v) \in \Sigma$,
2. there exists a scalar function f such that $\sigma_u^+ = f\sigma_u^-$ and $\sigma_v^+ = -f\sigma_v^-$, and
3. $\sigma_{uv}^+ \in \mathrm{span}\{\sigma_u^+, \sigma_v^+\}$, $\sigma_{uv}^- \in \mathrm{span}\{\sigma_u^-, \sigma_v^-\}$.

Thus, Λ is an Ω-surface, and $\sigma^\pm$ are called Christoffel dual lifts.

Proof Take $(u, v) \in \Sigma$ to be conformal coordinates for σ^+. From $\mathrm{d}\sigma^+ \wedge \mathrm{d}\sigma^- = 0$, we have

$$\sigma_u^+ \wedge \sigma_v^- - \sigma_v^+ \wedge \sigma_u^- = 0$$

and also

$$\sigma_u^- = \alpha\sigma_u^+ - \beta\sigma_v^+, \quad \sigma_v^- = \gamma\sigma_u^+ - \alpha\sigma_v^+$$

for some scalar functions α, β, γ. Then, because

$$(\sigma_u^-, \sigma_v^+) = (\sigma^-, \sigma_v^+)_u - (\sigma^-, \sigma_{uv}^+) = -(\sigma^-, \sigma_{uv}^+)$$
$$= -(\sigma^-, \sigma_u^+)_v + (\sigma_v^-, \sigma_u^+) = (\sigma_v^-, \sigma_u^+),$$

we have

$$\gamma = -\beta.$$

We can also now see that (u, v) are conformal coordinates for σ^- as well.

Set $E := (\sigma_u^+, \sigma_u^+)$. The compatibility condition $\sigma_{uv}^- = \sigma_{vu}^-$ gives $(\sigma_{uv}^- - \sigma_{vu}^-, \sigma_u^+) = (\sigma_{uv}^- - \sigma_{vu}^-, \sigma_v^+) = 0$, and thus

$$(\alpha_v + \beta_u)\sigma_u^+ + (\alpha_u - \beta_v)\sigma_v^+ + \beta(\sigma_{uu}^+ - \sigma_{vv}^+) + 2\alpha\sigma_{uv}^+ = 0 \qquad (6.8)$$

and

$$0 = (\alpha_v + \beta_u)E + \beta E_u + \alpha E_v = (\alpha_u - \beta_v)E - \beta E_v + \alpha E_u.$$

Therefore,

$$0 = (\beta E)_u + (\alpha E)_v = (-\beta E)_v + (\alpha E)_u,$$

implying that $E(\alpha + \sqrt{-1}\beta)$ is holomorphic with respect to $u + \sqrt{-1}v$.

Let Z be any smooth vector field so that $(Z, \sigma_u^+) = (Z, \sigma_v^+) = 0$. Then (6.8) implies

$$2\alpha(\sigma_{uv}^+, Z) + \beta(\sigma_{uu}^+ - \sigma_{vv}^+, Z) = 0. \qquad (6.9)$$

Define $a + \sqrt{-1}b$ with $a, b \in \mathbb{R}$, by

$$(a + \sqrt{-1}b)^2 = E(\alpha + \sqrt{-1}\beta),$$

so that $a + \sqrt{-1}b$ is holomorphic as well. It follows that

$$d(bdu + adv) = d(-adu + bdv) = 0,$$

so we can make a conformal change of coordinates $(u, v) \to (\tilde{u}, \tilde{v})$ so that

$$d\tilde{u} = bdu + adv, \quad d\tilde{v} = -adu + bdv.$$

Also,

$$\partial_{\tilde{u}} = \frac{b}{a^2 + b^2}\partial_u + \frac{a}{a^2 + b^2}\partial_v, \quad \partial_{\tilde{v}} = \frac{-a}{a^2 + b^2}\partial_u + \frac{b}{a^2 + b^2}\partial_v.$$

A direct computation gives

$$\sigma^+_{\tilde{u}\tilde{v}} = \frac{1}{(a^2 + b^2)^2}\left(-ab\sigma^+_{uu} - a^2\sigma^+_{uv} + b^2\sigma^+_{uv} + ab\sigma^+_{vv}\right)$$

$$- \frac{a}{a^2 + b^2}\left(\frac{b}{a^2 + b^2}\right)_u \sigma^+_u - \frac{a}{a^2 + b^2}\left(\frac{a}{a^2 + b^2}\right)_u \sigma^+_v$$

$$+ \frac{b}{a^2 + b^2}\left(\frac{b}{a^2 + b^2}\right)_v \sigma^+_u + \frac{b}{a^2 + b^2}\left(\frac{a}{a^2 + b^2}\right)_v \sigma^+_v,$$

and then it follows from (6.9) that $(\sigma^+_{\tilde{u}\tilde{v}}, Z) = 0$. In particular, for any $Y + s^+ \in \mathcal{N}s^+$ where $s^+ = \mathrm{span}\{\sigma^+\}$, we see that $(Y, \sigma^+_u) = (Y, \sigma^+_v) = 0$ so that $(\tilde{u}, \tilde{v})$ are isothermic coordinates for σ^+. Similarly, $(\sigma^-_{\tilde{u}\tilde{v}}, Z) = 0$, and $(\tilde{u}, \tilde{v})$ are isothermic coordinates for σ^- as well.

Now,

$$\sigma^-_{\tilde{u}} = \frac{b\sigma^-_u + a\sigma^-_v}{a^2 + b^2} = \frac{b(\alpha\sigma^+_u - \beta\sigma^+_v) - a(\beta\sigma^+_u + \alpha\sigma^+_v)}{a^2 + b^2}$$

$$= \frac{-b}{E}\sigma^+_u - \frac{a}{E}\sigma^+_v = -\frac{a^2 + b^2}{E}\sigma^+_{\tilde{u}},$$

and similarly

$$\sigma^-_{\tilde{v}} = \frac{a^2 + b^2}{E}\sigma^+_{\tilde{v}},$$

giving us the second claim.

Finally, because $\sigma^+_{\tilde{u}\tilde{v}}$ is perpendicular to the entire space $\mathrm{span}\{\sigma^+_u, \sigma^+_v\}^\perp$ at any point in the domain, and in particular $(\sigma^+_{\tilde{u}\tilde{v}}, Z) = 0$ even for vector fields Z in $\mathrm{span}\{\sigma^+_u, \sigma^+_v\}^\perp$ such that $Z \not\perp \sigma^+$, it follows that $\sigma^+_{\tilde{u}\tilde{v}} \in \mathrm{span}\{\sigma^+_u, \sigma^+_v\} = \mathrm{span}\{\sigma^+_{\tilde{u}}, \sigma^+_{\tilde{v}}\}$. Similarly, $\sigma^-_{\tilde{u}\tilde{v}} \in \mathrm{span}\{\sigma^-_{\tilde{u}}, \sigma^-_{\tilde{v}}\}$. $\qquad\square$

Remark 6.4.21 The converse of Lemma 6.4.20 also holds, namely, given an Ω-surface, one can scale sections representing isothermic sphere congruences to obtain Christoffel dual lifts. See, for example, [176].

6.5   Other Lie Applicable Surfaces

We have shown that Ω-surfaces (enveloped by a pair of (real) isothermic sphere congruences in $P(L^5)$) admit Calapso transformations, and hence are called Lie applicable surfaces. In this section, we look at two other cases of Lie applicable surfaces.

Revisiting the arguments about Ω-surfaces in Sect. 5.10, we could also consider the case that the metric in (5.13) is Lorentz conformal so that

$$g_{11}(\alpha - k_1)^2 = -g_{22}(\alpha - k_2)^2.$$

This gives

$$\alpha = \frac{k_1\sqrt{g_{11}} \mp \sqrt{-1}k_2\sqrt{g_{22}}}{\sqrt{g_{11}} \mp \sqrt{-1}\sqrt{g_{22}}},$$

resulting in $s^\pm$ being complex conjugate to each other.

Consideration of this case requires us to complexify $\mathbb{R}^{4,2}$: We use $(\cdot, \cdot)$ to denote the bilinear extension of the $\mathbb{R}^{4,2}$ metric to $\mathbb{R}^{4,2}\otimes\mathbb{C}$. Note that this bilinear extension is defined without using complex conjugation, so, for example, for a non-null vector Z in $\mathbb{R}^{4,2}$, we have

$$((1 + \sqrt{-1})Z, (1 + \sqrt{-1})Z) = 2\sqrt{-1}(Z, Z) \notin \mathbb{R},$$

and thus the bilinear extension $(\cdot, \cdot)$ is not a bona fide metric. (Rather than using a true metric combined with additional internal complex conjugation, it is notationally simpler to use this bilinear extension to obtain the same object.)

In this case, Demoulin's equation becomes

$$\left(\frac{\sqrt{g_{11}}}{\sqrt{g_{22}}}\frac{k_{1,u}}{k_1 - k_2}\right)_v - \left(\frac{\sqrt{g_{22}}}{\sqrt{g_{11}}}\frac{k_{2,v}}{k_1 - k_2}\right)_u = 0. \tag{6.10}$$

These surfaces are called Ω as well, and the $s^\pm$ are called complex isothermic sphere congruences.

Exercise 6.5.22 In this exercise, we will aim to recover the complex analogue of Lemma 5.12.65. To do this, we first extend the notion of cross ratios to the complexified line in $\mathbb{R}^{4,2}$, that is, for $\mathbf{v}_1, \mathbf{v}_2, \mathbf{v}_3, \mathbf{v}_4 \in \mathbb{R}^{4,2} \otimes \mathbb{C}$ so that

$$\mathbf{v}_j = \mathbf{v} + \alpha_j\mathbf{w}$$

for some $\mathbf{v}, \mathbf{w} \in \mathbb{R}^{4,2}$ and $\alpha_j \in \mathbb{C}$, let

$$\mathrm{cr}(\mathbf{v}_1, \mathbf{v}_2, \mathbf{v}_3, \mathbf{v}_4) = \frac{\alpha_2 - \alpha_1}{\alpha_3 - \alpha_2}\frac{\alpha_4 - \alpha_3}{\alpha_1 - \alpha_4}.$$

Using this notion of cross ratios, show that the complex pair of isothermic sphere congruences harmonically separate the curvature spheres.

Exercise 6.5.23 Prove the complex version of Lemma 5.12.66 and Corollary 5.12.67. That is, using the harmonic separation, show that for the curvature sphere congruences K_1 and K_2, $\mathrm{cr}(K_1, \alpha_1 K_1 + \sqrt{-1}\alpha_2 K_2, K_2, \beta_1 K_1 - \sqrt{-1}\beta_2 K_2) = -1$ implies that the 2-dimensional vectors (α_1, α_2) and (β_1, β_2) are parallel. Therefore, conclude that the complex isothermic sections $S^{\pm}$ can be chosen so that

$$S^+ = \alpha_1 K_1 + \sqrt{-1}\alpha_2 K_2 \quad \text{and} \quad S^- = \overline{S^+} = \alpha_1 K_1 - \sqrt{-1}\alpha_2 K_2.$$

Exercise 6.5.24 Prove the complex version of Lemma 6.1.3. That is, denoting the curvature sphere congruences of a Legendre immersion Λ by $s_1 = \mathrm{span}\{K_1\}$ and $s_2 = \mathrm{span}\{K_2\}$, show that if $\alpha K_1 + \sqrt{-1}\beta K_2$ is Moutard for some real functions α and β, then so is $\alpha K_1 - \sqrt{-1}\beta K_2$. Rescaling K_1 and K_2 appropriately so that $K_{1,u} \parallel K_2$ and $K_{2,v} \parallel K_1$, show that Moutard lifts can be given by

$$\varsigma^{\pm} := K_1 \pm \sqrt{-1}K_2.$$

Remark 6.5.25 Analogously to Lemma 6.4.20, if there exist complex sections $\sigma^{\pm}$ of Λ such that

$$d\sigma^+ \wedge d\sigma^- = 0,$$

then one can show that $\sigma^{\pm}$ are sections of complex conjugate isothermic sphere congruences for Λ.

We could also consider the case that the metric in (5.13) is degenerate, i.e.,

$$(\alpha - k_1)^2 (\alpha - k_2)^2 = 0.$$

Without loss of generality assume that $\alpha = k_1$; hence, we are now considering the case when the curvature sphere congruence is isothermic. To obtain Demoulin's equation in this case, we will use the Moutard lift characterization of isothermic sphere congruences.

Suppose that $s = \mathrm{span}\{S\}$ is an isothermic curvature sphere congruence with $S = k_1 X + N$. Taking the Moutard lift $\varsigma \in \Gamma s$, we have that $\varsigma_{uv} \parallel \varsigma$. However, we have that $\varsigma = \beta S$ for some real function β; thus,

$$\varsigma_{uv} = (\beta S)_{uv} \parallel S,$$

and this implies

$$S_{uv} = \gamma S - (\log \beta)_u S_v - (\log \beta)_v S_u$$

for some function γ. Since $(S, \mathfrak{p}) = -1$, taking the inner product with $\mathfrak{p}$ tells us that γ must be zero. Using Rodrigues equations, we have

$$S_u = k_{1,u} X, \quad S_v = k_{1,v} X + (k_1 - k_2) X_v$$

so that

$$k_{1,uv} X + k_{1,u} X_v = (S_u)_v = -(\log \beta)_u (k_{1,v} X + (k_1 - k_2) X_v) - (\log \beta)_v k_{1,u} X.$$

Comparing coefficients, we see

$$(\log \beta)_u = -\frac{k_{1,u}}{k_1 - k_2} = -(\log(k_1 - k_2))_u - \frac{k_{2,u}}{k_1 - k_2}$$

and

$$k_{1,uv} + \frac{k_{1,u}}{k_2 - k_1} k_{1,v} + (\log \beta)_v k_{1,u} = 0.$$

Now using the Codazzi equations (2.35) tells us

$$(\log \beta)_u = -(\log(k_1 - k_2))_u - \tfrac{1}{2}(\log g_{22})_u,$$

$$(\log \beta)_v = -\frac{k_{1,uv}}{k_{1,u}} - \frac{g_{11,v}}{2g_{11}}.$$

Therefore,

$$\log \beta = -\log(k_{1,u}) - \tfrac{1}{2} \log g_{11} + f_1(u) = -\log(k_1 - k_2) - \tfrac{1}{2} \log g_{22} + f_2(v),$$

where $f_1(u)$, resp. $f_2(v)$, is some function depending only on u, resp. v. We can reparametrize the v coordinate so that $f_2(v) = c_2$ is constant. Thus, taking the exponential of this equation, we have

$$\frac{k_{1,u}}{k_1 - k_2} \frac{\sqrt{g_{11}}}{\sqrt{g_{22}}} = f_3(u)$$

for some function $f_3(u)$ depending only on u, so

$$\left(\frac{\sqrt{g_{11}}}{\sqrt{g_{22}}} \frac{k_{1,u}}{k_1 - k_2} \right)_v = 0, \tag{6.11}$$

giving us Demoulin's equation for Ω_0-surfaces.

Definition 6.5.26 An Ω_0-*surface* is a surface where one of the curvature sphere congruences is isothermic.

The curvature sphere congruence is isothermic, and hence admits a 1-parameter of associated flat connections. Hence, one can define the Calapso transformations of

Ω_0 surfaces via the Calapso transformations of the isothermic sphere congruence. The union of all Ω-surfaces (including the complex conjugate case) and all Ω_0-surfaces are referred to as *Lie applicable surfaces* (see, for example, [14, 101, 165, 176]).

Exercise 6.5.27 Using Exercise 5.5.35, show that channel surfaces are Ω_0 surfaces. (Hint: use the Demoulin's equation (6.11).)

6.6 Guichard Surfaces

We have seen that isothermic surfaces (Lemma 5.11.60) and L-isothermic surfaces (Lemma 5.11.58) are subclasses of Ω-surfaces. We now introduce another important subclass of Ω-surfaces:

Definition 6.6.28 ([109]) A surface $x : \Sigma \to \mathbb{R}^3$ is called a *Guichard surface* if it admits a surface $x^* : \Sigma \to \mathbb{R}^3$ with parallel curvature directions to x such that

$$\frac{1}{k_1 k_2^*} + \frac{1}{k_2 k_1^*} = \text{constant} \ (\neq 0),$$

where k_1, k_2 and k_1^*, k_2^* are the principal curvatures of x and x^* respectively.

The characterization of Guichard surfaces can be done without the need for a second surface, as shown by [56]: A surface parametrized by curvature lines in $\mathbb{R}^3$ is Guichard if and only if it satisfies the Calapso's equation

$$cg_{11}g_{22}(k_1 - k_2)^2 = g_{22} - \epsilon^2 g_{11}, \tag{6.12}$$

for some choice of curvature line coordinates (u, v), where c is a non-zero constant and $\epsilon \in \{1, \sqrt{-1}\}$. In fact, Calapso's equation (6.12) allows for the definition of Guichard surfaces to be extended to surfaces in other 3-dimensional space forms:

Proposition 6.6.29 *Suppose a surface $x : \Sigma \to \mathbb{R}^3$ is Guichard, and let $L : \Sigma \to P(L^4)$ be its lift to Möbius geometry. If $X_\kappa : \Sigma \to M_\kappa \cong \mathfrak{R}_\kappa$ is any other space form projection of L, then X_κ also satisfies Calapso's equation.*

Proof Since we have that x is Guichard, we may assume that Calapso's equation (6.12) holds, where g and k_i denote the first fundamental form and the principal curvatures of x in the Euclidean space. Now suppose for the lift of x in Möbius geometry L, let $X_\kappa \in \Gamma L$ with $X_\kappa : \Sigma \to M_\kappa \cong \mathfrak{R}_\kappa$. Denoting the first fundamental form and the principal curvatures of X_κ with $\mathfrak{R}_\kappa$ by $\hat{g}$ and $\hat{k}_i$, we use (5.22) to verify that

$$c\hat{g}_{11}\hat{g}_{22}(\hat{k}_1 - \hat{k}_2)^2 = ct^{-4}g_{11}g_{22}(tk_1 - tk_2)^2 = t^{-2}(g_{22} - \epsilon^2 g_{11}) = \hat{g}_{22} - \epsilon^2 \hat{g}_{11}.$$

Thus X_κ also satisfies Calapso's equation. $\square$

Thus, we will call a surface in any space form a *Guichard surface* if it satisfies Calapso's equation.

Remark 6.6.30 In fact, the notion of Guichard surfaces is sensible in Möbius geometry, so one need never reduce to a 3-dimensional space form (see [39]).

With this in mind, we show the following:

Lemma 6.6.31 (cf. [88]) *Guichard surfaces in space forms are Ω.*

Proof We take a Guichard surface in any space form and give it curvature line coordinates so that Calapso's equation (6.12) holds; therefore, taking the logarithm on both sides, we have

$$\log(k_1 - k_2) = \tfrac{1}{2}\log(g_{22} - \epsilon^2 g_{11}) - \tfrac{1}{2}\log c - \tfrac{1}{2}\log g_{11} - \tfrac{1}{2}\log g_{22}. \qquad (6.13)$$

Since Demoulin's equation (5.19) does not depend on the choice of the space form by Lemma 5.11.63, we will show that Calapso's equation implies Demoulin's equation:

$$\left(\frac{\sqrt{g_{11}}}{\sqrt{g_{22}}} \frac{k_{1,u}}{k_1 - k_2} \right)_v + \epsilon^2 \left(\frac{\sqrt{g_{22}}}{\sqrt{g_{11}}} \frac{k_{2,v}}{k_1 - k_2} \right)_u = 0,$$

for either $\epsilon = 1$ or $\epsilon = \sqrt{-1}$.

We take up just the case $\epsilon = 1$ here. Using the Codazzi equations (2.35), the left-hand side of Demoulin's equation becomes

$$\left(\frac{\sqrt{g_{11}}}{\sqrt{g_{22}}} \left((\log(k_1 - k_2))_u + \tfrac{1}{2}(\log g_{22})_u \right) \right)_v$$
$$- \left(\frac{\sqrt{g_{22}}}{\sqrt{g_{11}}} \left((\log(k_1 - k_2))_v + \tfrac{1}{2}(\log g_{11})_v \right) \right)_u,$$

and using (6.13), we obtain

$$\tfrac{1}{2} \left(\frac{\sqrt{g_{11}}}{\sqrt{g_{22}}} (\log(g_{22} - g_{11}))_u \right)_v - \tfrac{1}{2} \left(\frac{\sqrt{g_{11}}}{\sqrt{g_{22}}} (\log g_{11})_u \right)_v$$
$$- \tfrac{1}{2} \left(\frac{\sqrt{g_{22}}}{\sqrt{g_{11}}} (\log(g_{22} - g_{11}))_v \right)_u + \tfrac{1}{2} \left(\frac{\sqrt{g_{22}}}{\sqrt{g_{11}}} (\log g_{22})_v \right)_u$$
$$= \tfrac{1}{2} \left(\frac{\sqrt{g_{11}}}{\sqrt{g_{22}}} \left(\log \left(\tfrac{g_{22}}{g_{11}} - 1 \right) \right)_u \right)_v - \tfrac{1}{2} \left(\frac{\sqrt{g_{22}}}{\sqrt{g_{11}}} \left(\log \left(1 - \tfrac{g_{11}}{g_{22}} \right) \right)_v \right)_u,$$

which can be directly checked to vanish. $\qquad\qquad\square$

6.7   Linear Weingarten Surfaces

In Möbius geometry, we have seen that constant mean curvature surfaces are a subclass of isothermic surfaces, characterized by the existence of a certain polynomial conserved quantity. In this section, we consider the analogous problem for Ω-surfaces in Lie sphere geometry, and show that linear Weingarten surfaces can be characterized via polynomial conserved quantities of the flat connections coming from the enveloping isothermic sphere congruences.

Linear Weingarten surfaces in a space form $\mathfrak{R}_\kappa \cong M_\kappa$ are those surfaces whose (extrinsic) Gauss and mean curvatures K and H (with respect to the choice of the space form) satisfy

$$\alpha K + 2\beta H + \gamma = 0$$

for some constants α, β, γ not all zero (Figs. 6.1 and 6.2). Also recall the analogous definition of polynomial conserved quantities of isothermic flat connections: For an isothermic sphere congruence $s : \Sigma \to P(L^5)$ with associated 1-parameter family of flat connections Γ^λ, $P(\lambda)$ defined as

$$P(\lambda) = \sum_{k=0}^{n} \lambda^k P_k,$$

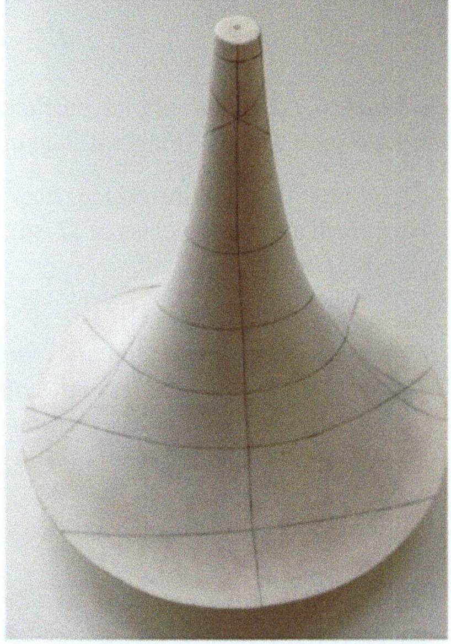

Fig. 6.1 Physical model of the pseudosphere in $\mathbb{R}^3$, which has constant negative Gaussian curvature, and hence is a linear Weingarten surface (owned by the geometry group at the TU Wien)

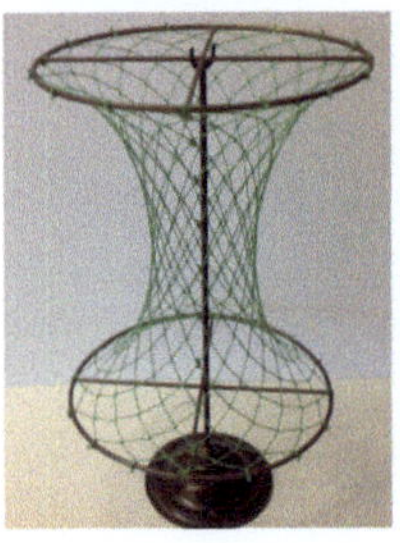

Fig. 6.2 Physical models of other surfaces of revolution in $\mathbb{R}^3$ with constant Gaussian curvature, which are again linear Weingarten surfaces (owned by the geometry group at the TU Wien)

for $P_k : \Sigma \to \mathbb{R}^{4,2}$ for $k = 0, \ldots, n$, is called a *polynomial conserved quantity of s of order n* if

$$\Gamma^\lambda P(\lambda) = 0$$

for all $\lambda \in \mathbb{R}$. We first treat a degenerate case:

Definition 6.7.32 A surface is *tubular* if one principal curvature is constant.

Thus a surface is tubular if $(k_1 - b)(k_2 - b) = 0$, where k_j are the principal curvatures of the surface and b is some constant. This amounts to the linear Weingarten condition:

$$K - 2bH + b^2 = 0.$$

Remark 6.7.33 For tubular surfaces in the Euclidean space, there is a parallel surface such that the image of the surface is a curve, called the *soul curve*.

The next theorem introduces how to identify linear Weingarten surfaces in some space form among the class of Ω-surfaces using the notion of polynomial conserved quantities:

Theorem 6.7.34 ([50, §3]) *Let* $\Lambda : \Sigma \to \mathcal{Z}$ *be a Legendre immersion. Then* Λ *projects to a non-tubular linear Weingarten surface in some space form if and only if* Λ *is an* Ω*-surface where the real pair (resp. complex pair) of enveloping isothermic sphere congruences admits constant conserved quantities* $Q^\pm \in \mathbb{R}^{4,2}$ *(resp.* $Q^+ = \overline{Q^-} \in \mathbb{R}^{4,2} \otimes \mathbb{C}$*) satisfying a non-degeneracy condition:*

$$\mathrm{span}\{Q^+, Q^-\} \ (\textit{resp. } \mathrm{span}\{\mathrm{Re}\ Q^+, \mathrm{Im}\ Q^+\}) \ \textit{is not a null 2-plane}. \qquad (6.14)$$

Proof First assume that Λ is an Ω-surface enveloped by a real pair of isothermic sphere congruences $s^\pm : \Sigma \to P(L^5)$ with associated flat connections $\Gamma^{\pm,\lambda}$. Denoting the constant conserved quantities of $\Gamma^{\pm,\lambda}$ as $Q^\pm$, we have that

$$(S^\pm, Q^\pm) = 0$$

for any sections $S^\pm \in \Gamma s^\pm$. Take the Moutard lifts $\varsigma^\pm \in \Gamma s^\pm$ as in Remark 6.1.4 so that

$$\varsigma^\pm = K_1 \pm K_2,$$

for $K_i \in \Gamma s_i$ where s_1 and s_2 denote the curvature sphere congruences.

Hence, the condition $(\varsigma^\pm, Q^\pm) = 0$ tells us that

$$(K_1, Q^\pm) = \mp(K_2, Q^\pm),$$

allowing us to obtain the relation

$$(K_1, Q^+)(K_2, Q^-) + (K_1, Q^-)(K_2, Q^+) = 0. \tag{6.15}$$

Now fix an orthogonal pair $\mathfrak{p}, \mathfrak{q} \in \mathrm{span}\{Q^+, Q^-\}$ with $(\mathfrak{p}, \mathfrak{p}) \neq 0$ and non-zero $\mathfrak{q}$ so that

$$Q^\pm = \alpha^\pm \mathfrak{p} + \beta^\pm \mathfrak{q}$$

for some $\alpha^\pm, \beta^\pm \in \mathbb{R} \setminus \{0\}$. Projecting Λ to the space form determined by $\mathfrak{p}$ and $\mathfrak{q}$, we have via Exercise 5.4.32 that

$$(K_i, Q^\pm) = (K_i, \alpha^\pm \mathfrak{p} + \beta^\pm \mathfrak{q}) = \alpha^\pm(K_i, \mathfrak{p}) + \beta^\pm(K_i, \mathfrak{q})$$

$$= (K_i, \mathfrak{p})\left(\alpha^\pm + \beta^\pm \frac{(K_i, \mathfrak{q})}{(K_i, \mathfrak{p})}\right) = (K_i, \mathfrak{p})(\alpha^\pm + \beta^\pm k_i).$$

Using this relation, (6.15) now implies in general that

$$(\alpha^+ + \beta^+ k_1)(\alpha^- + \beta^- k_2) + (\alpha^- + \beta^- k_1)(\alpha^+ + \beta^+ k_2) = 0,$$

or equivalently,

$$2\alpha^+\alpha^- + 2(\alpha^+\beta^- + \alpha^-\beta^+)\frac{k_1 + k_2}{2} + 2\beta^+\beta^- k_1 k_2 = 0. \tag{6.16}$$

As the choice for $\mathfrak{p}, \mathfrak{q} \in \mathrm{span}\{Q^+, Q^-\}$ were arbitrary, we see that Λ projects to a linear Weingarten surface in any space form determined by the choice of point sphere complex $\mathfrak{p}$ and space form vector $\mathfrak{q}$ in $\mathrm{span}\{Q^+, Q^-\}$. Finally, since $Q^+ \not\parallel Q^-$, we have

$$(\alpha^+\beta^- + \alpha^-\beta^+)^2 - 4\alpha^+\alpha^-\beta^+\beta^- = (\alpha^+\beta^- - \alpha^-\beta^+)^2 \neq 0$$

so that Λ is non-tubular. We leave the analogous proof of the complex case to an exercise.

To prove the converse, let X be the point sphere map of a Legendre immersion Λ with respect to some choice of point sphere complex $\mathfrak{p}$ and space form vector $\mathfrak{q}$ (such that $\mathrm{span}\{\mathfrak{p}, \mathfrak{q}\}$ is not null), and assume that X satisfies

$$\alpha K + 2\beta H + \gamma = 0$$

for some constant α, β, γ. Since the surface is non-tubular, we have that $\delta := \sqrt{\beta^2 - \alpha\gamma} \in \mathbb{R} \cup \sqrt{-1}\mathbb{R}$ is non-zero.

1. If $\alpha \neq 0$, define

$$Q^{\pm} = \alpha\mathfrak{q} + (\beta \mp \delta)\mathfrak{p}, \quad \sigma^{\pm} = \delta^{-1}((\delta \mp \beta)X \pm \alpha N).$$

When $\beta^2 - \alpha\gamma > 0$, the $Q^{\pm}$ and $\sigma^{\pm}$ will be real, and when $\beta^2 - \alpha\gamma < 0$, the $Q^{\pm}$ and $\sigma^{\pm}$ will be complex conjugate.
2. If $\alpha = 0$, define

$$Q^{+} = \beta\mathfrak{q} + \tfrac{1}{2}\gamma\mathfrak{p}, \quad Q^{-} = \mathfrak{p}, \quad \sigma^{+} = HX + N, \quad \sigma^{-} = X.$$

For case (1), we can verify that

$$(\sigma^{\pm}, Q^{\pm}) = -\alpha\delta^{-1}(\delta \mp \beta) \mp \alpha\delta^{-1}(\beta \mp \delta) = -\alpha\delta^{-1}(\delta \mp \beta \pm \beta - \delta) = 0,$$

while using

$$\sigma_u^{\pm} = \delta^{-1}(\delta \mp \beta \mp \alpha k_1)X_u =: \zeta_1^{\pm}X_u, \quad \sigma_v^{\pm} = \delta^{-1}(\delta \mp \beta \mp \alpha k_2)X_v =: \zeta_2^{\pm}X_v$$

we calculate that

$$\begin{aligned}
d\sigma^{+} \curlywedge d\sigma^{-} &= (\sigma_u^{+}du + \sigma_v^{+}dv) \curlywedge (\sigma_u^{-}du + \sigma_v^{-}dv) \\
&= (\sigma_u^{+} \wedge \sigma_v^{-} + \sigma_u^{-} \wedge \sigma_v^{+})(du \wedge dv) \\
&= (\zeta_1^{+}\zeta_2^{-} + \zeta_1^{-}\zeta_2^{+})(X_u \wedge X_v)(du \wedge dv) \\
&= -2\alpha\delta^{-2}\left(\alpha K + 2\beta H + \gamma\right)(X_u \wedge X_v)(du \wedge dv) = 0.
\end{aligned}$$

Also in case (2), it can be similarly verified that $(\sigma^{\pm}, Q^{\pm}) = 0$ and $d\sigma^{+} \curlywedge d\sigma^{-} = 0$.

Thus, $s^{\pm} := \mathrm{span}\{\sigma^{\pm}\}$ are isothermic sphere congruences enveloping $\Lambda := \mathrm{span}\{\sigma^{+}, \sigma^{-}\}$ (see Lemma 6.4.20 and Remark 6.5.25). $\square$

Exercise 6.7.35 Prove the necessary condition of Theorem 6.7.34 for complex case, analogously to the real case, as follows:

1. Use Exercise 6.5.24 to show that for some special sections $K_i \in \Gamma s_i$ of the curvature sphere congruences, we have

$$(K_1, Q_R^+)(K_2, Q_I^+) - (K_1, Q_I^+)(K_2, Q_R^+) = 0$$

where $Q^+ = Q_R^+ + \sqrt{-1}Q_I^+$.

2. Writing $Q^+ = \alpha \mathfrak{p} + \beta \mathfrak{q}$ for some complex functions $\alpha = \alpha_R + \sqrt{-1}\alpha_I$ and $\beta = \beta_R + \sqrt{-1}\beta_I$, show that

$$(K_i, Q_R^+) = (K_i, \mathfrak{p})(\alpha_R + \beta_R k_i) \quad \text{and} \quad (K_i, Q_I^+) = (K_i, \mathfrak{p})(\alpha_I + \beta_I k_i),$$

where $\mathfrak{p}, \mathfrak{q} \in \mathrm{span}\{\mathrm{Re}\, Q^+, \mathrm{Im}\, Q^+\}$ is an orthogonal pair.

3. Conclude that the surface is linear Weingarten in the space form determined by $\mathfrak{p}$ and $\mathfrak{q}$.

Proposition 6.7.36 ([50, §3]) *Let $\Lambda : \Sigma \to \mathcal{Z}$ be a Legendre immersion. Then Λ projects to a tubular surface in some space form if and only if Λ is an Ω_0-surface where the enveloping (isothermic) curvature sphere congruence admits a constant conserved quantity Q.*

Proof Suppose Λ is an Ω_0-surface with isothermic curvature sphere congruence $s : \Sigma \to P(L^5)$, and further suppose that

$$(S, Q) = 0$$

for any $S \in \Gamma s$. Choosing a point sphere complex $\mathfrak{p}$ and space form vector $\mathfrak{q}$ so that $Q = \alpha \mathfrak{p} + \beta \mathfrak{q}$, we assume without loss of generality that s is the curvature sphere corresponding to the u-direction. Then for $S = k_1 X + N$,

$$0 = (S, Q) = (k_1 X + N, \alpha \mathfrak{p} + \beta \mathfrak{q}) = -\beta k_1 - \alpha$$

where X and N are the point sphere map and tangent plane congruence with respect to $\mathfrak{p}$ and $\mathfrak{q}$. Therefore, we have k_1 is constant.

On the other hand, suppose that we have a tubular surface, and assume without loss of generality that k_1 is constant. Since tubular surfaces are channel surfaces, they are Ω_0-surfaces. Then for

$$\sigma := k_1 X + N,$$

we find that

$$Q := \mathfrak{q} - k_1 \mathfrak{p}$$

satisfies $(\sigma, Q) = 0$. This completes the proof of the theorem. $\qquad\square$

Among the class of linear Weingarten surfaces, there are surfaces that admit Weierstrass-type representations with the prototypical example being minimal surfaces in Euclidean spaces. We recall the following definition:

Definition 6.7.37 ([105, Equation (2)]) Non-tubular linear Weingarten surfaces in $\mathbb{H}^3 \cong M_{-1}$ that satisfy an equation of type

$$A(\pm H + 1) + B(K - 1) = 0$$

for some $A, B \in \mathbb{R}$ $(|A| + |B| > 0)$ are called *linear Weingarten surfaces of Bryant type* (Figs. 6.3 and 6.4).

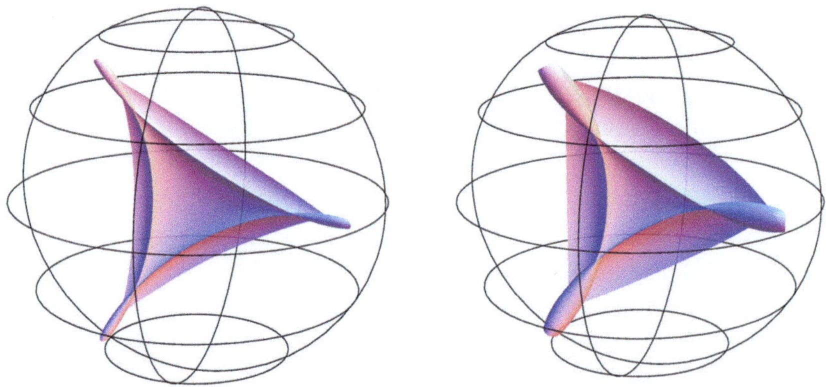

Fig. 6.3 Two surfaces in a deformation through a one-parameter family of linear Weingarten surfaces of Bryant type in $\mathbb{H}^3$, the first of which is a flat surface (see [126])

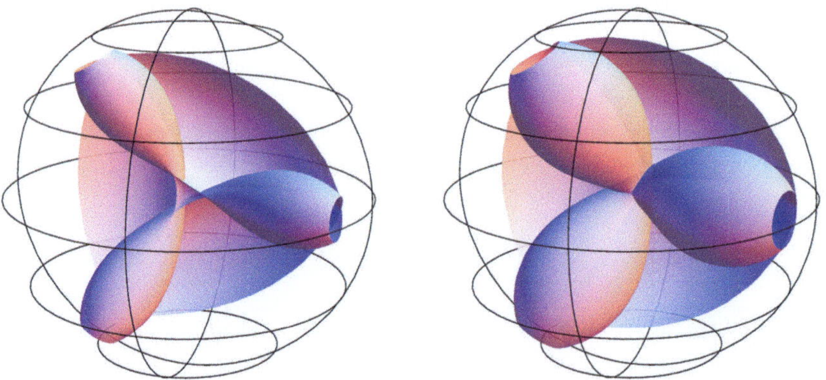

Fig. 6.4 Two further surfaces in that deformation starting in Fig. 6.3 through a one-parameter family of linear Weingarten surfaces of Bryant type in $\mathbb{H}^3$, the second of which is a CMC 1 surface (see [126])

Such surface classes admit Weierstrass-type representations; these distinguished subclass of linear Weingarten surfaces also admit a characterization via a condition on the constant conserved quantities:

Corollary 6.7.38 ([50, §4]) *Let $\Lambda : \Sigma \to \mathcal{Z}$ be a Legendre immersion. Then Λ projects to either*

1. a linear Weingarten surface of Bryant type in hyperbolic space, or
2. a minimal surface in Euclidean space

if and only if Λ is Ω with enveloping isothermic sphere congruences $s^\pm$ admitting constant conserved quantities $Q^\pm$ satisfying the following conditions:

1. one of $Q^\pm$ is lightlike, and
2. the signature of $\mathrm{span}\{Q^+, Q^-\}$ is either $(-\,0)$ or $(+-)$.

Proof Using the notations from the proof of Theorem 6.7.34, we first show the necessary condition: Assume without loss of generality that $(Q^+, Q^+) = 0$.

If $\mathrm{span}\{Q^+, Q^-\}$ has signature $(-\,0)$, then $(Q^+, Q^-) = 0$ with timelike Q^-, so choose $\mathfrak{q} = Q^+$ and $\mathfrak{p} = Q^-$. Hence, we have that $\alpha^+ = 0 = \beta^-$, and (6.16) now reads

$$2\alpha^- \beta^+ \frac{k_1 + k_2}{2} = 0,$$

that is, Λ projects to a minimal surface in the Euclidean space determined by $\mathfrak{p}$ and $\mathfrak{q}$.

If $\mathrm{span}\{Q^+, Q^-\}$ has signature $(+-)$, then let us choose orthogonal $\mathfrak{p}, \mathfrak{q} \in \mathrm{span}\{Q^+, Q^-\}$ so that $(\mathfrak{p}, \mathfrak{p}) = -1$ while $(\mathfrak{q}, \mathfrak{q}) = 1$. Then we can calculate that

$$0 = (Q^+, Q^+) = (\alpha^+ \mathfrak{p} + \beta^+ \mathfrak{q}, \alpha^+ \mathfrak{p} + \beta^+ \mathfrak{q}) = -(\alpha^+)^2 + (\beta^+)^2.$$

Thus using the relation $\beta^+ = \pm\alpha^+$, (6.16) now becomes

$$0 = 2\alpha^+ \alpha^- K + 2(\alpha^+ \beta^- \pm \alpha^- \alpha^+)H \pm 2\alpha^+ \beta^-$$

$$= 2\alpha^+ \alpha^- (K - 1) + 2(\alpha^+ \alpha^- \pm \alpha^+ \beta^-)(\pm H + 1).$$

Therefore, Λ projects to a linear Weingarten surface of Bryant type in the hyperbolic space determined by $\mathfrak{p}$ and $\mathfrak{q}$. The sufficiency proof is left as an exercise for the readers. $\qquad\square$

Exercise 6.7.39 Prove the sufficient direction of Corollary 6.7.38. (Hint: use the proof of Theorem 6.7.34 to see how the $\alpha^\pm, \beta^\pm$ of the necessary direction align with α, β, γ of the sufficient direction.)

Remark 6.7.40 The second condition on the conserved quantities in Corollary 6.7.38 is stronger than the non-degeneracy condition (6.14), as the condition allows us to choose a timelike point sphere complex $\mathfrak{p}$. If we assume the milder non-degeneracy condition (6.14) for Corollary 6.7.38, then we must also consider the cases when $\mathrm{span}\{Q^+, Q^-\}$ has signature $(+0)$ so that the point sphere complex is spacelike. It is known that this case corresponds to the projection being a *maximal surface* in Minkowski space, admitting its own Weierstrass-type representation first defined in [137, Theorem 1.1], and later refined to include singularities in [215, Theorem 2.6].

Remark 6.7.41 In the sufficiency proof of Theorem 6.7.34, the $\alpha = 0$ case corresponds to the projection being a cmc $H = -\frac{\gamma}{2\beta}$ surface. In fact, the two isothermic sphere congruences $\sigma^\pm$ enveloping Λ are the point spheres $\sigma^- = X$ and the mean curvature sphere congruence $\sigma^+ = HX + N$ allowing us to observe that all cmc surfaces are isothermic surfaces. In this setting, consider the constant conserved quantity Q^+ of $s^+ = \mathrm{span}\{\sigma^+\}$ so that $\Gamma^{+,\lambda}Q^+ = 0$. Then Theorem 6.2.15 tells us that

$$0 = \Gamma^{+,\lambda}Q^+ = g^\lambda \circ \Gamma^{-,\lambda}(g^{-\lambda}Q^+)$$

so that $g^{-\lambda}Q^+$ is a linear conserved quantity of $\Gamma^{-,\lambda}$ coming from the isothermic point sphere map. This linear conserved quantity was seen in Theorem 3.8.66.

For the non-cmc linear Weingarten surfaces, the following is known:

Lemma 6.7.42 ([50, §2]) *Non-tubular non-cmc linear Weingarten surfaces are Guichard.*

Proof By assumption, we have the linear Weingarten equation

$$\alpha K - 2\beta H + \gamma = 0$$

for some constants α, β and γ. Because the surface is not cmc, α is not zero, so without loss of generality we can take $\alpha = 1$, and the linear Weingarten equation becomes

$$(k_1 - \beta)(k_2 - \beta) = \rho := \beta^2 - \gamma,$$

and $\rho \neq 0$ because the surface is non-tubular. Suppose that (u, v) are curvature line coordinates of the surface.

The Case $\rho > 0$ Choose f so that

$$k_1 - \beta = \sqrt{\rho}\, \tanh f, \quad k_2 - \beta = \sqrt{\rho}\, \coth f.$$

Using the Codazzi equations (2.35), we have

$$\sqrt{\rho}\,\frac{f_v}{\cosh^2 f} = k_{1,v} = \frac{g_{11,v}}{2g_{11}}(k_2 - k_1) = \sqrt{\rho}(\coth f - \tanh f)\left(\tfrac{1}{2}(\log g_{11})_v\right),$$

implying

$$\frac{f_v}{\cosh f} = \frac{(\log g_{11})_v}{2 \sinh f}.$$

Thus,

$$\log(\cosh f) = \tfrac{1}{2} \log g_{11} + h(u),$$

where $h(u)$ is some function depending only on u. We can rescale the u coordinate so that $h(u) = c_1$ for some constant c_1. Thus

$$c_2 \cosh^2 f = g_{11}$$

for some constant c_2. We can then scale u by a constant scalar factor so that

$$g_{11} = \cosh^2 f.$$

Similarly, we can arrange that

$$g_{22} = \sinh^2 f.$$

Now

$$cg_{11}g_{22}(k_1 - k_2)^2 - g_{22} + \epsilon^2 g_{11}$$
$$= c \cosh^2 f \, \sinh^2 f \, (\sqrt{\rho} \tanh f - \sqrt{\rho} \coth f)^2 - \sinh^2 f + \epsilon^2 \cosh^2 f \, ,$$

and this is equal to zero when we set $c = -1/\rho$ and $\epsilon = 1$, thus Calapso's equation holds and the surface is Guichard.

The case $\rho < 0$: In this case we can similarly arrange that

$$k_1 - \beta = \sqrt{-\rho} \tan f, \quad k_2 - \beta = -\sqrt{-\rho} \cot f$$

and

$$g_{11} = \cos^2 f, \quad g_{22} = \sin^2 f.$$

Calapso's equation again holds, now using $\epsilon = \sqrt{-1}$.

$\square$

We also have this corollary concerning Calapso transforms of nontubular linear Weingarten surfaces, generalizing the Lawson correspondence for cmc surfaces.

Corollary 6.7.43 ([47, Theorem 6.18]) *Generically, Calapso transforms of nontubular linear Weingarten surfaces are again nontubular linear Weingarten in some space forms.*

Proof Take a nontubular linear Weingarten surface and its lift $\Lambda : \Sigma \to \mathcal{Z}$. Then by Theorem 6.7.34, Λ is Ω, and the enveloping isothermic sphere congruences $s^{\pm}$ admit constant conserved quantities $Q^{\pm}$. Denoting by $T^{\pm,\lambda}$ the Calapso transformation of $s^{\pm}$, we have by Definition 6.3.19 that the Calapso transform Λ^{μ} of Λ with parameter μ satisfies

$$\Lambda^{\mu} = T^{+,\mu}\Lambda = T^{-,\mu}\Lambda$$

with enveloping isothermic sphere congruences $s^{\pm,\mu} := T^{\pm,\mu}s^{\pm} : \Sigma \to P(L^5)$. The fact that $T^{\pm,\mu}$ takes values in $O_{4,2}$ then tells us

$$0 = (S^{\pm}, Q^{\pm}) = (T^{\pm,\mu}S^{\pm}, T^{\pm,\mu}Q^{\pm})$$

for any section $S^{\pm} \in \Gamma s^{\pm}$. Thus, $T^{\pm,\mu}Q^{\pm}$ are again conserved quantities of $s^{\pm,\mu}$, and an analogous argument to Theorem 3.7.58 tells us that these are constant conserved quantities. Then as long as the non-degeneracy condition (6.14) holds, Λ^{μ} projects to a linear Weingarten surface in some space form. $\qquad\square$

Recommended Further Readings for Chap. 5 and 6

The classical book by Blaschke [14] still stands as one of the most comprehensive treatments of sphere geometries of Möbius, Laguerre [145], and Lie [153]. A modern treatment of Lie sphere geometry is given in [65], exploring the basic notions of Lie sphere geometry and their subgeometries of any dimension and the notion of Legendre maps in this generalization.

Due to the nature of Lie sphere transformations preserving oriented contact, Lie sphere geometry provides a natural setting to study Legendre immersions, also known as fronts or wave fronts. The work [182] gave the first projective formulation for Legendre maps, and also shows that all Dupin hypersurfaces of a certain type are unique up to Lie sphere transformation.

Considering such contact structure also allows one to recover the geometry around singularities, as integral submanifolds of the contact structure on the space of lines in the Lie quadric. The work [180] does exactly this and looks at the behavior of singularities appearing on fronts under Lie sphere transformations.

Digressing from Lie sphere geometry temporarily, we recommend a few works examining singularities appearing on fronts: fronts, or wave fronts, have been considered since the early days of Huygens [130], as they have a wide range of application to physics—wave propagation, optics, diffraction phenomena, gravitational lensing, stability problems, thermodynamics and elasticity theory, to name a few. Such significance is reflected by the sheer number of works related to the problem including [63, 210, 219], but from a differential geometry perspective, the monograph [4] stands as one of the core references in this field.

One prominent research article in the study of front singularities from a differential geometric perspective is [193]. The work introduces new geometric invariants regarding front singularities, leading to generalizations of the Gauss-Bonnet formula to surfaces with singularities, expanding upon previous work on such formulas by [144, 148]. This article also characterizes the local behavior of surfaces at singularities in terms of the singular curvature, and finds relations between the singular curvature and the Gaussian curvature at nearby immersed points, making new connections between topological properties and global differential geometric properties. Also recommended is the textbook [212], where more attention is given to the necessary background materials.

Returning to topics related to Lie sphere geometry, Ribaucour transformations [187] relate two surfaces so that they envelop a common sphere congruence along the corresponding curvature lines. Thus, Lie sphere geometry provides a natural setting to consider Ribaucour transformations, and this is the work done in [43], giving a modern treatment of classical results including the permutability of Ribaucour transformations. We also mention a few works that provide an explicit formulation of Ribaucour transforms of surfaces in space forms: [79–81, 151, 172].

In a similar vein, channel surfaces are well-suited for discussions in Lie sphere geometry, and a modern account of such approach is given in [181]. The work also explores the Ω_0-structure of channel surfaces, characterizing channel surfaces as Ω_0-surfaces admitting polynomial conserved quantities, and discusses the various transformations of channel surfaces.

Now we switch our attention to Ω-surfaces defined by Demoulin [87–89]. Demoulin defined such surfaces in Euclidean 3-space via an equation, now known as Demoulin's equation, and proved the existence of a dual surface, akin to the Christoffel dual, for such surfaces. Furthermore, he noticed that isothermic surfaces and L-isothermic surfaces are all Ω in [87], and that Guichard surfaces are also Ω in [88] by showing that Calapso's equation for Guichard surfaces implies Demoulin's equation. Eisenhart then considered Darboux transformations of Ω-surfaces via Ribaucour transformations in [98, 99], showing that these transformations include the Eisenhart transformations of Guichard surfaces [100] and Darboux transformations of isothermic surfaces.

Analogous to isothermic surfaces in Möbius geometry, Ω- and Ω_0-surfaces constitute the second-order deformable surfaces of Lie sphere geometry, called *Lie applicable surfaces*, as noticed by Blaschke [14], and evidenced by the existence of Calapso transformations. This fact was revisited in [102, 165] employing a modern integrable systems technique.

The doctoral thesis [78] explores the gauge-theoretic approach to Lie applicable surfaces, in discussion of the more general notion of l-applicable maps. In doing so, the integrable structures of Lie applicable surfaces are shown to be induced by the pair of enveloping isothermic sphere congruences. A key difference between Lie applicable surfaces and isothermic surfaces is that they admit a gauge orbit's worth of closed 1-forms.

In [176], a special member of the gauge orbit is identified as the *middle potential* and describes the integrable structure of Lie applicable surfaces via 1-parameter family of flat connections coming from the middle potential. In addition, the work describes the transformation theory of Lie applicable surfaces, including Calapso transformation and Darboux transformation of Lie applicable surfaces, induced by the transformation theory of the enveloping isothermic sphere congruences. Also, Ω-surfaces are characterized by the existence of certain Combescure transformations, showing that the class of Ω-surfaces are O-surfaces in the sense of [200].

The middle potential and the associated flat connection, called the *middle pencil of connections* or the *middle pencil* for short, is further explored in [47]. By considering polynomial conserved quantities of the middle pencil, isothermic, Guichard, and L-isothermic surfaces are characterized as Ω-surfaces whose middle pencil admits linear conserved quantities of certain form. Furthermore, considering the behavior of polynomial conserved quantities under transformations, the work gives a unified description of Darboux transformations of isothermic surfaces, Eisenhart transformations of Guichard surfaces [100], and Bianchi-Darboux transformations of L-isothermic surfaces [168] as Darboux transformations of Ω-surfaces.

The integrable structure of isothermic surfaces can be interpreted as a curved flat characterization in the realm of Möbius geometry by considering the Darboux pair. Similar results have been derived for Ω-surfaces in [53], showing that curved flats in the Grassmannian of $(2, 2)$-planes are the same as Demoulin families of Ω-surfaces related by Darboux transformation.

The work [72] considers surfaces with spherical curvature lines following the works of Blaschke, showing that any such surface is generated by a spherical curve and a certain curve of Lie sphere transformations. In particular, by giving constrained elastic curves a Lie sphere geometric interpretation in terms of polynomial conserved quantities, the work also characterizes all Lie applicable surfaces with exactly one family of spherical curvature lines as Lie sphere transformations of constrained elastic curves in some space form.

Linear Weingarten surfaces, including those surfaces with Weierstrass-type representations such as minimal surfaces in Euclidean space or linear Weingarten surfaces of Bryant-type in hyperbolic space [105], are given a Lie sphere geometric characterization in [50] as Ω-surfaces with isothermic sphere congruences belonging to linear sphere complexes and hence admitting constant conserved quantities. Furthermore, the work efficiently shows that non-cmc linear Weingarten surfaces are Guichard.

Using the characterization of linear Weingarten surfaces in Lie sphere geometry, the work [118] explores channel linear Weingarten surfaces in all space forms. In particular, the work shows that any such surface must be a rotational surface in the space form, and by considering the parallel surfaces of constant Gauss curvature surfaces, obtains an explicit parameterization for a large class of these surfaces given in terms of Jacobi elliptic functions. Also, they recover a Lie sphere geometric version of Vessiot's theorem, namely, any Ω-surface that is additionally channel must be an isothermic surface in some space form. Finally, the work [184] obtains the explicit parametrizations of channel cmc surfaces in space forms, thereby completing the classification of all channel linear Weingarten surfaces in space forms given with explicit parametrization.

Within the class of linear Weingarten surfaces that admit a Weierstrass-type representation is the class of flat surfaces in hyperbolic space [106, 141], and these surface have been given a Lie sphere geometric characterization in [49]. In fact, much geometry is to be learned by considering these surfaces as *flat fronts* defined in [142] with a prototypical example being recognizing the type of singularities appearing on these surfaces as done in [139, 140].

The conserved quantities characterization of linear Weingarten surfaces and L-isothermic surfaces show that many linear Weingarten surfaces admitting Weierstrass-type representations are L-isothermic. In terms of surfaces in Euclidean space or hyperbolic space, this condition amounts to the isothermicity of the Gauss map or the hyperbolic Gauss map, respectively, and hence can be viewed as the inverse stereographic projection of a holomorphic function. Such is the starting observation for [178], where the work explores Ω-surfaces and L-isothermic surfaces in the realm of Laguerre geometry, giving a unified description of various Weierstrass-type representations via the Ω-dual construction.

Chapter 7
Discrete Ω-surfaces

7.1 Discrete Legendre Maps

We begin the discussion of discrete Ω-surfaces by introducing the discrete Legendre immersions.

Recall that for a smooth map $\Lambda : \Sigma \to \mathcal{Z}$ into the set of contact elements, the contact condition reads: for any two sections $S, \tilde{S} \in \Gamma\Lambda$, we have that

$$\mathrm{d}S \perp \tilde{S}.$$

Therefore, for $(u, v) \in \Sigma$, we have that

$$S(u, v), \tilde{S}(u, v) \perp S(u, v) + \epsilon S_u(u, v),$$
$$S(u, v), \tilde{S}(u, v) \perp \tilde{S}(u, v) + \epsilon \tilde{S}_v(u, v) \tag{7.1}$$

for $\epsilon \in \mathbb{R}$ close to zero. When (u, v) are curvature line coordinates, so that $S = K_1$ and $\tilde{S} = K_2$ are curvature spheres for the u and v directions, respectively, then Lemma 5.4.29 tells us that $S_u, \tilde{S}_v \in \Gamma\Lambda$. Then (7.1) is strengthened to $S + \epsilon S_u, \tilde{S} + \epsilon \tilde{S}_v \in \Gamma\Lambda$. Thus, because

$$S(u, v) + \epsilon S_u(u, v) = S(u + \epsilon, v) + \mathcal{O}(\epsilon^2),$$
$$\tilde{S}(u, v) + \epsilon \tilde{S}_V(u, v) = \tilde{S}(u, v + \epsilon) + \mathcal{O}(\epsilon^2) \tag{7.2}$$

for ϵ close to zero, $S(u + \epsilon, v), \tilde{S}(u, v + \epsilon)$ are first order approximations for sections in Λ.

Therefore, given a discrete map $\Lambda : \mathbb{Z}^2 \to \mathcal{Z}$, a (perhaps) naïve way to discretize the contact condition is to take $u \to m, v \to n, \epsilon \to 1$ and discard terms of at least second order in ϵ so that the conditions (7.1) and (7.2) now read for any elementary

© The Author(s), under exclusive license to Springer Nature Switzerland AG 2025   197
J. Cho et al., *Discrete Isothermic Surfaces in Lie Sphere Geometry*, Lecture Notes
in Mathematics 2375, https://doi.org/10.1007/978-3-031-95592-1_7

quadrilateral $ijk\ell$

$$S_i, \tilde{S}_i \perp S_i + \mathrm{d}S_{ij} = S_j \quad \text{and} \quad S_i, \tilde{S}_i \perp \tilde{S}_i + \mathrm{d}\tilde{S}_{i\ell} = \tilde{S}_\ell$$

for $S, \tilde{S} \in \Gamma\Lambda$, where we now view Λ as a discrete 2-dimensional null plane subbundle of the trivial bundle $\mathbb{Z}^2 \times \mathbb{R}^{4,2}$. Since $S_j, S_\ell \in L^5$, we see that $S_j, S_\ell \in \mathrm{span}\{S_i, \tilde{S}_i\} = \Lambda_i$. Therefore, we see that

$$S_j \in \Lambda_i \cap \Lambda_j \quad \text{and} \quad \tilde{S}_\ell \in \Lambda_i \cap \Lambda_\ell,$$

and switching the indices everywhere also tell us that

$$S_i \in \Lambda_i \cap \Lambda_j \quad \text{and} \quad \tilde{S}_i \in \Lambda_i \cap \Lambda_\ell.$$

Hence, we see that $S_i \parallel S_j \in \Lambda_i \cap \Lambda_j$ and $\tilde{S}_i \parallel \tilde{S}_\ell \in \Lambda_i \cap \Lambda_\ell$ providing the motivation to define the following:

Definition 7.1.1 ([27, Definition 18]) Let $\Lambda : \mathbb{Z}^2 \to \mathcal{Z}$ be a discrete map into the set of contact elements. We call Λ a *discrete Legendre map*, or a *principal contact element net*, if any pair of neighboring contact elements share a common sphere, that is, $\Lambda_i \cap \Lambda_j \neq \{0\}$.

Now, the immersion condition of the smooth Legendre map $\Lambda : \Sigma \to \mathcal{Z}$ required that if there is some $\mathbf{v} \in T_p\Sigma$ for some $p \in \Sigma$ such that for any section $S \in \Lambda$, $\mathrm{d}S(\mathbf{v}) \in \Lambda(p)$, then $\mathbf{v} = \mathbf{0}$. As in the consideration of the contact condition, we discretize the immersion condition and require that, on any edge ij, there is a section $S \in \Gamma\Lambda$ of a discrete Legendre map $\Lambda : \mathbb{Z}^2 \to \mathcal{Z}$ such that on any edge ij,

$$\mathrm{d}S_{ij} \notin \Lambda_i, \quad \text{and} \quad \mathrm{d}S_{ij} \notin \Lambda_j. \tag{7.3}$$

This implies that $\Lambda_i \neq \Lambda_j$, giving rise to the following definition:

Definition 7.1.2 ([27, Definition 18]) Let $\Lambda : \mathbb{Z}^2 \to \mathcal{Z}$ be a discrete Legendre map. We call Λ a *discrete Legendre immersion*, if any pair of neighboring contact elements share a unique common sphere, that is, $\dim(\Lambda_i \cap \Lambda_j) = 1$. The unique common sphere k_{ij} of the two neighboring contact elements, that is,

$$k_{ij} = \Lambda_i \cap \Lambda_j \in P(L^5),$$

is called the *curvature sphere* on an (unoriented) edge ij.

Remark 7.1.3 Recalling that Λ can be viewed as a discrete line bundle in the projective light cone $P(L^5)$, a discrete map Λ being a Legendre map is equivalent to requiring that Λ is a discrete line congruence in $P(L^5)$, in the sense of [94]. For non-degeneracy, we will always assume that a discrete Legendre immersion, viewed as a discrete line congruence in $P(L^5)$ is *generic*, (cf. [27, p. 13]), that is,

$$\Lambda_i + \Lambda_j + \Lambda_k + \Lambda_\ell = \mathrm{span}\{K_{ij}, K_{jk}, K_{k\ell}, K_{\ell i}\}$$

has dimension 4 for $K_{ij} \in \Gamma k_{ij}$.

The choice of a point sphere complex $\mathfrak{p}$ and space form vector $\mathfrak{q}$ allows us to define the *discrete point sphere map* $X : \mathbb{Z}^2 \to L^5$ and *discrete tangent plane congruence* $N : \mathbb{Z}^2 \to L^5$ as $X, N \in \Gamma \Lambda$ so that

$$(X, \mathfrak{p}) = (N, \mathfrak{q}) = 0 \quad \text{and} \quad (X, \mathfrak{q}) = (N, \mathfrak{p}) = -1.$$

Then for any $K_{ij} \in k_{ij}$, we see that $(X_i, K_{ij}) = (X_j, K_{ij}) = 0$ so that both vertices X_i and X_j lie on the common sphere k_{ij}; in this way, the sphere k_{ij} discretizes the property of smooth curvature spheres as the "best-fitting" (i.e. having the highest order of tangency) spheres in the curvature line directions.

On any edge ij, since k_{ij} is a map into the projective light cone, choose $K_{ij} \in k_{ij}$ so that

$$K_{ij} = N_i + \kappa_{ij} X_i = \alpha(N_j + \kappa_{ji} X_j) \tag{7.4}$$

for some scalars $\alpha, \kappa_{ij}, \kappa_{ji} \in \mathbb{R}$. Then taking the inner product of (7.4) with $\mathfrak{p}$, we see that $\alpha = 1$, and we can also see, by taking the inner product of (7.4) with $\mathfrak{q}$, that

$$\kappa_{ij} = \kappa_{ji}. \tag{7.5}$$

Then from (7.4) and (7.5), we have the discrete Rodrigues equation

$$\mathrm{d}N_{ij} = -\kappa_{ij}\mathrm{d}X_{ij}, \tag{7.6}$$

motivating us to call κ_{ij} the *principal curvatures*, defined on (unoriented) edges.

Exercise 7.1.4   Let $x : \mathbb{Z}^2 \to \mathbb{R}^3$ and $v : \mathbb{Z}^2 \to \mathbb{S}^2$ be defined as

$$x_{m,n} = (0, 0, n)^t, \quad v_{m,n} = \left(\cos \tfrac{2\pi m}{m_{per}}, \sin \tfrac{2\pi m}{m_{per}}, 0\right)^t,$$

and denote their lifts as $X, N : \mathbb{Z}^2 \to L^5$.

(1) Verify that

$$X_{m,n} = (\tfrac{1}{2}(1 + n^2), 0, 0, n, \tfrac{1}{2}(1 + n^2), 0)^t,$$

$$N_{m,n} = (0, \cos \tfrac{2\pi m}{m_{per}}, \sin \tfrac{2\pi m}{m_{per}}, 0, 0, 1)^t.$$

(2) For any elementary quadrilateral $ijk\ell$, show that $\mathrm{d}X_{ij} = 0$ and $\mathrm{d}N_{i\ell} = 0$, but $\mathrm{d}X_{i\ell} \neq 0$ and $\mathrm{d}N_{ij} \neq 0$, showing that $\Lambda := \mathrm{span}\{X, N\}$ satisfies the regularity condition (7.3).

Example 7.1.5 For the two different discretizations of the Clifford tori in Example 4.1.3, the Clifford torus with non-constant cross ratios lifts to

$$\Lambda = \text{span}\{X^{\mathbb{R}^3}, N^{\mathbb{R}^3}\} = \text{span}\{X^{\mathbb{S}^3}, N^{\mathbb{S}^3}\},$$

with

$$X^{\mathbb{R}^3} = \left(2 + \sqrt{2}\cos\tfrac{2\pi m}{m_{per}}, \cos\tfrac{2\pi n}{n_{per}}(\sqrt{2} + \cos\tfrac{2\pi m}{m_{per}}), \sin\tfrac{2\pi n}{n_{per}}(\sqrt{2} + \cos\tfrac{2\pi m}{m_{per}}),\right.$$
$$\left.\sin\tfrac{2\pi m}{m_{per}}, -1 - \sqrt{2}\cos\tfrac{2\pi m}{m_{per}}, 0\right)^t,$$

$$N^{\mathbb{R}^3} = \left(1 + \sqrt{2}\cos\tfrac{2\pi m}{m_{per}}, \cos\tfrac{2\pi n}{n_{per}}\cos\tfrac{2\pi m}{m_{per}}, \sin\tfrac{2\pi n}{n_{per}}\cos\tfrac{2\pi m}{m_{per}},\right.$$
$$\left.\sin\tfrac{2\pi m}{m_{per}}, -1 - \sqrt{2}\cos\tfrac{2\pi m}{m_{per}}, 1\right)^t,$$

$$X^{\mathbb{S}^3} = \frac{1}{2 + \sqrt{2}\cos\frac{2\pi m}{m_{per}}} X^{\mathbb{R}^3},$$

$$N^{\mathbb{S}^3} = -\frac{1 + \sqrt{2}\cos\frac{2\pi m}{m_{per}}}{2 + \sqrt{2}\cos\frac{2\pi m}{m_{per}}} X^{\mathbb{R}^3} + N^{\mathbb{R}^3},$$

while the one with constant cross ratios lifts to

$$\Lambda = \text{span}\{X^{\mathbb{R}^3}, N^{\mathbb{R}^3}\} = \text{span}\{X^{\mathbb{S}^3}, N^{\mathbb{S}^3}\},$$

with

$$X^{\mathbb{S}^3} = \left(1, \tfrac{1}{\sqrt{2}}\sin\tfrac{2\pi m}{m_{per}}, \tfrac{1}{\sqrt{2}}\cos\tfrac{2\pi n}{n_{per}}, \tfrac{1}{\sqrt{2}}\sin\tfrac{2\pi n}{n_{per}}, \tfrac{1}{\sqrt{2}}\cos\tfrac{2\pi m}{m_{per}}, 0\right)^t,$$

$$N^{\mathbb{S}^3} = \left(0, \tfrac{1}{\sqrt{2}}\sin\tfrac{2\pi m}{m_{per}}, -\tfrac{1}{\sqrt{2}}\cos\tfrac{2\pi n}{n_{per}}, -\tfrac{1}{\sqrt{2}}\sin\tfrac{2\pi n}{n_{per}}, \tfrac{1}{\sqrt{2}}\cos\tfrac{2\pi m}{m_{per}}, 1\right)^t,$$

$$X^{\mathbb{R}^3} = \frac{\sqrt{2}}{\sqrt{2} + \cos\frac{2\pi m}{m_{per}}} X^{\mathbb{S}^3},$$

$$N^{\mathbb{R}^3} = -\frac{\cos\frac{2\pi m}{m_{per}}}{\sqrt{2} + \cos\frac{2\pi m}{m_{per}}} X^{\mathbb{S}^3} + N^{\mathbb{S}^3}.$$

In both examples, one can check that Λ is a Legendre immersion.

7.2   Contact Elements of Discrete Legendre Maps

The geometry of contact elements of discrete Legendre maps can be better understood by considering its relationship to circular nets, introduced in Sect. 4.1 in the context of Euclidean geometry. For this, let us consider a discrete Legendre map $\Lambda : \mathbb{Z}^2 \to \mathcal{Z}$ into the set of contact elements, and let X be the point sphere map with respect to some choice of point sphere complex $\mathfrak{p}$ and space form vector $\mathfrak{q}$. Allowing for degenerate situations, the fact that X is the point sphere map implies $(X, \mathfrak{p}) = 0$ and $(X, \mathfrak{q}) = -1$ so that X takes values in an affine hyperplane; that is, on any elementary quadrilateral $ijk\ell$,

$$\dim(\mathrm{span}\{X_i, X_j, X_k, X_\ell\}) \leq 3. \tag{7.7}$$

Viewing X as a map into the light cone L^4 of $\mathrm{span}\{\mathfrak{p}\}^\perp \cong \mathbb{R}^{4,1}$, Remark 2.3.30 tells us that X_i, X_j, X_k, X_ℓ are contained in a circle. Thus, we have that discrete Legendre maps project to circular nets in Möbius geometry $\mathrm{span}\{\mathfrak{p}\}^\perp \cong \mathbb{R}^{4,1}$ (see [27, Theorem 25]).

On the other hand, suppose we have a circular net $x : \mathbb{Z}^2 \to \mathbb{R}^3$ in Euclidean 3-space with vertex normals v as defined in Sect. 4.1. As in the smooth case, at every vertex i, let us consider the "tangent" plane Π_i of x at x_i by letting Π_i be the plane perpendicular to the vertex normal v_i at the point x_i. With the notion of tangent plane available, we can now consider the pencil of spheres that are tangent to Π_i at x_i (with the same orientation); this defines a null 2-plane in Lie sphere geometry $\mathbb{R}^{4,2}$. Thus for every circular net $x : \mathbb{Z}^2 \to \mathbb{R}^3$ with given normals v, we have a corresponding discrete map $\Lambda : \mathbb{Z}^2 \to \mathcal{Z}$ into the set of contact elements.

To see that so defined Λ is discrete Legendre, we note that for any edge ij, the normals v_i and v_j are symmetric with respect to the perpendicular bisector of dx_{ij}. Denoting the normal line at vertex i by N_i, that is, the line through x_i in the direction of v_i, we then have N_i and N_j must intersect at some point (generically), and thus there is a common sphere that is tangent to both tangent planes Π_i and Π_j. Thus, the contact element Λ_i and Λ_j must have a common sphere, i.e. we have $\Lambda_i \cap \Lambda_j \neq \{0\}$. Therefore, Λ is a discrete Legendre map, and we call Λ a *discrete Legendre lift*. (See also the discussion in [27].)

Remark 7.2.6 In fact, (7.7) implies that X is a Q-net (see Remark 4.1.1).

In particular, we now have a notion of discrete sphere congruence enveloping a discrete Legendre map in Lie sphere geometry (and hence for any space form projections under consideration):

Definition 7.2.7 Let $\Lambda : \mathbb{Z} \to \mathcal{Z}$ be a discrete Legendre map, and suppose that $s : \mathbb{Z}^2 \to P(L^5)$ is a sphere congruence. Viewing Λ as a 2-dimensional subbundle of trivial bundle $\mathbb{Z}^2 \times \mathbb{R}^{4,2}$, if s is a subbundle of Λ, then s is a sphere congruence enveloping Λ.

7.3   Discrete Curvatures via Mixed Area: Revisited

We have seen that the *mixed areas* defined for Euclidean 3-space give rise to the notion of discrete Gauss and mean curvatures in Sect. 4.2 via normal vectors. Having a notion of contact for circular nets in any space form projection now allows us to define curvatures; to achieve this, we first generalize the notion of mixed areas to $\mathbb{R}^{4,2}$.

From Exercise 4.2.5, recall that for two quadrilaterals $P = (p_i, p_j, p_k, p_\ell)$, $Q = (q_i, q_j, q_k, q_\ell)$ with parallel corresponding edges, the mixed area of two polygons in a plane with unit normal $\mathbf{n}$ satisfies

$$A(P, Q)_{ijk\ell} = \frac{1}{2}\det(\mathrm{d}p_{ik}, \mathrm{d}q_{j\ell}, \mathbf{n}) = \frac{1}{2}\det(\mathrm{d}q_{ik}, \mathrm{d}p_{j\ell}, \mathbf{n}).$$

On the other hand, for any two vectors $v, w \in \mathbb{R}^3$, with unit length $\mathbf{n}$ normal to the plane $\mathrm{span}\{v, w\}$ so that $\{v, w, \mathbf{n}\}$ are positively oriented, we have that

$$v \times w = \det(v, w, \mathbf{n})\mathbf{n},$$

where $\times$ denotes the standard Euclidean cross product. We can consider $v \times w$ as the vector encoding two pieces of information: the area of the parallelogram spanned by v and w as its length, and the normal $\mathbf{n}$ to the plane spanned by the parallelogram as its direction. Therefore, we may replace $\det(v, w, \mathbf{n})$ with $v \times w$, and view the mixed area as a vector in $\mathbb{R}^3$ via

$$A(P, Q)_{ijk\ell} = \frac{1}{2}\mathrm{d}p_{ik} \times \mathrm{d}q_{j\ell} = \frac{1}{2}\mathrm{d}q_{ik} \times \mathrm{d}p_{j\ell},$$

where we used the fact that the corresponding edges are parallel.

The wedge product is a generalization of the Euclidean cross product; in the Euclidean 3-space, they can be explicitly identified via

$$\mathbb{R}^3 \ni \alpha\mathbf{e}_1 + \beta\mathbf{e}_2 + \gamma\mathbf{e}_3 = v \times w \sim v \wedge w = \alpha\mathbf{e}_2 \wedge \mathbf{e}_3 + \beta\mathbf{e}_3 \wedge \mathbf{e}_1 + \gamma\mathbf{e}_1 \wedge \mathbf{e}_2 \in \wedge^2\mathbb{R}^3$$

for any $v, w \in \mathbb{R}^3$. Therefore, the mixed area can also be viewed as a vector in $\wedge^2 \mathbb{R}^3$, where

$$A(P, Q)_{ijk\ell} = \frac{1}{2}\mathrm{d}p_{ik} \wedge \mathrm{d}q_{j\ell} = \frac{1}{2}\mathrm{d}q_{ik} \wedge \mathrm{d}p_{j\ell}.$$

This formulation of mixed area in terms of wedge product allows for a generalization to other vector spaces including $\mathbb{R}^{4,2}$.

Let P_*, Q_* for $* = i, j, k, \ell$ be the vertices of two edge-parallel planar quadrilaterals in $\mathbb{R}^{4,2}$ such that the edges are non-null. Then we define the mixed area of P and Q as a vector in $\wedge^2 \mathbb{R}^{4,2}$ via

$$\mathcal{A}(P, Q)_{ijk\ell} := \frac{1}{2}\mathrm{d}P_{ik} \wedge \mathrm{d}Q_{j\ell} = \frac{1}{2}\mathrm{d}Q_{ik} \wedge \mathrm{d}P_{j\ell}. \tag{7.8}$$

With the notion of mixed areas available, we now define the Gaussian and mean curvatures of the space form projection of discrete Legendre immersions. For this, let $\Lambda : \mathbb{Z}^{4,2} \to \mathcal{Z}$ be a Legendre immersion, and denote by $X, N \in \Gamma\Lambda$ the point sphere map and tangent plane congruence of Λ with respect to some choice of point sphere complex and space form vector. Then the Rodrigues equation (7.6) tells us that X and N are edge-parallel, allowing us to define the following, with extra regularity assumptions that none of the edges of X and N are null and that $\mathcal{A}(X, X) \neq 0$:

Definition 7.3.8 ([48, Lemma and Definition 2.3]) Let $\Lambda : \mathbb{Z}^{4,2} \to \mathcal{Z}$ be a Legendre immersion, and denote by $X, N \in \Gamma\Lambda$ the point sphere map and tangent plane congruence of Λ with respect to some choice of point sphere complex and space form vector. Real-valued functions $\mathcal{H}$ and $\mathcal{K}$ defined on faces via the relation

$$\mathcal{A}(X, X)_{ijk\ell}\mathcal{H}_{ijk\ell} := -\mathcal{A}(X, N)_{ijk\ell}, \tag{7.9}$$

$$\mathcal{A}(X, X)_{ijk\ell}\mathcal{K}_{ijk\ell} := \mathcal{A}(N, N)_{ijk\ell}. \tag{7.10}$$

are called the *Gauss curvature* and *mean curvature*, respectively, of X.

Example 7.3.9 Computations show that the lifts of the discrete Clifford tori in Example 7.1.5 satisfy

$$H^{\mathbb{R}^3} = \frac{\sqrt{2} + \cos\frac{2\pi m}{m_{per}} + \cos\frac{2\pi(m+1)}{m_{per}}}{2\sqrt{2} + \cos\frac{2\pi m}{m_{per}} + \cos\frac{2\pi(m+1)}{m_{per}}},$$

$$K^{\mathbb{R}^3} = \frac{\sqrt{2}\left(\cos\frac{2\pi m}{m_{per}} + \cos\frac{2\pi(m+1)}{m_{per}}\right)}{4 + \sqrt{2}\cos\frac{2\pi m}{m_{per}} + \sqrt{2}\cos\frac{2\pi(m+1)}{m_{per}}},$$

$$H^{\mathbb{S}^3} = 0 , \quad K^{\mathbb{S}^3} = -1$$

in the non-constant cross ratio case ($K^{\mathbb{S}^3}$ is extrinsic Gauss curvature), while

$$H^{\mathbb{R}^3} = \frac{1 - (1 + \sqrt{2}\cos\frac{2\pi m}{m_{per}})(1 + \sqrt{2}\cos\frac{2\pi(m+1)}{m_{per}})}{4 + \sqrt{2}\cos\frac{2\pi m}{m_{per}} + \sqrt{2}\cos\frac{2\pi(m+1)}{m_{per}}},$$

$$K^{\mathbb{R}^3} = \frac{-4 - 3\sqrt{2}(\cos\frac{2\pi m}{m_{per}} + \cos\frac{2\pi(m+1)}{m_{per}}) - 4\cos\frac{2\pi m}{m_{per}}\cos\frac{2\pi(m+1)}{m_{per}}}{4 + \sqrt{2}\cos\frac{2\pi m}{m_{per}} + \sqrt{2}\cos\frac{2\pi(m+1)}{m_{per}}},$$

$$H^{\mathbb{S}^3} = 0 , \quad K^{\mathbb{S}^3} = -1$$

in the constant cross ratio case (again, $K^{\mathbb{S}^3}$ is extrinsic Gauss curvature).

The next theorem shows that the definition of Gaussian curvature $\mathcal{K}$ and mean curvature $\mathcal{H}$ within the Lie sphere geometric setup (Definition 7.3.8) is consistent with that of the Euclidean space K, H as defined in Definition 4.2.6.

Theorem 7.3.10 *Let $x : \mathbb{Z}^2 \to \mathbb{R}^3$ be a circular net with Gauss map $v : \mathbb{Z}^2 \to \mathbb{S}^2$ with Gaussian curvature K and mean cuvature H. Denoting the lifts of x, v as $X, N : \mathbb{Z}^2 \to L^5$, respectively, let $\Lambda := \mathrm{span}\{X, N\}$. Then we have $\mathcal{K} = K$ and $\mathcal{H} = H$.*

Proof Since by Remark 7.2.6, X has planar quadrilaterals, and N is edge-parallel to X, there are some p_{11}, p_{12}, p_{21}, p_{22} such that

$$\mathrm{d}N_{ik} = p_{11}\mathrm{d}X_{ik} + p_{12}\mathrm{d}X_{j\ell}, \quad \mathrm{d}N_{j\ell} = p_{21}\mathrm{d}X_{ik} + p_{22}\mathrm{d}X_{j\ell}.$$

Accordingly, we have that

$$\mathrm{d}v_{ik} = p_{11}\mathrm{d}x_{ik} + p_{12}\mathrm{d}x_{j\ell}, \quad \mathrm{d}v_{j\ell} = p_{21}\mathrm{d}x_{ik} + p_{22}\mathrm{d}x_{j\ell}.$$

From $\mathcal{A}(N, N)_{ijk\ell} = \mathcal{K}_{ijk\ell}\mathcal{A}(X, X)_{ijk\ell}$ we have

$$\mathcal{K}_{ijk\ell} = p_{11}p_{22} - p_{12}p_{21}.$$

We have that

$$A(v, v)_{ijk\ell} = (p_{11}p_{22} - p_{12}p_{21})\mathrm{d}x_{ik} \wedge \mathrm{d}x_{j\ell} = K_{ijk\ell}A(x, x)_{ijk\ell},$$

and thus $\mathcal{K} = K$ on every face.

Similarly, we can verify that

$$\mathcal{H} = H = \frac{1}{2}(p_{11} + p_{22}),$$

giving us the claim. $\qquad\qquad\square$

Because the mean and the Gauss curvatures are defined on faces, while the principal curvatures are defined on edges, the relationships between them are not immediately apparent. The actual relations between the mean curvature, Gauss curvature, and principal curvatures have been determined as follows: let

$$\mathrm{d}X_{ik} = a\mathrm{d}X_{ij} + b\mathrm{d}X_{i\ell}$$

for some a and b. Using the equations

$$\mathrm{d}N_{ij} = -\kappa_{ij}\mathrm{d}X_{ij}, \quad \mathrm{d}N_{i\ell} = -\kappa_{i\ell}\mathrm{d}X_{i\ell},$$

$$\mathrm{d}N_{jk} = -\kappa_{jk}\mathrm{d}X_{jk} = -\kappa_{jk}(\mathrm{d}X_{ik} - \mathrm{d}X_{ij}) = -\kappa_{jk}(a\mathrm{d}X_{ij} + b\mathrm{d}X_{i\ell} - \mathrm{d}X_{ij}),$$

$$\mathrm{d}N_{ik} = p_{11}\mathrm{d}X_{ik} + p_{12}\mathrm{d}X_{j\ell}, \quad \mathrm{d}N_{j\ell} = p_{21}\mathrm{d}X_{ik} + p_{22}\mathrm{d}X_{j\ell},$$

$$\mathrm{d}N_{ik} = \mathrm{d}N_{ij} + \mathrm{d}N_{jk}, \quad \mathrm{d}N_{j\ell} = \mathrm{d}N_{i\ell} - \mathrm{d}N_{ij}, \quad \mathrm{d}X_{j\ell} = \mathrm{d}X_{i\ell} - \mathrm{d}X_{ij},$$

we have

$$\alpha_1 \mathrm{d}X_{ij} + \alpha_2 \mathrm{d}X_{i\ell} = 0, \quad \alpha_3 \mathrm{d}X_{ij} + \alpha_4 \mathrm{d}X_{i\ell} = 0$$

for

$$\alpha_1 = -ap_{11} + p_{12} - \kappa_{ij} + \kappa_{jk} - a\kappa_{jk}, \quad \alpha_2 = -bp_{11} - p_{12} - b\kappa_{jk},$$

$$\alpha_3 = -ap_{21} + p_{22} + \kappa_{ij}, \quad \alpha_4 = -bp_{21} - p_{22} - \kappa_{i\ell}.$$

It follows that all α_* must be zero, and hence

$$p_{11} = -\frac{\kappa_{ij} - \kappa_{jk} + a\kappa_{jk} + b\kappa_{jk}}{a+b}, \quad p_{12} = -\frac{-b\kappa_{ij} + b\kappa_{jk}}{a+b},$$

$$p_{21} = \frac{\kappa_{ij} - \kappa_{i\ell}}{a+b}, \quad p_{22} = -\frac{b\kappa_{ij} + a\kappa_{i\ell}}{a+b},$$

where $a+b \neq 0$ due to the non-degeneracy condition of non-null edges. Continuing,

$$\mathrm{d}X_{\ell k} = \mathrm{d}X_{ik} - \mathrm{d}X_{i\ell} = a\mathrm{d}X_{ij} + b\mathrm{d}X_{i\ell} - \mathrm{d}X_{i\ell},$$

$$\mathrm{d}N_{k\ell} = \mathrm{d}N_{i\ell} - \mathrm{d}N_{ij} - \mathrm{d}N_{jk}.$$

Then $\mathrm{d}N_{k\ell} = -\kappa_{k\ell}\mathrm{d}X_{k\ell}$ leads to the two equations

$$\kappa_{ij} - \kappa_{jk} + a\kappa_{jk} - a\kappa_{k\ell} = -\kappa_{i\ell} + b\kappa_{jk} + \kappa_{k\ell} - b\kappa_{k\ell} = 0.$$

Thus,

$$a = \frac{-\kappa_{ij} + \kappa_{jk}}{\kappa_{jk} - \kappa_{k\ell}}, \quad b = \frac{-\kappa_{i\ell} + \kappa_{k\ell}}{-\kappa_{jk} + \kappa_{k\ell}},$$

and it is now straightforward to prove the following lemma:

Lemma 7.3.11 ([24, Proposition 11]) *We have*

$$\mathcal{H}_{ijk\ell} = \frac{\kappa_{i\ell}\kappa_{jk} - \kappa_{ij}\kappa_{k\ell}}{\kappa_{ij} - \kappa_{jk} + \kappa_{k\ell} - \kappa_{i\ell}},$$

$$\mathcal{K}_{ijk\ell} = \frac{\kappa_{ij}\kappa_{jk}\kappa_{k\ell} - \kappa_{ij}\kappa_{i\ell}\kappa_{jk} + \kappa_{ij}\kappa_{i\ell}\kappa_{k\ell} - \kappa_{i\ell}\kappa_{jk}\kappa_{k\ell}}{\kappa_{ij} - \kappa_{jk} + \kappa_{k\ell} - \kappa_{i\ell}}.$$

7.4 Discrete Isothermic Sphere Congruences

Now we will consider discrete isothermic sphere congruences. First, we define discrete isothermic sphere congruences using the analogous Moutard lifts characterization seen in Theorem 4.6.33.

Definition 7.4.12 Let $s : \mathbb{Z}^2 \to P(L^5)$ be a discrete sphere congruence. If there is some $\varsigma \in \Gamma s$ such that ς is Moutard, that is,

$$\varsigma_\ell - \varsigma_j \parallel \varsigma_k - \varsigma_i$$

on every elementary quadrilateral $ijk\ell$, then we call s a *discrete isothermic sphere congruence*.

We would like to now recover the discrete analogue of the smooth result seen in Lemma 6.4.20; namely, a pair of sphere congruences admitting Koenigs dual lifts imply that the sphere congruence is isothermic. For this, suppose that we have two distinct discrete sphere congruences $s^\pm : \mathbb{Z}^2 \to P(L^5)$ admitting Koenigs dual lifts $\sigma^\pm \in \Gamma s^\pm$ defined analogously as in Definition 4.4.18 so that on every edge ij, we have

$$d\sigma_{ij}^+ = \alpha_{ij} d\sigma_{ij}^- \tag{7.11}$$

for some real $\alpha_{ij} = \alpha_{ji}$, and

$$d\sigma_{ik}^\pm \parallel d\sigma_{j\ell}^\mp, \text{ (or equivalently, } \mathcal{A}(\sigma^+, \sigma^-) = 0).$$

Using a completely analogous argument to Lemma 4.4.20, one can show the following:

Lemma 7.4.13 *Koenigs dual $\sigma^\pm : \mathbb{Z}^2 \to L^5$ have planar quadrilaterals.*

Lemma 7.4.14 *For Koenigs dual $\sigma^\pm : \mathbb{Z}^2 \to L^5$ with $d\sigma_{ij}^+ = \alpha_{ij} d\sigma_{ij}^-$, we have*

$$\alpha_{ij}\,\alpha_{k\ell} = \alpha_{i\ell}\alpha_{jk}.$$

Proof Using the compatibility condition of σ^+ and (7.11), we have

$$d\sigma_{j\ell}^+ \parallel (\alpha_{jk} - \alpha_{k\ell})d\sigma_{ik}^- = (\alpha_{jk} - \alpha_{ij})d\sigma_{ij}^- + (\alpha_{i\ell} - \alpha_{k\ell})d\sigma_{i\ell}^-$$

$$= (\alpha_{ij}^{-1}\alpha_{jk} - 1)d\sigma_{ij}^+ + (1 - \alpha_{i\ell}^{-1}\alpha_{k\ell})d\sigma_{i\ell}^+$$

$$= A\sigma_i^+ + B\sigma_j^+ + C\sigma_\ell^+$$

for

$$A = \alpha_{i\ell}^{-1}\alpha_{k\ell} - \alpha_{ij}^{-1}\alpha_{jk}, \quad B = \alpha_{ij}^{-1}\alpha_{jk} - 1, \quad C = 1 - \alpha_{i\ell}^{-1}\alpha_{k\ell}.$$

It follows that $A = 0$ and $B = -C$, implying the result. $\square$

Hence, we can argue similarly as in Lemma 4.4.17 to obtain the Christoffel formula:

Lemma 7.4.15 ([48, Equation (3.1)]) *There exists a function* $r : \mathbb{Z}^2 \to \mathbb{R}$ *such that*

$$\mathrm{d}\sigma_{ij}^- = r_i r_j \mathrm{d}\sigma_{ij}^+. \tag{7.12}$$

Using an analogous definition of discrete Moutard lifts as in Definition 4.6.31, we show how Moutard lifts can be recovered from Koenigs dual lifts:

Theorem 7.4.16 ([48]) *Let* $s^\pm : \mathbb{Z}^2 \to P(L^5)$ *be sphere congruences admitting Koenigs dual lifts* $\sigma^\pm \in \Gamma s^\pm$. *If* $\varsigma^\pm \in \Gamma s^\pm$ *are defined via*

$$\varsigma^\pm := r^{\pm 1}\sigma^\pm, \tag{7.13}$$

then $\varsigma^\pm$ *are Moutard lifts of the sphere congruence* $s^\pm$.

Proof Calculations analogous to (4.10) in the proof of Theorem 4.4.22 imply that

$$(r_k - r_i)(r_\ell \sigma_\ell^+ - r_j \sigma_j^+) = (r_\ell - r_j)(r_k \sigma_k^+ - r_i \sigma_i^+).$$

Thus $(\varsigma_\ell^+ - \varsigma_j^+) \parallel (\varsigma_k^+ - \varsigma_i^+)$. The fact that ς^- is Moutard lift is shown similarly. $\square$

Thus, we conclude:

Theorem 7.4.17 *Let* $s^\pm : \mathbb{Z}^2 \to P(L^5)$ *be sphere congruences admitting Koenigs dual lifts* $\sigma^\pm \in \Gamma s^\pm$. *Then* $s^\pm$ *are isothermic sphere congruences.*

We can also recover the edge-labeling from the Moutard lifts:

Lemma 7.4.18 ([48, Lemma 3.5]) *We have* $a_{ij}^\pm := (\varsigma_i^\pm, \varsigma_j^\pm)$ *is an edge-labeling, that is,* $a_{ij}^\pm = a_{k\ell}^\pm$ *and* $a_{i\ell}^\pm = a_{jk}^\pm$. *Furthermore, we have that* $a_{ij}^+ = a_{ij}^-$, *that is, the two discrete isothermic sphere congruences share the same edge-labeling.*

Proof The fact that $a_{ij}^\pm$ is an edge-labeling can be shown using an analogous argument to the necessity proof of Theorem 4.6.33. Now, we have

$$a_{ij}^+ = (\varsigma_i^+, \varsigma_j^+) = r_i r_j (\sigma_i^+, \sigma_j^+) = -\frac{1}{2} r_i r_j (\mathrm{d}\sigma_{ij}^+, \mathrm{d}\sigma_{ij}^+)$$

$$= -\frac{1}{2} \frac{1}{r_i r_j} (\mathrm{d}\sigma_{ij}^-, \mathrm{d}\sigma_{ij}^-) = \frac{1}{r_i r_j} (\sigma_i^-, \sigma_j^-) = (\varsigma_i^-, \varsigma_j^-) = a_{ij}^-.$$

$\square$

Remark 7.4.19 Even though the notion of cross ratios is unavailable in Lie sphere geometry in general, Lemma 7.4.13 tells us that every elementary quadrilateral of $\sigma^{\pm}$ takes values in an affine plane. Thus, a notion of cross ratios can be defined for every elementary quadrilateral of $\sigma^{\pm}$, and it can be checked that

$$\mathrm{cr}(\sigma_i^{\pm}, \sigma_j^{\pm}, \sigma_k^{\pm}, \sigma_\ell^{\pm}) = \frac{a_{ij}}{a_{i\ell}}. \tag{7.14}$$

For example, if an elementary quadrilateral of $\sigma^{\pm}$ spans a subspace $V \cong \mathbb{R}^{2,1}$, then one can use Comment 2 in Sect. 2.5 to verify (7.14). See [48, Lemma 3.5], for details.

7.5 Discrete Ω-surfaces

For discrete sphere congruences $s^{\pm} : \mathbb{Z}^2 \to P(L^5)$ admitting Koenigs dual lifts $\sigma^{\pm} \in \Gamma s^{\pm}$, further assume that $(\sigma^+, \sigma^-) = 0$ so that $\Lambda := \mathrm{span}\{\sigma^+, \sigma^-\} : \mathbb{Z}^2 \to \mathcal{Z}$. Then (7.12) implies that

$$K_{ij} := \sigma_j^- - r_i r_j \sigma_j^+ = \sigma_i^- - r_i r_j \sigma_i^+ \in \Lambda_i \cap \Lambda_j. \tag{7.15}$$

Therefore, we have that Λ is a discrete Legendre map.

Motivated by the behavior of infinitesimal quadrilaterals of Christoffel dual lifts for smooth Ω-surfaces considered in Lemma 6.4.20, we have the following definition of discrete Ω-surfaces:

Definition 7.5.20 ([48, Definition 3.1]) A discrete Legendre map $\Lambda : \mathbb{Z}^2 \to \mathcal{Z}$ is called a *discrete Ω-surface* if there are either

(1) two sections $\sigma^{\pm} \in \Gamma\Lambda$, or
(2) complex conjugate sections $\sigma^{\pm} = \sigma_1 \pm \sqrt{-1}\sigma_2$ for $\sigma_1, \sigma_2 \in \Gamma\Lambda$,

such that $\sigma^{\pm}$ are Koenigs duals of each other.

Remark 7.5.21 Our discussions leading up to the Definition 7.5.20 only considered real sections of Λ. As in the case smooth Ω-surfaces, seen in Sect. 6.5, we expand the definition of discrete Ω-surfaces to include complex conjugate isothermic sphere congruences. We leave checking isothermicity of complex conjugate Koenigs nets to the readers, which is also done in [48].

Remark 7.5.22 Here we consider two nonzero vectors $\mathbf{v}, \mathbf{w}$ in $\mathbb{R}^{4,2} \otimes \mathbb{C}$ to be parallel if there exists $f \in \mathbb{C} \setminus \{0\}$ such that $\mathbf{v} = f\mathbf{w}$. In particular, if $\mathbf{v}$ and $\mathbf{w}$ are complex conjugate, then f will be unitary.

Using the definition of discrete Ω-surfaces, we can show that discrete minimal surfaces and constant mean curvature surfaces in some space form are discrete Ω-surfaces as follows: Denoting the point sphere map of Λ as X for some choice of point sphere complex and space form vector, if $\mathcal{H} \equiv 0$, that is, $\mathcal{A}(X, N) = 0$, then $\sigma^- := X$ and $\sigma^+ := N$ are Koenigs dual, and Λ is an Ω-surface.

Similarly, if $\mathcal{H} \equiv c \neq 0$, then we have that $\mathcal{A}(X, \mathcal{H}X + N) = 0$ since the mixed area is bilinear. Thus, with Koenigs dual $\sigma^- := X$ and $\sigma^+ := \mathcal{H}X + N$, we have that Λ is an Ω-surface. This is the discrete analogue of the smooth result stated in Remark 6.7.41. (Also compare these results with the Euclidean formulation given in Corollaries 4.10.59 and 4.10.62 with Corollary 4.4.23.)

Example 7.5.23 For the discrete Clifford tori with constant cross ratios in Example 7.1.5, we have that the Königs dual lifts are

$$\sigma^- = X^{\mathbb{S}^3}, \quad \sigma^+ = N^{\mathbb{S}^3}.$$

We can also directly compute that

$$X^{\mathbb{S}^3}_{m+1,n} - X^{\mathbb{S}^3}_{m,n} = N^{\mathbb{S}^3}_{m+1,n} - N^{\mathbb{S}^3}_{m,n},$$

$$X^{\mathbb{S}^3}_{m,n+1} - X^{\mathbb{S}^3}_{m,n} = N^{\mathbb{S}^3}_{m,n} - N^{\mathbb{S}^3}_{m,n+1},$$

$$X^{\mathbb{S}^3}_{m+1,n+1} - X^{\mathbb{S}^3}_{m,n} = N^{\mathbb{S}^3}_{m+1,n} - N^{\mathbb{S}^3}_{m,n+1},$$

$$X^{\mathbb{S}^3}_{m,n+1} - X^{\mathbb{S}^3}_{m+1,n} = N^{\mathbb{S}^3}_{m,n} - N^{\mathbb{S}^3}_{m+1,n+1},$$

and come to the same conclusion. This can be done for the Clifford tori with non-constant cross ratios as well.

7.6 Isothermic Sphere Congruences of Discrete Ω-surfaces

In the case of smooth Ω-surfaces, generically, the pair of isothermic sphere congruences is unique. However, this is not the case for discrete Ω-surfaces as we see in the following theorem:

Theorem 7.6.24 ([48, Lemma 3.3, Corollary 3.4]) *Let $\Lambda : \mathbb{Z}^2 \to \mathcal{Z}$ be a discrete Ω-surface, spanned by Koenigs dual lifts of isothermic sphere congruences. Then there exists infinitely many Königs dual lifts spanning Λ.*

Proof Let $\Lambda = \mathrm{span}\{\sigma^+, \sigma^-\}$ with Königs dual lifts $\sigma^\pm \in \Gamma\Lambda$. For any constants $c^\pm$ $(c^+ \neq c^-)$, set

$$\tilde{\sigma}^\pm = \frac{1}{r + c^\mp}\{\sigma^- + c^\pm r\sigma^+\}.$$

Then we have that

$$(r_i + c^-)(r_j + c^-)d\tilde{\sigma}_{ij}^+ = (r_i + c^-)(\sigma_j^- + c^+ r_j \sigma_j^+) - (r_j + c^-)(\sigma_i^- + c^+ r_i \sigma_i^+)$$
$$= (r_i + c^+)(\sigma_j^- + c^- r_j \sigma_j^+) - (r_j + c^-)(\sigma_i^- + c^- r_i \sigma_i^+)$$
$$= (r_i + c^+)(r_j + c^+)d\tilde{\sigma}_{ij}^-,$$

so that

$$\mathrm{d}\tilde{\sigma}_{ij}^- = \tilde{r}_i \tilde{r}_j \mathrm{d}\tilde{\sigma}_{ij}^+,$$

where

$$\tilde{r} = \frac{r + c^-}{r + c^+}.$$

Then an analogous argument to the proof of Theorem 4.4.22 tells us that the opposite diagonals are parallel. $\qquad\square$

Remark 7.6.25 In fact, the constants $c^\pm$ can be chosen as complex conjugate numbers to obtain a pair of complex conjugate Koenigs dual lifts from a pair of real Koenigs duals, and vice versa. Therefore, in discussing discrete Ω-surfaces, one only needs to consider a real pair of Koenigs dual sections. For more information, see [48].

In Sect. 4.5, we have seen that discrete isothermic surfaces in $P(L^4)$ admit a 1-parameter families of flat connections, and we now recover the analogous result in the realm of Lie sphere geometry, without using the notion of cross ratios. First, we analogously define the orthogonal transformation in Definition 3.6.40 for $\mathbb{R}^{4,2}$: for $s_1, s_2 \in P(L^5)$ such that $s_1 \oplus s_2$ span a 2-plane of signature $(+-)$, let

$$\Gamma_{s_2}^{s_1}(t)Y = \begin{cases} tY, & \text{if } Y \in s_1, \\ t^{-1}Y, & \text{if } Y \in s_2, \\ Y, & \text{if } Y \in (s_1 \oplus s_2)^\perp, \end{cases}$$

for any $Y \in \mathbb{R}^{4,2}$. Then as in Lemma 3.6.42, we have that

$$\Gamma_{s_2}^{s_1}(t)Y = Y + (V, W)^{-1}\{(t-1)(Y, W)V + (t^{-1} - 1)(Y, V)W\} \qquad (7.16)$$

for any choice of non-zero $V \in s_1$ and $W \in s_2$.

Analogous to isothermic surfaces in Möbius geometry seen in Sect. 4.5, we can show that a discrete isothermic sphere congruence $s : \mathbb{Z}^2 \to P(L^5)$, satisfying a mild non-degeneracy condition, $s_i \not\perp s_k$ and $s_j \not\perp s_\ell$, gives rise to a 1-parameter family of discrete flat connections:

Lemma 7.6.26 ([48, Lemma 3.7]) *Let $s : \mathbb{Z}^2 \to P(L^5)$ be discrete isothermic with edge-labeling a. Then the connection Γ^λ defined on the trivial bundle $\mathbb{Z}^2 \times \mathbb{R}^{4,2}$*

by

$$\Gamma_{ij}^{\lambda} := \Gamma_{s_j}^{s_i}(1 - \lambda a_{ij})$$

on any edge ij is flat for any $\lambda \in \mathbb{R}$.

Proof If we show that

$$\Gamma_{s_j}^{s_i}(1 - \lambda a_{ij})\Gamma_{s_k}^{s_j}(1 - \lambda a_{jk}) = \Gamma_{s_\ell}^{s_j}\left(\tfrac{1-\lambda a_{i\ell}}{1-\lambda a_{ij}}\right) = \Gamma_{s_\ell}^{s_i}(1 - \lambda a_{i\ell})\Gamma_{s_k}^{s_\ell}(1 - \lambda a_{\ell k}),$$

$$\tag{7.17}$$

then this would imply the desired conclusion. To show the second equality, we will show both sides hold true for $Y \in s_j$, $Y \in s_\ell$, and $Y \in (s_j \oplus s_\ell)^{\perp}$. Keeping in mind that a is an edge-labeling, take the Moutard lift $\varsigma \in \Gamma s$, and see immediately that

$$\Gamma_{s_\ell}^{s_i}(1 - \lambda a_{i\ell})\Gamma_{s_k}^{s_\ell}(1 - \lambda a_{\ell k})\varsigma_\ell = (1 - \lambda a_{\ell k})\Gamma_{s_\ell}^{s_i}(1 - \lambda a_{i\ell})\varsigma_\ell$$

$$= \frac{1 - \lambda a_{ij}}{1 - \lambda a_{i\ell}}\varsigma_\ell = \Gamma_{s_\ell}^{s_j}\left(\tfrac{1-\lambda a_{i\ell}}{1-\lambda a_{ij}}\right)\varsigma_\ell.$$

On the other hand, writing the flat connection explicitly using Moutard lifts via (7.16), we have that

$$\Gamma_{s_j}^{s_i}(1 - \lambda a_{ij})Y = Y - \lambda(\varsigma_j, Y)\varsigma_i + \tfrac{\lambda}{1-\lambda a_{ij}}(\varsigma_i, Y)\varsigma_j.$$

Therefore, we can calculate that

$$\Gamma_{s_\ell}^{s_i}(1-\lambda a_{i\ell})\Gamma_{s_k}^{s_\ell}(1-\lambda a_{\ell k})\varsigma_j = \Gamma_{s_\ell}^{s_i}(1-\lambda a_{i\ell})\left((1 - \lambda a_{i\ell})\varsigma_\ell - \alpha\varsigma_i + \alpha\tfrac{1-\lambda a_{i\ell}}{1-\lambda a_{ij}}\varsigma_k\right),$$

where we have used the fact that

$$\varsigma_\ell - \varsigma_j = \alpha(\varsigma_i - \varsigma_k)$$

since ς is Moutard. Verifying that

$$\Gamma_{s_\ell}^{s_i}(1 - \lambda a_{i\ell})\varsigma_k = \varsigma_k - \lambda a_{ij}\varsigma_i + \tfrac{\lambda}{1-\lambda a_{i\ell}}\tfrac{a_{ij}-a_{i\ell}}{\alpha}\varsigma_\ell,$$

we can conclude that

$$\Gamma_{s_\ell}^{s_i}(1 - \lambda a_{i\ell})\Gamma_{s_k}^{s_\ell}(1 - \lambda a_{\ell k})\varsigma_j = \tfrac{1-\lambda a_{i\ell}}{1-\lambda a_{ij}}\left(\varsigma_\ell - \alpha(\varsigma_i - \varsigma_k)\right)$$

$$= \tfrac{1-\lambda a_{i\ell}}{1-\lambda a_{ij}}\varsigma_j = \Gamma_{s_\ell}^{s_j}\left(\tfrac{1-\lambda a_{i\ell}}{1-\lambda a_{ij}}\right)\varsigma_j.$$

Finally, for $Y \in (s_j \oplus s_\ell)^\perp$, we see that

$$\Gamma_{s_\ell}^{s_i}(1 - \lambda a_{i\ell})\Gamma_{s_k}^{s_\ell}(1 - \lambda a_{\ell k})Y = \Gamma_{s_\ell}^{s_i}(1 - \lambda a_{i\ell})\left(Y - \lambda(\varsigma_k, Y)\varsigma_\ell\right)$$

$$= Y + \tfrac{\lambda}{1-\lambda a_{i\ell}}(\varsigma_i, Y)\varsigma_\ell - \tfrac{\lambda}{1-\lambda a_{i\ell}}(\varsigma_k, Y)\varsigma_\ell$$

$$= Y + \tfrac{\lambda}{1-\lambda a_{i\ell}}\tfrac{1}{\alpha}(\varsigma_\ell - \varsigma_j, Y)\varsigma_\ell = Y.$$

The first equality can be shown using a symmetric argument by switching the indices appropriately. $\square$

Remark 7.6.27 Essentially, the statement of Lemma 3.6.49 is identical to the condition (7.17).

Now let $\Lambda : \mathbb{Z}^2 \to \mathcal{Z}$ be a discrete Ω-surface with some choice of Koenigs dual sections $\sigma^\pm \in \Gamma\Lambda$, and denote by $s^\pm : \mathbb{Z}^2 \to P(L^5)$ the isothermic sphere congruences determined by the Koenigs dual sections, i.e., $s^\pm = \operatorname{span}\{\sigma^\pm\}$. Suppose that $\varsigma^\pm \in \Gamma s^\pm$ are Moutard lifts, so that $a_{ij} = (\varsigma_i^+, \varsigma_j^+) = (\varsigma_i^-, \varsigma_j^-)$ is an edge-labeling. From (7.15), we then see that

$$a_{ij} = (\sigma_i^+, \sigma_j^-) = (\sigma_j^+, \sigma_i^-).$$

We now recover a discrete version of Theorem 6.2.15.

Theorem 7.6.28 ([48]) *Denoting by $\Gamma^{\pm,\lambda}$ the isothermic flat connections associated with $s^\pm$, we have that*

$$g^\lambda \bullet \Gamma^{-,\lambda} = \Gamma^{+,\lambda},$$

where

$$g^\lambda := 1 + \lambda\sigma^- \wedge \sigma^+ = \exp(\lambda\sigma^- \wedge \sigma^+).$$

Proof On some fixed edge ij, it is enough to show that both sides agree for $Y \in s_i^+$, $Y \in s_j^+$, and $Y \in (s_i^+ \oplus s_j^+)^\perp$. Verifying the following using (7.15),

$$g_i^\lambda\sigma_j^+ = \sigma_j^+ + \lambda a_{ij}\sigma_i^+ - \tfrac{\lambda a_{ij}}{r_i r_j}\sigma_i^- = \sigma_j^+ - \tfrac{\lambda a_{ij}}{r_i r_j}(\sigma_j^- - r_i r_j\sigma_j^+),$$

$$g_i^\lambda\sigma_j^- = \sigma_j^- + \lambda a_{ij}r_i r_j\sigma_i^+ - \lambda a_{ij}\sigma_i^- = \sigma_j^- - \lambda a_{ij}(\sigma_j^- - r_i r_j\sigma_j^+),$$

we can calculate that

$$g_i^\lambda\Gamma_{ij}^{-,\lambda}g_j^{-\lambda}\sigma_j^+ = g_i^\lambda\Gamma_{ij}^{-,\lambda}\sigma_j^+ = g_i^\lambda\left(\sigma_j^+ + \tfrac{\lambda a_{ij}}{1-\lambda a_{ij}}\tfrac{1}{r_i r_j}\sigma_j^-\right)$$

$$= \tfrac{1}{1-\lambda a_{ij}}\sigma_j^+ = \Gamma_{ij}^{+,\lambda}\sigma_j^+.$$

Similar arguments can be used to show the rest of the claim. $\square$

Exercise 7.6.29 Show that for K_{ij} as in (7.15), one has that $\Gamma_{ij}^{\pm,\lambda} K_{ij} = K_{ij}$.

7.7 Calapso Transformations of Discrete Ω-surfaces

Calapso transformations for discrete isothermic surfaces were defined in Definition 4.7.35 via trivializing gauge transformations. Analogously, we define the Calapso transformations of isothermic sphere congruence $s : \mathbb{Z}^2 \to P(L^5)$ with associated flat connections Γ^λ using $T^\lambda : \mathbb{Z}^2 \to O_{4,2}$ such that

$$T^\lambda \bullet \Gamma^\lambda = id.$$

Now suppose that $\Lambda : \mathbb{Z}^2 \to \mathcal{Z}$ is a discrete Ω-surface admitting Koenigs dual lifts $\sigma^\pm \in \Gamma\Lambda$. Consider the Calapso transformations $T^{\pm,\lambda}$ of $s^\pm = \operatorname{span}\{\sigma^\pm\}$. Then from Theorem 7.6.28, we have that

$$T^{-,\lambda} \bullet \Gamma^{-,\lambda} = id = T^{+,\lambda} \bullet \Gamma^{+,\lambda} = T^{+,\lambda} \bullet (g^\lambda \bullet \Gamma^{-,\lambda}) = (T^{+,\lambda} g^\lambda) \bullet \Gamma^{-,\lambda},$$

so that with correct choice of initial conditions, we have that

$$T^{-,\lambda} = T^{+,\lambda} g^\lambda.$$

For any section $S \in \Gamma\Lambda$, we easily see that $g^\lambda S = S$ so that $T^{-,\lambda} S = T^{+,\lambda} S$. In particular, we have that on every vertex

$$T^{-,\lambda} \Lambda = T^{+,\lambda} g^\lambda \Lambda = T^{+,\lambda} \Lambda.$$

Therefore, define $\Lambda^\mu : \mathbb{Z}^2 \to \mathcal{Z}$ by

$$\Lambda^\mu := T^{+,\mu} \Lambda = T^{-,\mu} \Lambda.$$

Defining

$$\sigma^{\pm,\mu} := T^{\pm,\mu} \sigma^\pm,$$

and using Exercise 7.6.29, we compute that

$$r_i r_j \mathrm{d}(T^{+,\mu}\sigma^+)_{ij} = r_i r_j \mathrm{d}(T^{-,\mu}\sigma^+)_{ij} = T_i^{-,\mu}(\Gamma_{ij}^{-,\mu} r_i r_j \sigma_j^+ - r_i r_j \sigma_i^+)$$

$$= T_i^{-,\mu}(\Gamma_{ij}^{-,\mu}(r_i r_j \sigma_j^+ + K_{ij}) - (r_i r_j \sigma_i^+ + K_{ij}))$$

$$= T_i^{-,\mu}(\Gamma_{ij}^{-,\mu}\sigma_j^- - \sigma_i^-) = T_j^{-,\mu}\sigma_j^- - T_i^{-,\mu}\sigma_i^- = \mathrm{d}(T^{-,\mu}\sigma^-)_{ij},$$

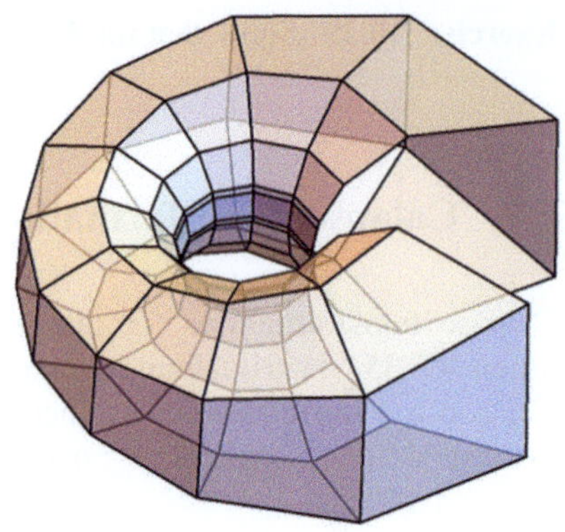

Fig. 7.1 A Calapso transform, using $\mu = -0.1$, of the Clifford torus with constant cross ratios and $m_{per} = n_{per} = 10$. Note that this transform is not periodic with respect to the domain that the Clifford torus is periodic on

where K_{ij} is as defined in (7.15). Therefore, $\sigma^{\pm,\mu}$ satisfy the Christoffel formula (7.11), and calculations analogous to the proof of Theorem 4.4.22 tells us that

$$(r_\ell - r_j)\mathrm{d}\sigma_{ik}^{-,\mu} = r_j r_\ell (r_k - r_i)\mathrm{d}\sigma_{j\ell}^{+,\mu},$$

allowing us to conclude that $\sigma^{\pm,\mu} \in \Gamma \Lambda^\mu$ are Koenigs duals:

Definition 7.7.30 Let $\Lambda : \mathbb{Z}^2 \to \mathcal{Z}$ be a discrete Ω-surface admitting Koenigs dual lifts $\sigma^\pm \in \Gamma \Lambda$. Denote by $T^{\pm,\lambda}$ the respective Calapso transformations of isothermic sphere congruences $s^\pm : \mathbb{Z}^2 \to P(L^5)$ coming from $\sigma^\pm$. Then we call $\Lambda^\mu := T^{+,\mu}\Lambda = T^{-,\mu}\Lambda$ a *Calapso transform* of the discrete Ω-surface Λ with parameter μ (Fig. 7.1).

Exercise 7.7.31 Let $\Lambda : \mathbb{Z}^2 \to \mathcal{Z}$ be a discrete Ω-surface admitting Moutard lifts $\varsigma^\pm \in \Gamma \Lambda$. Show that $\varsigma^{\pm,\mu} := T^{\pm,\mu}\varsigma^\pm$ are Moutard sections of Λ^μ.

7.8 Discrete Linear Weingarten Surfaces

With the notions of Gauss and mean curvatures defined, we define discrete linear Weingarten surfaces analogously to the smooth case:

Definition 7.8.32 Denoting by X and N the point sphere map and the tangent plane congruence of a discrete Legendre map $\Lambda : \mathbb{Z}^2 \to \mathcal{Z}$ for some choice of point sphere complex and space form vector, we say Λ projects to a *discrete linear Weingarten surface* if the mean curvature $\mathcal{H}$ and the Gauss curvature $\mathcal{K}$ satisfy

$$\alpha \mathcal{K} + 2\beta \mathcal{H} + \gamma = 0$$

for some constants α, β, γ not all zero. Furthermore, when $\beta^2 - \alpha\gamma = 0$, we say that the discrete linear Weingarten surface is *tubular*.

Recalling the analogous definition of polynomial conserved quantities from Definition 4.9.48, for a 1-parameter family of flat connections Γ^λ of an isothermic sphere congruence $s : \mathbb{Z}^2 \to P(L^5)$, an analogous result of Lemma 4.9.52 says that Q is a constant conserved quantity of Γ^λ if Q is constant, and $(S, Q) = 0$ for any $S \in \Gamma s$. The next theorem characterizes discrete linear Weingarten surfaces in space forms as discrete Ω-surfaces with enveloping isothermic sphere congruences admitting constant conserved quantities:

Theorem 7.8.33 ([48, Theorem 2.8]) *Suppose* $\Lambda : \mathbb{Z}^2 \to \mathcal{Z}$ *is a discrete* Ω-*surface admitting Koenigs dual lifts* $\sigma^\pm$ *in* $\mathbb{R}^{4,2}$ *(resp.* $\sigma^+ = \overline{\sigma^-} \in \mathbb{R}^{4,2} \otimes \mathbb{C}$*). Then* Λ *projects to a non-tubular discrete linear Weingarten surface in some space form if and only if the isothermic sphere congruences determined by the Koenigs dual lifts admit constant conserved quantities* $Q^\pm \in \mathbb{R}^{4,2}$ *(resp.* $Q^+ = \overline{Q^-} \in \mathbb{R}^{4,2} \otimes \mathbb{C}$*) such that*

(1) $V = \mathrm{span}\{Q^+, Q^-\}$ *(resp.* $V = \mathrm{span}\{\mathrm{Re}\, Q^+, \mathrm{Im}\, Q^+\}$*) is not a null 2-plane, and*

(2) $(\sigma^\pm, Q^\mp) \neq 0$.

Proof If $Q^\pm$ are constant conserved quantities of $s^\pm = \mathrm{span}\{\sigma^\pm\}$, then we have that $(\sigma^\pm, Q^\pm) = 0$ so that $(d\sigma^\mp_{ij}, Q^\pm) = 0$. Then the second condition on the constant conserved quantities allows us to assume without loss of generality that

$$(\sigma^\pm, Q^\mp) \equiv 1.$$

Choose some orthogonal pair $\mathfrak{p}, \mathfrak{q} \in V$ so that

$$\mathfrak{p} = a^+ Q^+ + a^- Q^- \quad \text{and} \quad \mathfrak{q} = b^+ Q^+ + b^- Q^-,$$

for some $a^\pm, b^\pm$ with $(\mathfrak{p}, \mathfrak{p}) \neq 0$. Defining $\vartheta = a^- b^+ - a^+ b^- \neq 0$, we find that

$$\vartheta X = a^+ \sigma^+ - a^- \sigma^- \quad \text{and} \quad \vartheta N = -b^+ \sigma^+ + b^- \sigma^-,$$

where X and N are the point sphere map and tangent plane congruence of Λ with respect to point sphere complex $\mathfrak{p}$ and space form vector $\mathfrak{q}$. Therefore, we find that

$$\sigma^+ = -b^- X - a^- N \quad \text{and} \quad \sigma^- = -b^+ X - a^+ N,$$

and the bilinearity of the mixed area tells us

$$0 = \mathcal{A}(\sigma^+, \sigma^-) = b^+ b^- \mathcal{A}(X, X) + (a^- b^+ + a^+ b^-) \mathcal{A}(X, N) + a^+ a^- \mathcal{A}(N, N).$$

Note that even in the complex conjugate case, we have that the coefficients of the above equation are all real constants. Thus, Λ projects to a linear Weingarten surface in the space form determined by the point sphere complex $\mathfrak{p}$ and space form vector $\mathfrak{q}$. Finally, we note that

$$(a^- b^+ + a^+ b^-)^2 - 4 a^+ a^- b^+ b^- = (a^- b^+ - a^+ b^-)^2 = \vartheta^2 \neq 0$$

so that the space form projection is non-tubular.

Conversely, suppose that Λ projects to a non-tubular discrete linear Weingarten surface in the space form determined by $\mathfrak{p}$ and $\mathfrak{q}$. Then there exist real constants α, β, γ not all zero so that

$$\gamma \mathcal{A}(X, X) - 2\beta \mathcal{A}(X, N) + \alpha \mathcal{A}(N, N) = 0. \tag{7.18}$$

Choosing δ so that

$$\delta^2 = \beta^2 - \alpha\gamma,$$

we have that $\delta \in \mathbb{R} \cup i\mathbb{R}$, and δ is not zero because we have assumed the linear Weingarten net is non-tubular. We choose $Q^{\pm}$ and $\sigma^{\pm}$ as follows in two different cases:

- **Case $\alpha \neq 0$:** we set $Q^{\pm} = \frac{1}{2\alpha}(\alpha \mathfrak{q} + \beta \mathfrak{p} \pm \delta \mathfrak{p})$, $\sigma^{\pm} = X \pm \frac{1}{\delta}(\beta X - \alpha N)$.
- **Case $\alpha = 0$:** noting that $\beta \neq 0$, we set $Q^{-} = \mathfrak{p}$, $Q^{+} = \mathfrak{q} + \frac{\gamma}{2\beta}\mathfrak{p} = \mathfrak{q} - \mathcal{H}\mathfrak{p}$, $\sigma^{-} = X$, $\sigma^{+} = N + \mathcal{H}X$.

In either case it is straightforward to check that Q^{-} and Q^{+} are independent and $(\sigma^{\pm}, Q^{\pm}) = 0$. Also, $\mathrm{d}\sigma_{ij}^{+} \parallel \mathrm{d}\sigma_{ij}^{-}$ on all edges ij, by (7.6). Finally, using (7.18), we can compute that

$$\mathcal{A}(\sigma^{+}, \sigma^{-}) = 0,$$

telling us that $\sigma^{\pm}$ are Königs dual sections of Λ. $\qquad\qquad\qquad\qquad\qquad\square$

Finally, the Calapso transformation of discrete Ω-surfaces gives a Lawson-type correspondence for discrete linear Weingarten surfaces in space forms by considering the analogous behavior of polynomial conserved quantities under Calapso transformations in Theorem 4.9.54:

Theorem 7.8.34 ([48, Theorem and Definition 4.1]) *Generically, if Λ projects to a non-tubular linear Weingarten net, then the Calapso transform Λ^{μ} of Λ does as well in certain space forms.*

Certain special cases occur as follows (note that the case of CGC $\mathcal{K} = 0$ surfaces is excluded as this is tubular):

Exercise 7.8.35 Check that the following classes of discrete surfaces in space forms determined by $\mathfrak{p}$ and $\mathfrak{q}$ lift to Ω-surfaces with constant conserved quantities $Q^{\pm}$:

- minimal surfaces in space forms for $Q^{-} = \mathfrak{p}$, $Q^{+} = \mathfrak{q}$. See Fig. 4.6.
- cmc $\mathcal{H}$ surfaces for $Q^{-} = \mathfrak{p}$, $Q^{+} = \mathfrak{q} - \mathcal{H}\mathfrak{p}$. See Figs. 4.7, 7.2, and 7.4.
- flat surfaces in $\mathbb{H}^3$ for $(\mathfrak{q}, \mathfrak{q}) = -(\mathfrak{p}, \mathfrak{p}) = 1$ and $Q^{\pm} = \frac{1}{2}(\mathfrak{q} \pm \mathfrak{p})$. Note that then $(Q^{+}, Q^{+}) = (Q^{-}, Q^{-}) = 0$. See Fig. 7.5.
- flat surfaces in $\mathbb{S}^{2,1}$ for $(\mathfrak{q}, \mathfrak{q}) = -(\mathfrak{p}, \mathfrak{p}) = -1$ and $Q^{\pm} = \frac{1}{2}(\mathfrak{q} \pm \mathfrak{p})$. Note that then $(Q^{+}, Q^{+}) = (Q^{-}, Q^{-}) = 0$. See Fig. 7.5.

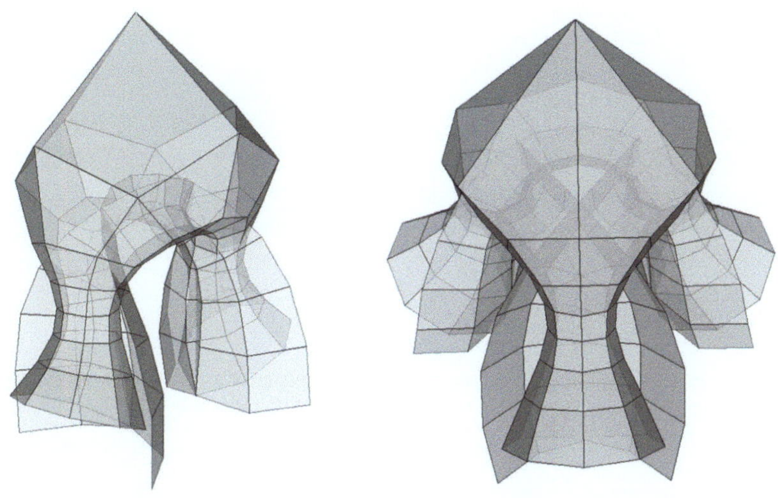

Fig. 7.2 Discrete 2-legged and 3-legged cmc Smyth surfaces in $\mathbb{R}^3$

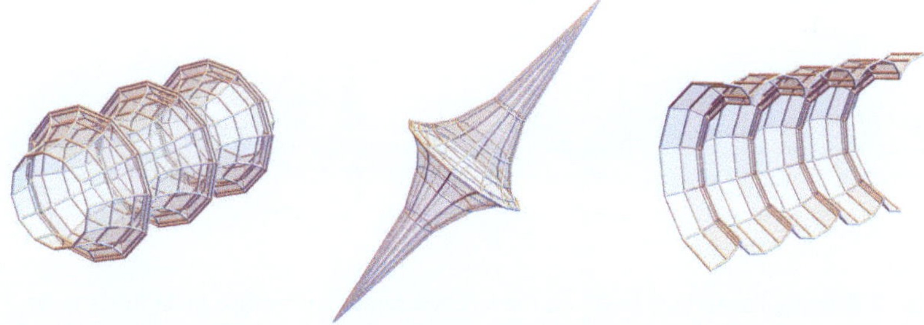

Fig. 7.3 Discrete surfaces with constant negative Gauss curvature in $\mathbb{R}^3$ on the left and the middle; a discrete surface with constant positive Gauss curvature in $\mathbb{R}^3$ on the right

- constant Gauss curvature $\mathcal{K} > 0$ surfaces for $\alpha = 1$ and $\gamma < 0$ so that $Q^{\pm} = \frac{1}{2}(\mathfrak{q} \pm \sqrt{-\gamma}\,\mathfrak{p}) \in \mathbb{R}^{4,2}$. See Fig. 7.3.
- discrete K-surfaces for $\alpha = 1$ and $\gamma > 0$. Then $Q^{\pm} = \frac{1}{2}(\mathfrak{q} \pm \sqrt{-\gamma}\,\mathfrak{p})$ are complex conjugate. See Fig. 7.3.

Example 7.8.36 The discrete Clifford tori, both with non-constant and constant cross ratios, are examples of minimal surfaces in $\mathbb{S}^3$, where by Examples 7.1.5 and 7.3.9, we have $\sigma^- = X^{\mathbb{S}^3}$, $\sigma^+ = N^{\mathbb{S}^3}$, and $Q^+ = (1, 0, 0, 0, 0, 0)^t = \mathfrak{q}$, $Q^- = (0, 0, 0, 0, 0, 1)^t = \mathfrak{p}$, so that $(\sigma^{\pm}, Q^{\pm}) = 0$. On the other hand, the discrete Clifford tori can also be seen as discrete K-surfaces, when using $\alpha = \gamma = 1$, $\beta = 0$, $\delta = \sqrt{-1}$, $Q^{\pm} = \frac{1}{2}(\mathfrak{q} \pm \sqrt{-1}\,\mathfrak{p})$ and $\sigma^{\pm} = X^{\mathbb{S}^3} \pm \sqrt{-1}N^{\mathbb{S}^3}$. This phenomenon is due to the non-uniqueness of Koenigs dual lifts.

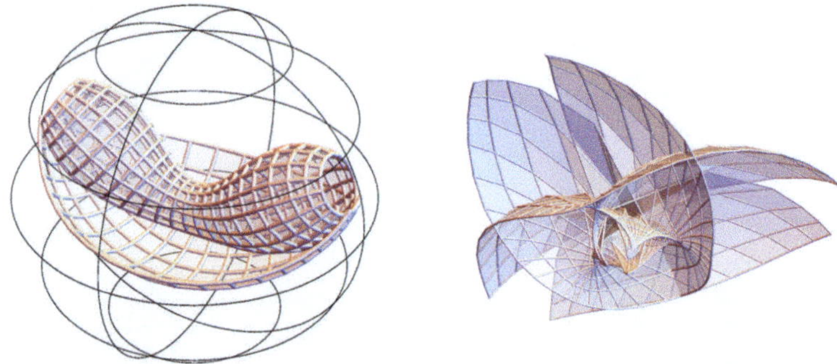

Fig. 7.4 A discrete cmc $\mathcal{H} = 1$ surface in $\mathbb{H}^3$, the Enneper cousin on the left, shown in the Poincaré ball model; a discrete cmc $\mathcal{H} = 1$ surface in de Sitter space $\mathbb{S}^{2,1}$ on the right, shown in the hollow ball model

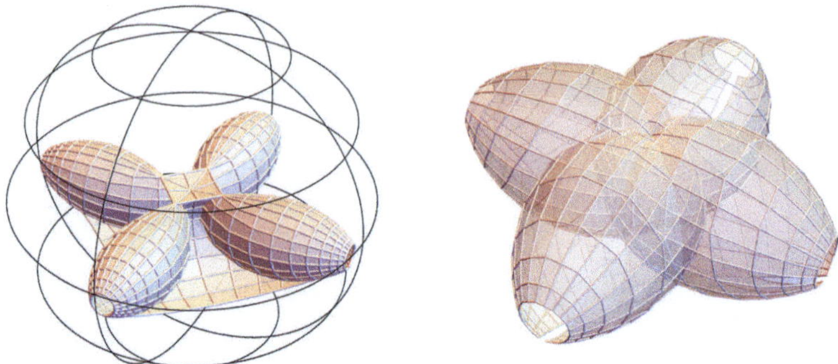

Fig. 7.5 A discrete flat surface in $\mathbb{H}^3$ on the left; a discrete flat surface in de Sitter space $\mathbb{S}^{2,1}$ on the right

Recommended Further Readings for Chap. 7

Discrete Legendre maps, also known as *principal contact-element nets*, are defined in [27] in the realm of Lie sphere geometry. Via reduction to subgeometries, the work shows that discrete Legendre maps in Lie sphere geometry are equivalent to circular nets in Möbius geometry or conical nets in Laguerre geometry defined in [156]. The work also explores a discrete version of Ribaucour transformations, called discrete R-congruences, a concept also considered in [92].

Considering Ribaucour transformations of curves, [44] shows that one can interpolate between any two curves via successive Ribaucour transformations. Applying this technique to the semi-discrete Legendre maps, the work obtains a "smoothing" of semi-discrete circular surfaces via channel surfaces, analogous to the smoothing of circular nets via Dupin cyclides obtained in [15, 21].

Building on the concept of discrete Ribaucour transformations, the work [189] explores the enveloping surfaces of a given discrete R-congruence. In the smooth case, one can generically find a pair of surfaces enveloping a sphere congruence; the work shows that in the discrete case, one finds a 2-parameter family of discrete enveloping surfaces.

The work [119] considers discrete channel surfaces, a notion that needs careful examination due to the principal curvatures being defined on edges. Using the Lie cyclides characterization of smooth channel surfaces in the realm of Lie sphere geometry, the work offers a convincing discretization of channel surfaces, recovering the discrete version of Vessiot's theorem [217]: a discrete isothermic channel surface is Möbius equivalent to a surface of revolution, cylinder, or a cone.

Discrete Ω-surfaces are a fairly new topic, with the first definition appearing in [78], employing a gauge-theoretic description akin to the characterization of discrete isothermic surfaces via 1-parameter family of discrete flat connections. This idea is then discussed in depth in the work [48] which most of the material in Chapter 7 is based on. The work formally introduces discrete Ω-surfaces using Koenigs duality characterization of isothermicity and recovers the flat connections. Then this notion is applied to the theory of discrete linear Weingarten surfaces, characterizing them in terms of discrete Ω-surfaces with isothermic sphere congrueces belonging to linear sphere complexes and hence admits constant conserved quantities.

A comprehensive discretization of Ω-surfaces is given in [40]. The discrete integrable structure of Ω-surfaces is explained in terms of the 3-dimensional permutability of transformations of isothermic sphere congruences. Such observation immediately leads to a gauge-theoretic description of Ω-surfaces, capturing various characterizations of discrete Ω-surfaces that the smooth counterparts enjoy, including the concept of dual Ω-surfaces. Also, via polynomial conserved quantities, discrete isothermic, Guichard, and L-isothermic surfaces are characterized, while the Ω-dual construction yields a Euclidean characterization of the aforementioned subclasses, making connection with discrete O-surfaces of [199].

Applying the Lie sphere geometric characterization of linear Weingarten surfaces, [190] takes an in-depth look at each case of linear Weingarten surfaces admitting discrete Weierstrass representations via Lie sphere geometry. A notable consideration is given to the detection of singularities on discrete surfaces: observing the behavior of principal curvatures near singularities in the smooth case [208], the work examines the discrete principal curvatures to detect possible singularities on discrete surfaces. This approach to the detection of singularities is utilized to study singularities appearing on semi-discrete linear Weingarten surfaces admitting Weierstrass-type representations in [223], where the representations are derived independently of Lie sphere geometry.

Another application of the characterization of linear Weingarten surfaces is the discrete analogue [179] of the work [178]. As in the smooth case, the work takes advantage of the discrete L-isothermicity of certain linear Weingarten surfaces, and obtains various discrete Weierstrass-type representations in the realm of Laguerre geometry.

Among those discrete linear Weingarten surfaces admitting Weierstrass-type representations are discrete flat surfaces in hyperbolic space. The work [126] recovers the representation for discrete flat surfaces in hyperbolic space, and considers how they are related to discrete cmc-1 surfaces and linear Weingarten surfaces of Bryant type. Furthermore, discrete focal surfaces, or *caustics*, are explored, making connections to singularity theory.

The cyclic systems of Ribaucour [187, 188] treat 1-parameter families of surfaces that are orthogonal to circle congruences, and include the transformations of K-surfaces given by Bianchi [11] and parallel families of flat fronts in hyperbolic space as special cases. The work [120] discretizes the cyclic systems in the realm of Lie sphere geometry, and examines how the discrete cyclic systems are related to the parallel families of discrete flat fronts in hyperbolic space.

References

1. Agafonov, S.I., Bobenko, A.I.: Discrete Z^γ and Painlevé equations. Internat. Math. Res. Notices **2000**(4), 165–193 (2000). http://doi.org/10.1155/S1073792800000118
2. Akivis, M.A., Goldberg, V.V.: Conformal differential geometry and its generalizations. Pure and Applied Mathematics. John Wiley & Sons, Inc., New York (1996). http://doi.org/10.1002/9781118032633
3. Ando, H., Hay, M., Kajiwara, K., Masuda, T.: An explicit formula for the discrete power function associated with circle patterns of Schramm type. Funkcial. Ekvac. **57**(1), 1–41 (2014). http://doi.org/10.1619/fesi.57.1
4. Arnol'd, V.I.: Singularities of caustics and wave fronts. Mathematics and Its Applications (Soviet Series), vol. 62. Kluwer Academic Publishers Group, Dordrecht (1990). http://doi.org/10.1007/978-94-011-3330-2
5. Arnol'd, V.I., Guseĭn-Zade, S.M., Varchenko, A.N.: Singularities of differentiable maps. vol. I. Monographs in Mathematics, vol. 82. Birkhäuser Boston, Inc., Boston (1985). http://doi.org/10.1007/978-1-4612-5154-5
6. Bäcklund, A.V.: Om ytor med konstant negativ krökning. Lunds Universitets Årsskrift **19**, 1–48 (1883)
7. Bianchi, L.: Complementi alle ricerche sulle superficie isoterme. Ann. Mat. Pura Appl. (3) **12**(1), 19–54 (1906). http://doi.org/10.1007/BF02419495
8. Bianchi, L.: Lezioni di Geometria Differenziale, vol. II, 2nd edn. Enrico Spoerri, Pisa (1903)
9. Bianchi, L.: Ricerche sulle superficie elicoidali e sulle superficie a curvatura costante. Ann. Scuola Norm. Sup. Pisa Cl. Sci. **2**, 285–341 (1879)
10. Bianchi, L.: Ricerche sulle superficie isoterme e sulla deformazione delle quadriche. Ann. Mat. Pura Appl. (3) **11**(1), 93–157 (1905). http://doi.org/10.1007/BF02419963
11. Bianchi, L.: Sopra i sistemi tripli ortogonali di Weingarten. Ann. Mat. Pura Appl. (2) **13**(1), 177–234 (1885). http://doi.org/10.1007/BF02420799
12. Bianchi, L.: Sulla trasformazione di Bäcklund per le superficie pseudosferiche. Rend. Lincei **5**(1), 3–12 (1892)
13. Björling, E.G.: In integrationem aequationis Derivatarum partialium superficiei, cujus in puncto unoquoque principales ambo radii curvedinis aequales sunt signoque contrario. Arch. Math. Phys. **4**, 290–315 (1844)
14. Blaschke, W.: Vorlesungen über Differentialgeometrie und Geometrische Grundlagen von Einsteins Relativitätstheorie III: Differentialgeometrie der Kreise und Kugeln. Springer, Berlin (1929)
15. Bo, P., Pottmann, H., Kilian, M., Wang, W., Wallner, J.: Circular arc structures. ACM Trans. Graph. **30**(4), 101:1–101:12 (2011). http://doi.org/10.1145/2010324.1964996

16. Bobenko, A.I.: Discrete conformal maps and surfaces. In: Clarkson, P.A., Nijhoff, F.W. (eds.) Symmetries and Integrability of Difference Equations (Canterbury, 1996). London Mathematical Society Lecture Note series, vol. 255, pp. 97–108. Cambridge University Press, Cambridge (1999). http://doi.org/10.1017/CBO9780511569432.009

17. Bobenko, A.I.: Integrable surfaces. Funktsional. Anal. i Prilozhen. **24**(3), 68–69 (1990). http://doi.org/10.1007/BF01077966

18. Bobenko, A.I., Eitner, U.: Painlevé Equations in the Differential Geometry of Surfaces. Lecture Notes in Mathematics, vol. 1753. Springer-Verlag, Berlin (2000). http://doi.org/10.1007/b76883

19. Bobenko, A.I., Hertrich-Jeromin, U., Lukyanenko, I.: Discrete constant mean curvature nets in space forms: Steiner's formula and Christoffel duality. Discrete Comput. Geom. **52**(4), 612–629 (2014). http://doi.org/10.1007/s00454-014-9622-5

20. Bobenko, A.I., Hoffmann, T., Sageman-Furnas, A.O.: Compact Bonnet pairs: isometric tori with the same curvatures (2021). arXiv: 2110.06335

21. Bobenko, A.I., Huhnen-Venedey, E.: Curvature line parametrized surfaces and orthogonal coordinate systems: discretization with Dupin cyclides. Geom. Dedicata **159**, 207–237 (2012). http://doi.org/10.1007/s10711-011-9653-5

22. Bobenko, A.I., Pinkall, U.: Discrete isothermic surfaces. J. Reine Angew. Math. **475**, 187–208 (1996). http://doi.org/10.1515/crll.1996.475.187

23. Bobenko, A.I., Pinkall, U.: Discrete surfaces with constant negative Gaussian curvature and the Hirota equation. J. Differ. Geom. **43**(3), 527–611 (1996). http://doi.org/10.4310/jdg/1214458324

24. Bobenko, A.I., Pottmann, H., Wallner, J.: A curvature theory for discrete surfaces based on mesh parallelity. Math. Ann. **348**(1), 1–24 (2010). http://doi.org/10.1007/s00208-009-0467-9

25. Bobenko, A.I., Suris, Y.B.: Discrete Differential Geometry. Graduate Studies in Mathematics, vol. 98. American Mathematical Society, Providence (2008)

26. Bobenko, A.I., Suris, Y.B.: Isothermic surfaces in sphere geometries as Moutard nets. Proc. R. Soc. Lond. Ser. A Math. Phys. Eng. Sci. **463**(2088), 3171–3193 (2007). http://doi.org/10.1098/rspa.2007.1902

27. Bobenko, A.I., Suris, Y.B.: On organizing principles for discrete differential geometry. Geometry of spheres. Russ. Math. Surv. **62**(1), 1–43 (2007). http://doi.org/10.1070/RM2007v062n01ABEH004380

28. Bonnet, O.: Mémoire sur la théorie des surfaces applicables sur une surface donnée. J. Éc. Polytech. **42**, 1–151 (1867)

29. Boothby, W.M.: An Introduction to Differentiable Manifolds and Riemannian Geometry. Academic Press, New York-London (1975)

30. Bor, G., Levi, M., Perline, R., Tabachnikov, S.: Tire tracks and integrable curve evolution. Int. Math. Res. Not. IMRN **2020**(9), 2698–2768 (2020). http://doi.org/10.1093/imrn/rny087

31. Bour, E.: Théorie de la déformation des surfaces. J. Éc. Polytech. **39**, 1–148 (1862)

32. Boussinesq, J.: Théorie de l'intumescence liquide appelée onde solitaire ou de translation, se propageant dans un canal rectangulaire. C. R. Acad. Sci. Paris **72**, 755–759 (1871)

33. Boussinesq, J.: Théorie des ondes et des remous qui se propagent le long d'un canal rectangulaire horizontal, en communiquant au liquide contenu dans ce canal des vitesses sensiblement pareilles de la surface au fond. J. Math. Pures Appl. (2) **17**, 55–108 (1872)

34. Bryant, R.L.: Surfaces of mean curvature one in hyperbolic space. Astérisque **154–155**, 321–347 (1987)

35. Bücking, U., Matthes, D.: Constructing solutions to the Björling problem for isothermic surfaces by structure preserving discretization. In: Bobenko, A.I. (ed.) Advances in Discrete Differential Geometry, pp. 309–345. Springer, Berlin (2016). http://doi.org/10.1007/978-3-662-50447-5_10

36. Burstall, F.E.: Isothermic surfaces: conformal geometry, Clifford algebras and integrable systems. In: Terng, C.-L. (ed.) Integrable Systems, Geometry, and Topology. AMS/IP Studies in Advanced Mathematics, vol. 36, pp. 1–82. American Mathematical Society, Providence (2006) http://doi.org/10.1090/amsip/036/01

37. Burstall, F.E.: Notes on transformations in integrable geometry. In: Chiossi, S.G., Fino, A., Musso, E., Podestà, F., Vezzoni, L. (eds.) Special Metrics and Group Actions in Geometry. Springer INdAM Series, vol. 23, pp. 59–80. Springer, Cham (2017) http://doi.org/10.1007/978-3-319-67519-0_3

38. Burstall, F.E., Calderbank, D.M.J.: Conformal submanifold geometry I-III (2010). arXiv: 1006.5700

39. Burstall, F.E., Calderbank, D.M.J.: Conformal submanifold geometry IV-V. In progress.

40. Burstall, F.E., Cho, J., Hertrich-Jeromin, U., Pember, M., Rossman, W.: Discrete Ω-nets and Guichard nets via discrete Koenigs nets. Proc. Lond. Math. Soc. (3) **126**(2), 790–836 (2023). http://doi.org/10.1112/plms.12499

41. Burstall, F.E., Donaldson, N.M., Pedit, F., Pinkall, U.: Isothermic submanifolds of symmetric R-spaces. J. Reine Angew. Math. **660**, 191–243 (2011). http://doi.org/10.1515/crelle.2011.075

42. Burstall, F.E., Ferus, D., Leschke, K., Pedit, F., Pinkall, U.: Conformal Geometry of Surfaces in S^4 and Quaternions. Lecture Notes in Mathematics, vol. 1772. Springer-Verlag, Berlin (2002) http://doi.org/10.1007/b82935

43. Burstall, F.E., Hertrich-Jeromin, U.: The Ribaucour transformation in Lie sphere geometry. Differ. Geom. Appl. **24**(5), 503–520 (2006). http://doi.org/10.1016/j.difgeo.2006.04.007

44. Burstall, F.E., Hertrich-Jeromin, U., Lara Miro, M.: Ribaucour coordinates. Beitr. Algebra Geom. **60**(1), 39–55 (2019). http://doi.org/10.1007/s13366-018-0391-9

45. Burstall, F.E., Hertrich-Jeromin, U., Müller, C., Rossman, W.: Semi-discrete isothermic surfaces. Geom. Dedicata **183**, 43–58 (2016). http://doi.org/10.1007/s10711-016-0143-7

46. Burstall, F.E., Hertrich-Jeromin, U., Pedit, F., Pinkall, U.: Curved flats and isothermic surfaces. Math. Z. **225**(2), 199–209 (1997). http://doi.org/10.1007/PL00004308

47. Burstall, F.E., Hertrich-Jeromin, U., Pember, M., Rossman, W.: Polynomial conserved quantities of Lie applicable surfaces. Manuscripta Math. **158**(3–4), 505–546 (2019). http://doi.org/10.1007/s00229-018-1033-0

48. Burstall, F.E., Hertrich-Jeromin, U., Rossman, W.: Discrete linear Wein-garten surfaces. Nagoya Math. J. **231**, 55–88 (2018). http://doi.org/10.1017/nmj.2017.11

49. Burstall, F.E., Hertrich-Jeromin, U., Rossman, W.: Lie geometry of flat fronts in hyperbolic space. C. R. Math. Acad. Sci. Paris **348**(11–12), 661–664 (2010). http://doi.org/10.1016/j.crma.2010.04.018

50. Burstall, F.E., Hertrich-Jeromin, U., Rossman, W.: Lie geometry of linear Weingarten surfaces. C. R. Math. Acad. Sci. Paris **350**(7–8), 413–416 (2012). http://doi.org/10.1016/j.crma.2012.03.018

51. Burstall, F.E., Hertrich-Jeromin, U., Rossman, W., Santos, S.D.: Discrete special isothermic surfaces. Geom. Dedicata **174**, 1–11 (2015). http://doi.org/10.1007/s10711-014-0001-4

52. Burstall, F.E., Hertrich-Jeromin, U., Rossman, W., Santos, S.D.: Discrete surfaces of constant mean curvature. In: Kobayashi, S.-P. (ed.) Development in Differential Geometry of Submanifolds. RIMS Kôkyûroku, vol. 1880, pp. 133–179. Research Institute for Mathematical Sciences (RIMS), Kyoto (2014). arXiv: 0804.2707

53. Burstall, F.E., Pember, M.: Lie applicable surfaces and curved flats. Manuscripta Math. **168**(3–4), 525–533 (2022). http://doi.org/10.1007/s00229-021-01304-8

54. Burstall, F.E., Santos, S.D.: Formal conserved quantities for isothermic surfaces. Geom. Dedicata **172**, 191–205 (2014). http://doi.org/10.1007/s10711-013-9915-5

55. Burstall, F.E., Santos, S.D.: Special isothermic surfaces of type d. J. Lond. Math. Soc. (2) **85**(2), 571–591 (2012). http://doi.org/10.1112/jlms/jdr050

56. Calapso, P.: Alcune superficie di Guichard e le relative trasformazioni. Ann. Mat. Pura Appl. (3) **11**(1), 201–251 (1905). http://doi.org/10.1007/BF02419966

57. Calapso, P.: Sulla superficie a linee di curvatura isoterme. Rend. Circ. Mat. Palermo **17**(2), 275–286 (1903). http://doi.org/10.1007/BF03012748

58. Calapso, P.: Sulle trasformazioni delle superficie isoterme. Ann. Mat. Pura Appl. (3) **24**, 11–48 (1915). http://doi.org/10.1007/BF02419671

59. Carl, W.: On semidiscrete constant mean curvature surfaces and their associated families. Monatsh. Math. **182**(3), 537–563 (2017). http://doi.org/10.1007/s00605-016-0929-6

60. Cartan, E.: Les espaces à connexion conforme. Ann. Soc. Pol. Math. **2**, 171–221 (1923).

61. Case, K.M.: On discrete inverse scattering problems. II. J. Math. Phys. **14**(7), 916–920 (1973). http://doi.org/10.1063/1.1666417

62. Case, K.M., Kac, M.: A discrete version of the inverse scattering problem. J. Math. Phys. **14**(5), 594–603 (1973). http://doi.org/10.1063/1.1666364

63. Cayley, A.: On evolutes and parallel curves. Q. J. Pure Appl. Math. **11**, 183–200 (1871)

64. Cayley, A.: On the surfaces divisible into squares by their curves of curvature. Proc. Lond. Math. Soc. **4**, 120–121 (1871). http://doi.org/10.1112/plms/s1-4.1.120

65. Cecil, T.E.: Lie Sphere Geometry. Universitext, 2nd edn. Springer, New York (2008)

66. Cho, J.: Zero mean curvature surfaces of Bonnet-type. Ph.D. Thesis. Kobe University, 2019

67. Cho, J., Leschke, K., Ogata, Y.: Discrete constant mean curvature cylinders and isothermic tori. Discrete Comput. Geom. (2024, to appear). http://doi.org/10.1007/s00454-024-00697-z

68. Cho, J., Leschke, K., Ogata, Y.: Generalised Bianchi permutability for isothermic surfaces. Ann. Global Anal. Geom. **61**(4), 799–829 (2022). http://doi.org/10.1007/s10455-022-09833-5

69. Cho, J., Leschke, K., Ogata, Y.: New explicit CMC cylinders and same-lobed CMC multibubbletons (2022). arXiv: 2205.14675

70. Cho, J., Leschke, K., Ogata, Y.: Periodic discrete Darboux transforms. Differ. Geom. Appl. **91**, 102065:1–25 (2023) http://doi.org/10.1016/j.difgeo.2023.102065

71. Cho, J., Ogata, Y.: Simple factor dressings and Bianchi–Bäcklund transformations. Ill. J. Math. **63**(4), 619–631 (2019). http://doi.org/10.1215/00192082-7988989

72. Cho, J., Pember, M., Szewieczek, G.: Constrained elastic curves and surfaces with spherical curvature lines. Indiana Univ. Math. J. **72**(5), 2059–2099 (2023). http://doi.org/10.1512/iumj. 2023.72.9487

73. Cho, J., Rossman, W., Seno, T.: Discrete mKdV equation via Darboux transformation. Math. Phys. Anal. Geom. **24**(3), 25:1–11 (2021). http://doi.org/10.1007/s11040-021-09398-y

74. Cho, J., Rossman, W., Seno, T.: Infinitesimal Darboux transformation and semi-discrete mKdV equation. Nonlinearity **35**(4), 2134–2146 (2022). http://doi.org/10.1088/1361-6544/ ac591f

75. Cho, J., Rossman, W., Yang, S.-D.: Discrete minimal nets with symmetries. In: Hoffmann, T., Kilian, M., Leschke, K., Martin, F. (eds.) Minimal Surfaces: Integrable Systems and Visualisation. Springer Proceedings in Mathematics & Statistics, vol. 349, pp. 35–50. Springer International Publishing, Cham (2021). http://doi.org/10.1007/978-3-030-68541-6_3

76. Christoffel, E.B.: Ueber einige allgemeine Eigenschaften der Minimumsflächen. J. Reine Angew. Math. **67**, 218–228 (1867). http://doi.org/10.1515/crll.1867.67.218

77. Cieśliński, J., Goldstein, P., Sym, A.: Isothermic surfaces in $\mathbf{E}^3$ as soliton surfaces. Phys. Lett. A **205**(1), 37–43 (1995). http://doi.org/10.1016/0375-9601(95)00504-V

78. Clarke, D.: Integrability in submanifold geometry. Ph.D. Thesis. University of Bath, 2012

79. Corro, A.V., Ferreira, W., Tenenblat, K.: Minimal surfaces obtained by Ribaucour transformations. Geom. Dedicata **96**, 117–150 (2003). http://doi.org/10.1023/A:1022139119383

80. Corro, A.V., Ferreira, W., Tenenblat, K.: Ribaucour transformations for constant mean curvature and linear Weingarten surfaces. Pac. J. Math. **212**(2), 265–296 (2003). http://doi. org/10.2140/pjm.2003.212.265

81. Corro, A.V., Martínez, A., Tenenblat, K.: Ribaucour transformations for flat surfaces in the hyperbolic 3-space. J. Math. Anal. Appl. **412**(2), 720–743 (2014). http://doi.org/10.1016/j. jmaa.2013.10.080

82. Darboux, G.: Leçons sur la Théorie générale des Surfaces et les Applications Géométriques du Calcul Infinitéstimal, Première Partie. Gauthier-Villars, Paris (1887)

83. Darboux, G.: Sur les surfaces à courbure constante. C. R. Acad. Sci. Paris **97**, 892–894 (1883)

84. Darboux, G.: Sur les surfaces isothermiques. Ann. Sci. Éc. Norm. Supér. (3) **16**, 491–508 (1899). http://doi.org/10.24033/asens.473

85. Darboux, G.: Sur les surfaces isothermiques. C. R. Acad. Sci. Paris **128**, 1299–1305 (1899)

86. Delaunay, C.: Sur la surface de révolution dont la courbure moyenne est constante. J. Math. Pures Appl. 309–320 (1841)
87. Demoulin, A.: Sur les surfaces Ω. C. R. Acad. Sci. Paris **153**, 927–929 (1911)
88. Demoulin, A.: Sur les surfaces R et les surfaces Ω. C. R. Acad. Sci. Paris **153**, 705–707 (1911)
89. Demoulin, A.: Sur les surfaces R et les surfaces Ω. C. R. Acad. Sci. Paris **153**, 593 (1911)
90. Demoulin, A.: Sur les systèmes et les congruences K. C. R. Acad. Sci. Paris **150**, 156–159 (1910)
91. Desbrun, M., Hirani, A.N., Leok, M., Marsden, J.E.: Discrete exterior calculus (2005). arXiv: math/0508341
92. Doliwa, A.: Quadratic reductions of quadrilateral lattices. J. Geom. Phys. **30**(2), 169–186 (1999). http://doi.org/10.1016/S0393-0440(98)00053-9
93. Doliwa, A., Santini, P.M.: Multidimensional quadrilateral lattices are integrable. Phys. Lett. A **233**(4–6), 365–372 (1997). http://doi.org/10.1016/S0375-9601(97)00456-8
94. Doliwa, A., Santini, P.M., Mañas, M.: Transformations of quadrilateral lattices. J. Math. Phys. **41**(2), 944–990 (2000). http://doi.org/10.1063/1.533175
95. Donaldson, N.M.: Symmetric R-spaces: submanifold geometry and transformation theory. Ph.D. Thesis. University of Bath, 2005
96. Dorfmeister, J., Pedit, F., Wu, H.: Weierstrass type representation of harmonic maps into symmetric spaces. Commun. Anal. Geom. **6**(4), 633–668 (1998). http://doi.org/10.4310/CAG.1998.v6.n4.a1
97. Dupin, C.: Applications de Géométrie et de Méchanique. Bachelier, Paris (1822)
98. Eisenhart, L.P.: Surfaces Ω and their transformations. Trans. Am. Math. Soc. **16**(3), 275–310 (1915). http://doi.org/10.2307/1988993
99. Eisenhart, L.P.: Transformations of surfaces Ω. II. Trans. Am. Math. Soc. **17**(1), 53–99 (1916). http://doi.org/10.2307/1988826
100. Eisenhart, L.P.: Transformations of surfaces of guichard and surfaces applicable to quadrics. Ann. Mat. Pura Appl. (3) **22**(1), 191–247 (1914). http://doi.org/10.1007/BF02419557
101. Ferapontov, E.V.: Analog of Wilczynski's projective frame in Lie sphere geometry: Lie-applicable surfaces and commuting Schrödinger operators with magnetic fields. Int. J. Math. **13**(9), 959–985 (2002). http://doi.org/10.1142/S0129167X0200154X
102. Ferapontov, E.V.: Lie sphere geometry and integrable systems. Tohoku Math. J. (2) **52**(2), 199–233 (2000). http://doi.org/10.2748/tmj/1178224607
103. Ferus, D., Pedit, F.: Curved flats in symmetric spaces. Manuscripta Math. **91**(4), 445–454 (1996). http://doi.org/10.1007/BF02567965
104. Fujimori, S., Saji, K., Umehara, M., Yamada, K.: Singularities of maximal surfaces. Math. Z. **259**(4), 827–848 (2008). http://doi.org/10.1007/s00209-007-0250-0
105. Gálvez, J.A., Martínez, A., Milán, F.: Complete linear Weingarten surfaces of Bryant type. A Plateau problem at infinity. Trans. Am. Math. Soc. **356**(9), 3405–3428 (2004). http://doi.org/10.1090/S0002-9947-04-03592-5
106. Gálvez, J. A., Martínez, A., Milán, F.: Flat surfaces in the hyperbolic 3-space. Math. Ann. **316**(3), 419–435 (2000). http://doi.org/10.1007/s002080050337
107. Gardner, C.S., Greene, J.M., Kruskal, M.D., Miura, R M.: KortewegdeVries equation and generalization. VI. Methods for exact solution. Commun. Pure Appl. Math. **27**, 97–133 (1974). http://doi.org/10.1002/cpa.3160270108
108. Gardner, C.S., Greene, J.M., Kruskal, M.D., Miura, R.M.: Method for solving the Korteweg-deVries equation. Phys. Rev. Lett. **19**(19), 1095–1097 (1967). http://doi.org/10.1103/PhysRevLett.19.1095
109. Guichard, C.: Sur les surfaces isothermiques. C. R. Acad. Sci. Paris **130**, 159–162 (1900)
110. Hertrich-Jeromin, U.: Integrable systems in Möbius geometry. In: Mladenov, I.M., de León, M. (eds.) Geometry, Integrability and Quantization, pp. 11–47. Softex, Sofia (2007). http://doi.org/10.7546/giq-8-2007-11-47
111. Hertrich-Jeromin, U.: Introduction to Möbius Differential Geometry. London Mathematical Society Lecture Note Series, vol. 300. Cambridge University Press, Cambridge (2003)

112. Hertrich-Jeromin, U.: Supplement on curved flats in the space of point pairs and isothermic surfaces: a quaternionic calculus. Doc. Math. **2**, 335–350 (1997). http://doi.org/10.4171/DM/33

113. Hertrich-Jeromin, U.: Transformations of discrete isothermic nets and discrete cmc-1 surfaces in hyperbolic space. Manuscripta Math. **102**(4), 465–486 (2000). http://doi.org/10.1007/s002290070037

114. Hertrich-Jeromin, U., Hoffmann, T., Pinkall, U.: A discrete version of the Darboux transform for isothermic surfaces. In: Bobenko, A.I., Seiler, R. (eds.) Discrete integrable geometry and physics (Vienna, 1996). Oxford Lecture Series in Mathematics and Its Applications, vol. 16, pp. 59–81. Oxford University Press, New York (1999)

115. Hertrich-Jeromin, U., Honda, A.: Minimal Darboux transformations. Beitr. Algebra Geom. **58**(1), 81–91 (2017). http://doi.org/10.1007/s13366-016-0301-y

116. Hertrich-Jeromin, U., Musso, E., Nicolodi, L.: Möbius geometry of surfaces of constant mean curvature 1 in hyperbolic space. Ann. Global Anal. Geom. **19**(2), 185–205 (2001). http://doi.org/10.1023/A:1010738712475

117. Hertrich-Jeromin, U., Pedit, F.: Remarks on the Darboux transform of isothermic surfaces. Doc. Math. **2**, 313–333 (1997). http://doi.org/10.4171/DM/32

118. Hertrich-Jeromin, U., Pember, M., Polly, D.: Channel linear Weingarten surfaces in space forms. Beitr. Algebra Geom. **64**(4), 969–1009 (2023). http://doi.org/10.1007/s13366-022-00664-w

119. Hertrich-Jeromin, U., Rossman, W., Szewieczek, G.: Discrete channel surfaces. Math. Z. **294**, 747–767 (2020). http://doi.org/10.1007/s00209-019-02389-4

120. Hertrich-Jeromin, U., Szewieczek, G.: Discrete cyclic systems and circle congruences. Ann. Mat. Pura Appl. (4) **201**(6), 2797–2824 (2022). http://doi.org/10.1007/s10231-022-01219-5

121. Hirani, A.N.: Discrete exterior calculus. Ph.D. Thesis. California Institute of Technology, 2003

122. Hirota, R.: Nonlinear partial difference equations. I. A difference analogue of the Korteweg-de Vries equation. J. Phys. Soc. Jpn. **43**(4), 1424–1433 (1977). http://doi.org/10.1143/JPSJ.43.1424

123. Hirota, R.: Nonlinear partial difference equations. III. Discrete sine-Gordon equation. J. Phys. Soc. Jpn. **43**(6), 2079–2086 (1977). http://doi.org/10.1143/JPSJ.43.2079

124. Hoffmann, T.: Discrete cmc surfaces and discrete holomorphic maps. In: Bobenko, A.I., Seiler, R. (eds.). Discrete Integrable Geometry and Physics (Vienna, 1996). Oxford Lecture Series in Mathematics and Its Applications, vol. 16, pp. 97–112. Oxford University Press, New York (1999)

125. Hoffmann, T., Kobayashi, S., Ye, Z.: Discrete constant mean curvature surfaces on general graphs. Geom. Dedicata **216**(6), 71:1–34 (2022). http://doi.org/10.1007/s10711-022-00733-3

126. Hoffmann, T., Rossman, W., Sasaki, T., Yoshida, M.: Discrete flat surfaces and linear Weingarten surfaces in hyperbolic 3-space. Trans. Am. Math. Soc. **364**(11), 5605–5644 (2012). http://doi.org/10.1090/S0002-9947-2012-05698-4

127. Hoffmann, T., Sageman-Furnas, A.O., Wardetzky, M.: A discrete parametrized surface theory in $\mathbb{R}^3$. Int. Math. Res. Not. IMRN **2017**(14), 4217–4258 (2017). http://doi.org/10.1093/imrn/rnw015

128. Hoffmann, T., Szewieczek, G.: Isothermic nets with spherical parameter lines from discrete holomorphic maps (2024). arXiv: 2403.13476.

129. Hoffmann, T., Ye, Z.: A discrete extrinsic and intrinsic Dirac operator. Exp. Math. **31**(3), 920–935 (2022). http://doi.org/10.1080/10586458.2020.1727798

130. Huygens, C.: Traité de la Lumière. Chez Pierre vander Aa, Leyde (1690)

131. Inoguchi, J.-I., Kajiwara, K., Matsuura, N., Ohta, Y.: Explicit solutions to the semi-discrete modified KdV equation and motion of discrete plane curves. J. Phys. A **45**(4), 045206:1–16 (2012). http://doi.org/10.1088/1751-8113/45/4/045206

132. Jensen, G.R., Musso, E., Nicolodi, L.: Surfaces in Classical Geometries. Universitext. Springer, Cham (2016). http://doi.org/10.1007/978-3-319-27076-0

133. Jorge, L.P., Meeks, III, W.H.: The topology of complete minimal surfaces of finite total Gaussian curvature. Topology **22**(2), 203–221 (1983). http://doi.org/10.1016/0040-9383(83)90032-0

134. Kaji, S., Kajiwara, K., Park, H.: Linkage mechanisms governed by integrable deformations of discrete space curves. In: Euler, N., Nucci, M.C. (eds.) Nonlinear Systems and Their Remarkable Mathematical Structures, vol. 2, pp. 356–381. Chapman and Hall/CRC, New York (2019). http://doi.org/10.1201/9780429263743

135. Kamberov, G., Pedit, F., Pinkall, U.: Bonnet pairs and isothermic surfaces. Duke Math. J. **92**(3), 637–644 (1998). http://doi.org/10.1215/S0012-7094-98-09219-5

136. Kilian, M.: Bubbletons are not embedded. Osaka J. Math. **49**(3), 653–663 (2012). http://doi.org/10.18910/23143

137. Kobayashi, O.: Maximal surfaces in the 3-dimensional Minkowski space $\mathbf{L}^3$. Tokyo J. Math. **6**(2), 297–309 (1983). http://doi.org/10.3836/tjm/1270213872

138. Kobayashi, S.-P., Inoguchi, J.: Characterizations of Bianchi-Bäcklund transformations of constant mean curvature surfaces. Int. J. Math. **16**(2), 101–110 (2005). http://doi.org/10.1142/S0129167X05002801

139. Kokubu, M., Rossman, W., Saji, K., Umehara, M., Yamada, K.: Addendum: Singularities of flat fronts in hyperbolic space. Pac. J. Math. **294**(2), 505–509 (2018). http://doi.org/10.2140/pjm.2018.294.505

140. Kokubu, M., Rossman, W., Saji, K., Umehara, M., Yamada, K.: Singularities of flat fronts in hyperbolic space. Pac. J. Math. **221**(2), 303–351 (2005). http://doi.org/10.2140/pjm.2005.221.303

141. Kokubu, M., Umehara, M., Yamada, K.: An elementary proof of Small's formula for null curves in PSL(2, $\mathbf{C}$) and an analogue for Legendrian curves in PSL(2, $\mathbf{C}$). Osaka J. Math. **40**(3), 697–715 (2003)

142. Kokubu, M., Umehara, M., Yamada, K.: Flat fronts in hyperbolic 3-space. Pac. J. Math. **216**(1), 149–175 (2004). http://doi.org/10.2140/pjm.2004.216.149

143. Korteweg, D. J., de Vries, G.: On the change of form of long waves advancing in a rectangular canal, and on a new type of long stationary waves. Philos. Mag. (5) **39**(240), 422–443 (1895). http://doi.org/10.1080/14786449508620739

144. Kossowski, M.: The Boy-Gauss-Bonnet theorems for C^∞-singular surfaces with limiting tangent bundle. Ann. Global Anal. Geom. **21**(1), 19–29 (2002). http://doi.org/10.1023/A:1014239727720

145. Laguerre, E.: Sur la transformation par directions réciproques. C. R. Acad. Sci. Paris **92**, 71–73 (1881)

146. Lam, W.Y.: Discrete minimal surfaces: critical points of the area functional from integrable systems. Int. Math. Res. Not. IMRN **2018**(6), 1808–1845 (2018). http://doi.org/10.1093/imrn/rnw267

147. Lam, W.Y., Yasumoto, M.: Trivalent maximal surfaces in Minkowski 3-space. In: Cañadas-Pinedo, M.A., Flores, J.L., Palomo, F.J. (eds.) Lorentzian Geometry and Related Topics. Springer Proceedings in Mathematics & Statistics, vol. 211, pp. 169–184. Springer International Publishing, Cham (2017). http://doi.org/10.1007/978-3-319-66290-9_10

148. Langevin, R., Levitt, G., Rosenberg, H.: Classes d'homotopie de surfaces avec rebroussements et queues d'aronde dans $\mathbf{R}^3$. Can. J. Math. **47**(3), 544–572 (1995). http://doi.org/10.4153/CJM-1995-030-5

149. Lawson H., Blaine, J.: Complete minimal surfaces in S^3. Ann. Math. (2) **92**, 335–374 (1970). http://doi.org/10.2307/1970625

150. Lax, P.D.: Integrals of nonlinear equations of evolution and solitary waves. Commun. Pure Appl. Math. **21**, 467–490 (1968). http://doi.org/10.1002/cpa.3160210503

151. Lemes, M.V., Roitman, P., Tenenblat, K., Tribuzy, R.: Lawson correspondence and Ribaucour transformations. Trans. Am. Math. Soc. **364**(12), 6229–6258 (2012). http://doi.org/10.1090/S0002-9947-2012-05440-7

152. Levi, D., Benguria, R.: Bäcklund transformations and nonlinear differential difference equations. Proc. Nat. Acad. Sci. U.S.A. **77**, 5025–5027 (1980). http://doi.org/10.1073/pnas.77.9.5025

153. Lie, S.: Ueber Complexe, insbesondere Linien- und Kugel-Complexe, mit Anwendung auf die Theorie partieller Differential-Gleichungen. Math. Ann. **5**(1), 145–208 (1872). http://doi.org/10.1007/BF01446331

154. Lie, S.: Ueber Flächen, deren Krümmungsradien durch eine Relation verknüpft sind. Arch. Math. Naturvid. **4**, 507–512 (1879)

155. Lie, S.: Untersuchungen über Differentialgleichungen. IV. Christ. Forh. **18**, 1–5 (1883)

156. Liu, Y., Pottmann, H., Wallner, J., Yang, Y.-L., Wang, W.: Geometric modeling with conical meshes and developable surfaces. ACM Trans. Graph. **25**(3), 681–689 (2006). http://doi.org/10.1145/1141911.1141941

157. Mahler, A.: Bianchi-Bäcklund transformations for constant mean curvature surfaces with umbilics - theory and applications. Ph.D. Thesis. University of Toledo, 2002

158. Matsuura, N.: Discrete KdV and discrete modified KdV equations arising from motions of planar discrete curves. Int. Math. Res. Not. IMRN **2012**(8), 1681–1698 (2012). http://doi.org/10.1093/imrn/rnr080

159. Moutard, M.: Sur la construction des équations de la forme $\frac{1}{z}\frac{d^2z}{dxdy} = \lambda(x, y)$, qui admettent une intégrale générale explicite. J. Éc. Polytech. **45**, 1–11 (1878)

160. Müller, C.: Planar discrete isothermic nets of conical type. Beitr. Algebra Geom. **57**(2), 459–482 (2016). http://doi.org/10.1007/s13366-015-0256-4

161. Müller, C.: Semi-discrete constant mean curvature surfaces. Math. Z. **279**(1–2), 459–478 (2015). http://doi.org/10.1007/s00209-014-1377-4

162. Müller, C., Wallner, J.: Semi-discrete isothermic surfaces. Results Math. **63**(3-4), 1395–1407 (2013). http://doi.org/10.1007/s00025-012-0292-4

163. Musso, E.: Deformation of surfaces in Möbius space. Rend. Istit. Mat. Univ. Trieste **27**(1-2), 25–45 (1995)

164. Musso, E., Nicolodi, L.: A variational problem for surfaces in Laguerre geometry. Trans. Am. Math. Soc. **348**(11), 4321–4337 (1996). http://doi.org/10.1090/S0002-9947-96-01698-4

165. Musso, E., Nicolodi, L.: Deformation and applicability of surfaces in Lie sphere geometry. Tohoku Math. J. (2) **58**(2), 161–187 (2006). http://doi.org/10.2748/tmj/1156256399

166. Musso, E., Nicolodi, L.: Isothermal surfaces in Laguerre geometry. Boll. Un. Mat. Ital. B (7) **11**, 125–144 (1997)

167. Musso, E., Nicolodi, L.: On the equation defining isothermic surfaces in Laguerre geometry. In: Szenthe, J. (ed.) New Developments in Differential Geometry, Budapest 1996, pp. 285–294. Kluwer Academic Publishers, Dordrecht (1999). http://doi.org/10.1007/978-94-011-5276-1_20

168. Musso, E., Nicolodi, L.: The Bianchi-Darboux transform of L-isothermic surfaces. Int. J. Math. **11**(7), 911–924 (2000). http://doi.org/10.1142/S0129167X00000465

169. Nijhoff, F.W., Quispel, G.R.W., Capel, H.W.: Direct linearization of nonlinear difference-difference equations. Phys. Lett. A **97**(4), 125–128 (1983). http://doi.org/10.1016/0375-9601(83)90192-5

170. Nimmo, J.J.C., Schief, W.K.: Superposition principles associated with the Moutard transformation: an integrable discretization of a $(2 + 1)$-dimensional sine-Gordon system. Proc. R. Soc. Lond. Ser. A **453**(1957), 255–279 (1997). http://doi.org/10.1098/rspa.1997.0015

171. Nutbourne, A.W., Martin, R.R.: Differential Geometry Applied to Curve and Surface Design, vol. 1. Ellis Horwood Ltd., Chichester (1988)

172. Ogata, Y.: Ribaucour transformations and their singularities. J. Geom. **113**(1), Paper No. 3, 14 (2022). http://doi.org/10.1007/s00022-021-00618-y

173. Ogata, Y., Teramoto, K.: Duality between cuspidal butterflies and cuspidal S_1^- singularities on maximal surfaces. Note Mat. **38**(1), 115–130 (2018). http://doi.org/10.1285/i15900932v38n1p115

174. Ogata, Y., Yasumoto, M.: Construction of discrete constant mean curvature surfaces in Riemannian spaceforms and applications. Differ. Geom. Appl. **54**, 264–281 (2017). http://doi.org/10.1016/j.difgeo.2017.04.010

175. Pacheco, R.: Bianchi-Bäcklund transforms and dressing actions, revisited. Geom. Dedicata **146**, 85–99 (2010). http://doi.org/10.1007/s10711-009-9427-5

176. Pember, M.: Lie applicable surfaces. Commun. Anal. Geom. **28**(6), 1407–1450 (2020). http://doi.org/10.4310/CAG.2020.v28.n6.a5

177. Pember, M.: Special surface classes. Ph.D. Thesis. University of Bath, 2015

178. Pember, M.: Weierstrass-type representations. Geom. Dedicata **204**(1), 299–309 (2020). http://doi.org/10.1007/s10711-019-00456-y

179. Pember, M., Polly, D., Yasumoto, M.: Discrete Weierstrass-type representations. Discrete Comput. Geom. **70**(3), 816–844 (2023). http://doi.org/10.1007/s00454-022-00439-z

180. Pember, M., Rossman, W., Saji, K., Teramoto, K.: Characterizing singularities of a surface in Lie sphere geometry. Hokkaido Math. J. **48**(2), 281–308 (2019). http://doi.org/10.14492/hokmj/1562810509

181. Pember, M., Szewieczek, G.: Channel surfaces in Lie sphere geometry. Beitr. Algebra Geom. **59**(4), 779–796 (2018). http://doi.org/10.1007/s13366-018-0394-6

182. Pinkall, U.: Dupin hypersurfaces. Math. Ann. **270**(3), 427–440 (1985). http://doi.org/10.1007/BF01473438

183. Pinkall, U., Polthier, K.: Computing discrete minimal surfaces and their conjugates. Experiment. Math. **2**(1), 15–36 (1993). http://doi.org/10.1080/10586458.1993.10504266

184. Polly, D.: Rotational cmc surfaces in terms of Jacobi elliptic functions (2023). arXiv: 2305.15514

185. Polthier, K., Rossman, W.: Discrete constant mean curvature surfaces and their index. J. Reine Angew. Math. **549**, 47–77 (2002). http://doi.org/10.1515/crll.2002.066

186. Pottmann, H., Liu, Y., Wallner, J., Bobenko, A.I., Wang, W.: Geometry of multi-layer freeform structures for architecture. ACM Trans. Graph. (TOG) **26**(3), 65-1–65-11 (2007). http://doi.org/10.1145/1276377.1276458

187. Ribaucour, A.: Sur la déformation des surfaces. C. R. Acad. Sci. Paris **70**, 330–333 (1870)

188. Ribaucour, A.: Sur les systèmes cycliques. C. R. Acad. Sci. Paris **76**, 478–482 (1873)

189. Rörig, T., Szewieczek, G.: The Ribaucour families of discrete R-congruences. Geom. Dedicata **214**, 251–275 (2021). http://doi.org/10.1007/s10711-021-00614-1

190. Rossman, W., Yasumoto, M.: Discrete linear Weingarten surfaces with singularities in Riemannian and Lorentzian spaceforms. In: Izumiya, S., Ishikawa, G., Yamamoto, M., Saji, K., Yamamoto, T., Takahashi, M. (eds.) Singularities in Generic Geometry. Advanced Studies in Pure Mathematics, vol. 78, pp. 383–410. The Mathematical Society of Japan, Tokyo (2018). http://doi.org/10.2969/aspm/07810383

191. Rossman, W., Yasumoto, M.: Weierstrass representation for semi-discrete minimal surfaces, and comparison of various discretized catenoids. J. Math-for-Ind. **4B**, 109–118 (2012)

192. Russell, J.S.: Report on waves. In: Report of the Fourteenth Meeting of the British Association for the Advancement of Science. John Murray, London (1845)

193. Saji, K., Umehara, M., Yamada, K.: The geometry of fronts. Ann. Math. (2) **169**(2), 491–529 (2009). http://doi.org/10.4007/annals.2009.169.491

194. Santos, S.D.: Special isothermic surfaces. Ph.D. Thesis. University of Bath, 2008

195. Sauer, R.: Differenzengeometrie. Berlin-New York: Springer-Verlag (1970)

196. Sauer, R.: Parallelogrammgitter als Modelle pseudosphärischer Flächen. Math. Z. **52**, 611–622 (1950). http://doi.org/10.1007/BF02230715

197. Sauer, R., Graf, H.: Über Flächenverbiegung in Analogie zur Verknickung offener Facettenflache. Math. Ann. **105**(1), 499–535 (1931). http://doi.org/10.1007/BF01455828

198. Schief, W.K.: On a maximum principle for minimal surfaces and their integrable discrete counterparts. J. Geom. Phys. **56**(9), 1484–1495 (2006). http://doi.org/10.1016/j.geomphys.2005.07.007

199. Schief, W.K.: On the unification of classical and novel integrable surfaces. II. Difference geometry. R. Soc. Lond. Proc. Ser. A Math. Phys. Eng. Sci. **459**(2030), 373–391 (2003). http://doi.org/10.1098/rspa.2002.1008

200. Schief, W.K., Konopelchenko, B.G.: On the unification of classical and novel integrable surfaces. I. Differential geometry. R. Soc. Lond. Proc. Ser. A Math. Phys. Eng. Sci. **459**(2029), 67–84 (2003). http://doi.org/10.1098/rspa.2002.1007

201. Schneider, R.: Convex bodies: the Brunn-Minkowski theory. Encyclopedia of Mathematics and Its Applications, vol. 44. Cambridge University Press, Cambridge (1993). http://doi.org/10.1017/CBO9780511526282

202. Schramm, O.: Circle patterns with the combinatorics of the square grid. Duke Math. J. **86**(2), 347–389 (1997). http://doi.org/10.1215/S0012-7094-97-08611-7

203. Stephenson, K.: Introduction to circle packing. Cambridge University Press, Cambridge (2005)

204. Sterling, I., Wente, H.C.: Existence and classification of constant mean curvature multibubbletons of finite and infinite type. Indiana Univ. Math. J. **42**(4), 1239–1266 (1993). http://doi.org/10.1512/iumj.1993.42.42057

205. Stillwell, J.: The Four Pillars of Geometry. Springer, New York (2005). http://doi.org/10.1007/0-387-29052-4

206. Tabachnikov, S.: On the bicycle transformation and the filament equation: results and conjectures. J. Geom. Phys. **115**, 116–123 (2017). http://doi.org/10.1016/j.geomphys.2016.05.013

207. Tabachnikov, S., Tsukerman, E.: On the discrete bicycle transformation. Publ. Mat. Urug. **14**, 201–219 (2013)

208. Teramoto, K.: Parallel and dual surfaces of cuspidal edges. Differential Geom. Appl. **44**, 52–62 (2016) http://doi.org/10.1016/j.difgeo.2015.10.005

209. Terng, C.-L., Uhlenbeck, K.: Bäcklund transformations and loop group actions. Commun. Pure Appl. Math. **53**(1), 1–75 (2000). http://doi.org/10.1002/(SICI)1097-0312(200001)53:1<1::AID-CPA1>3.3.CO;2-L

210. Thom, R.: Stabilité Structurelle et Morphogénèse. Mathematical Physics Monograph Series. W. A. Benjamin, Inc., Reading (1972)

211. Uhlenbeck, K.: Harmonic maps into Lie groups: classical solutions of the chiral model. J. Differ. Geom. **30**(1), 1–50 (1989). http://doi.org/10.4310/jdg/1214443286

212. Umehara, M., Saji, K., Yamada, K.: Differential Geometry of Curves and Surfaces with Singularities. Series in Algebraic and Differential Geometry, vol. 1. World Scientific Publishing Co. Pte. Ltd., Hackensack (2022). http://doi.org/10.1142/12284

213. Umehara, M., Yamada, K.: A duality on CMC-1 surfaces in hyperbolic space, and a hyperbolic analogue of the Osserman inequality. Tsukuba J. Math. **21**(1), 229–237 (1997). http://doi.org/10.21099/tkbjm/1496163174

214. Umehara, M., Yamada, K.: A parametrization of the Weierstrass formulae and perturbation of complete minimal surfaces in $\mathbf{R}^3$ into the hyperbolic 3-space. J. Reine Angew. Math. **432**, 93–116 (1992). http://doi.org/10.1515/crll.1992.432.93

215. Umehara, M., Yamada, K.: Maximal surfaces with singularities in Minkowski space. Hokkaido Math. J. **35**(1), 13–40 (2006). http://doi.org/10.14492/hokmj/1285766302

216. Umehara, M., Yamada, K.: Surfaces of constant mean curvature c in $\mathbf{H}^3(-c^2)$ with prescribed hyperbolic Gauss map. Math. Ann. **304**(2), 203–224 (1996). http://doi.org/10.1007/BF01446291

217. Vessiot, E.: Contribution à la géométrie conforme. Théorie des surfaces. I. Bull. Soc. Math. Fr. **54**, 139–179 (1926)

218. Weierstrass, K.T.: Untersuchungen über die Flächen, deren mittlere Krümmung überall gleich Null ist. Monatsber. Berliner Akad., 612–625 (1866)

219. Whitney, H.: The singularities of a smooth n-manifold in $(2n-1)$-space. Ann. Math. (2) **45**, 247–293 (1944). http://doi.org/10.2307/1969266

220. Wunderlich, W.: Zur Differenzengeometrie der Flächen konstanter negativer Krümmung. Österreich. Akad. Wiss. Math.-Nat. Kl. S.-B. IIa. **160**, 39–77 (1951)

221. Yasumoto, M.: Discrete maximal surfaces with singularities in Minkowski space. Differ. Geom. Appl. **43**, 130–154 (2015). http://doi.org/10.1016/j.difgeo.2015.09.006

222. Yasumoto, M.: Semi-discrete maximal surfaces with singularities in Minkowski space. In: Hoffmann, T., Kilian, M., Leschke, K., Martin, F. (eds.). Minimal Surfaces: Integrable Systems and Visualisation. Springer Proceedings in Mathematics & Statistics, vol. 349, pp. 263–280. Springer International Publishing, Cham (2021). http://doi.org/10.1007/978-3-030-68541-6_16

223. Yasumoto, M., Rossman, W.: Semi-discrete linear Weingarten surfaces with Weierstrass-type representations and their singularities. Osaka J. Math. **57**(1), 169–185 (2020)
224. Zabusky, N.J., Kruskal, M.D.: Interaction of "solitons" in a collisionless plasma and the recurrence of initial states. Phys. Rev. Lett. **15**(6), 240–243 (1965). http://doi.org/10.1103/PhysRevLett.15.240
225. Zakharov, V.E., Šabat, A.B.: Integration of the nonlinear equations of mathematical physics by the method of the inverse scattering problem. II. Funktsional. Anal. i Prilozhen. **13**(3), 13–22 (1979). http://doi.org/10.1007/BF01077483

Index

forwarded to the LNM Editorial Board, this is very helpful. If no reports are forwarded or if other questions remain unclear in respect of homogeneity etc, the series editors may wish to consult external referees for an overall evaluation of the volume.

5. Manuscripts should in general be submitted in English. Final manuscripts should contain at least 100 pages of mathematical text and should always include

   - a table of contents;
   - an informative introduction, with adequate motivation and perhaps some historical remarks: it should be accessible to a reader not intimately familiar with the topic treated;
   - a subject index: as a rule this is genuinely helpful for the reader.
   - For evaluation purposes, manuscripts should be submitted as pdf files.

6. Careful preparation of the manuscripts will help keep production time short besides ensuring satisfactory appearance of the finished book in print and online. After acceptance of the manuscript authors will be asked to prepare the final LaTeX source files (see LaTeX templates online: https://www.springer.com/gb/authors-editors/book-authors-editors/manuscriptpreparation/5636) plus the corresponding pdf- or zipped ps-file. The LaTeX source files are essential for producing the full-text online version of the book, see http://link.springer.com/bookseries/304 for the existing online volumes of LNM). The technical production of a Lecture Notes volume takes approximately 12 weeks. Additional instructions, if necessary, are available on request from lnm@springer.com.

7. Authors receive a total of 30 free copies of their volume and free access to their book on SpringerLink, but no royalties. They are entitled to a discount of 33.3 % on the price of Springer books purchased for their personal use, if ordering directly from Springer.

8. Commitment to publish is made by a *Publishing Agreement*; contributing authors of multiauthor books are requested to sign a *Consent to Publish form*. Springer-Verlag registers the copyright for each volume. Authors are free to reuse material contained in their LNM volumes in later publications: a brief written (or e-mail) request for formal permission is sufficient.

Addresses:
Professor Jean-Michel Morel, CMLA, École Normale Supérieure de Cachan, France
E-mail: moreljeanmichel@gmail.com

Professor Bernard Teissier, Equipe Géométrie et Dynamique,
Institut de Mathématiques de Jussieu – Paris Rive Gauche, Paris, France
E-mail: bernard.teissier@imj-prg.fr

Springer: Ute McCrory, Mathematics, Heidelberg, Germany,
E-mail: lnm@springer.com